Nontuberculous Mycobacteria (NTM)

Nontuberculous Mycobacteria (NTM)

Microbiological, Clinical and Geographical Distribution

Edited by

Ali Akbar Velayati
Mycobacteriology Research Centre (MRC), National Research Institute of Tuberculosis and Lung Disease (NRITLD), Shahid Beheshti University of Medical Sciences, Tehran, Iran

Parissa Farnia
Mycobacteriology Research Centre (MRC), National Research Institute of Tuberculosis and Lung Disease (NRITLD), Shahid Beheshti University of Medical Sciences, Tehran, Iran

ELSEVIER

ACADEMIC PRESS
An imprint of Elsevier

Academic Press is an imprint of Elsevier
125 London Wall, London EC2Y 5AS, United Kingdom
525 B Street, Suite 1650, San Diego, CA 92101, United States
50 Hampshire Street, 5th Floor, Cambridge, MA 02139, United States
The Boulevard, Langford Lane, Kidlington, Oxford OX5 1GB, United Kingdom

British Library Cataloguing-in-Publication Data

A catalogue record for this book is available from the British Library

Library of Congress Cataloging-in-Publication Data

A catalog record for this book is available from the Library of Congress

ISBN: 978-0-12-814692-7

For Information on all Academic Press publications
visit our website at https://www.elsevier.com/books-and-journals

Publisher: Stacy Masucci
Acquisition Editor: Katie Chan
Editorial Project Manager: Pat Gonzalez
Production Project Manager: Kiruthika Govindaraju
Cover Designer: Miles Hitchen

Typeset by MPS Limited, Chennai, India

Contents

List of Contributors

Imran Ahmed Department of Pathology and Laboratory Medicine, Aga Khan University, Karachi, Pakistan

Rutger Bennet Astrid Lindgren Children's Hospital, Karolinska University Hospital, Stockholm, Sweden

Mridula Bose Department of Microbiology, Vallabhbhai Patel Chest Institute, University of Delhi, Delhi, India

Margareta Eriksson Astrid Lindgren Children's Hospital, Karolinska University Hospital, Stockholm, Sweden

Parissa Farnia Mycobacteriology Research Centre (MRC), National Research Institute of Tuberculosis and Lung Disease (NRITLD), Shahid Beheshti University of Medical Sciences, Tehran, Iran

Poopak Farnia Department of Biotechnology, School of Advanced Technologies in Medicine, Shahid Beheshti University of Medical Sciences, Tehran, Iran

Jalaledin Ghanavi Mycobacteriology Research Centre (MRC), National Research Institute of Tuberculosis and Lung Disease (NRITLD), Shahid Beheshti University of Medical Sciences, Tehran, Iran

Rumina Hasan Department of Pathology and Laboratory Medicine, Aga Khan University, Karachi, Pakistan

Sven Hoffner Department of Public Health Sciences, Karolinska Institute, Stockholm, Sweden

Seema Irfan Department of Pathology and Laboratory Medicine, Aga Khan University, Karachi, Pakistan

Hamidreza Jamaati Chronic Respiratory Diseases Research Center, National Research Institute of Tuberculosis and Lung Diseases (NRITLD), Shahid Beheshti University of Medical Sciences, Tehran, Iran

Bodil Jönsson Department of Infectious Medicine, Institute of Biomedicine, University of Gothenburg, Gothenburg, Sweden

Lars-Olof. Larsson Division of Respiratory Medicine, Department of Medicine, Karolinska University Hospital, Stockholm, Sweden

Payam Mehrian Pediatric Respiratory Diseases Research Centre (PRDRC), National Research Institute of Tuberculosis and Lung Disease (NRITLD), Shahid Beheshti University of Medical Sciences, Tehran, Iran

Diane Ordway Mycobacteria Research Laboratory, Department of Microbiology, Immunology and Pathology, Colorado State University, Fort Collins, CO, United States

Maria Owais Ziauddin Medical University, Karachi, Pakistan

Malin Ridell Department of Microbiology and Immunology, Institute of Biomedicine, University of Gothenburg, Gothenburg, Sweden

Ruxana T. Sadikot Section of Pulmonary, Critical Care Medicine, Emory University, Atlanta, GA, United States

Shima Saif Mycobacteriology Research Centre (MRC), National Research Institute of Tuberculosis and Lung Disease (NRITLD), Shahid Beheshti University of Medical Sciences, Tehran, Iran

Sadia Shakoor Department of Pathology and Laboratory Medicine, Aga Khan University, Karachi, Pakistan

Payam Tabarsi Clinical Tuberculosis and Epidemiology Research Center, National Research Institute of Tuberculosis and Lung Diseases (NRITLD), Shahid Beheshti University of Medical Sciences, Tehran, Iran

Enrico Tortoli Emerging Bacterial Pathogens Unit, IRCCS San Raffaele Scientific Institute, Milan, Italy

Mandira Varma-Basil Department of Microbiology, Vallabhbhai Patel Chest Institute, University of Delhi, Delhi, India

Ali Akbar Velayati Mycobacteriology Research Centre (MRC), National Research Institute of Tuberculosis and Lung Disease (NRITLD), Shahid Beheshti University of Medical Sciences, Tehran, Iran

Preface

There has been a recent dramatic interest in the incidence of pulmonary nontuberculous Mycobacteria (NTM). Twenty years ago, NTM were regarded as nonvirulent bacteria that were not appreciated widely as human pathogens. Today, they are considered to be in every environmental factor, that is, soil, water, and air. Recently, it was shown that NTM can cause diseases from various environmental conditions rather than from person-to-person "in very rare exceptional cases." Ernest Runyon (1959) put these human disease-associated Mycobacteria into four main groups. The number of identified and cataloged NTM species has been increasing rapidly, from about 50 in 1997 to over 160 by January, 2016. The surge is mainly due to improved isolation and identification techniques. However, even with these new techniques, the Runyon classification is still sometimes used to organize the Mycobacteria into categories. The most common clinical manifestation of NTM disease is lung disease, but lymphatic, skin/soft tissue, and disseminated diseases are also important. Pulmonary disease caused by NTM is most often seen in postmenopausal women and patients with underlying lung disease such as cystic fibrosis (CF), bronchiectasis, and prior tuberculosis. It is not uncommon for alpha 1-antitrypsin deficiency, Marfan syndrome, or primary ciliary dyskinesia patients to have pulmonary NTM colonization and/or infection. Pulmonary NTM can also be found in individuals with AIDS and malignant disease. It can be caused by many NTM species which depends on region, but most frequently MAC and M. kansasii. Clinical symptoms vary in scope and intensity, but commonly include chronic cough, often with purulent sputum. Hemoptysis may also be present. Systemic symptoms include malaise, fatigue, and weight loss in advanced stages of disease. The diagnosis of NTM pulmonary infection requires the presence of symptoms, radiologic abnormalities, and microbiologic cultures. In this book, we give an overview of the taxonomy and identification of NTM. Then, the susceptibility of these species are discussed. Additionally to clinical symptoms and radiological patterns, the geographical distribution of NTM are presented. The authors hope that by going through the chapters of this book, clinicians and researchers will give more attention to NTM.

Although one should keep in mind that NTM study needs intellectual trends, funds, and governmental support for high tech-laboratories to detect and report NTM infection within suspected cases.

Ali Akbar Velayati and Parissa Farnia

Mycobacteriology Research Centre (MRC), National Research Institute of Tuberculosis and Lung Disease (NRITLD), Shahid Beheshti University of Medical Sciences, Tehran, Iran

REFERENCE

Runyon, E.H., 1959. Anonymous mycobacteria in pulmonary disease. Med. Clin. North Am. 43 (1), 273–290. PubMed 3612432.

Chapter 1

The Taxonomy of the Genus *Mycobacterium*

Enrico Tortoli
Emerging Bacterial Pathogens Unit, IRCCS San Raffaele Scientific Institute, Milan, Italy

The genus *Mycobacterium* includes more than 190 species and belongs to the family of *Mycobacteriaceae*, class *Corynobacteriales*, type *Actinobacteria*, and kingdom *Bacteria*. It was first proposed in 1896 (Lehmann and Neuman, 1896) to host organisms considered at that time to be halfway between fungi and bacteria.

The Swedish botanist Carl Linnaeus is considered the founder of taxonomy: a hierarchical classification of plants (Linnaeus, 1735), subsequently extended to animals, based on similarity of phenotypic characters. A similarly important innovation introduced by Linnaeus is the binomial nomenclature, still in use for naming species.

Following the discovery of DNA (Watson and Crick, 1953), genetic characters started to be investigated and to be used for taxonomic purposes. The determination of DNA base composition (guanine + cytosine%) represented the first step (Barbu et al., 1956). A few years later Wayne et al. proposed to measure the DNA relatedness among strains in a paper aiming to reconcile the competing taxonomic approaches: phenotypic and genotypic (Wayne et al., 1987). He suggested that in members of different species this parameter, measured by DNA−DNA hybridization (DDH), should be lower than 70%. The 70% threshold is still considered a gold standard for species circumscription, the DDH test is, however, hardly performed in modern laboratories (Chan et al., 2012).

The genetic sequencing introduced by Sanger (Sanger et al., 1977), a milestone in biological sciences, had an enormous impact on taxonomy and represented significant progress toward a phylogeny-based classification. The rRNA, which is highly conserved because of the essential role of ribosomes in the protein synthesis, soon became the primary target. Among its three subunits, the 16S has been by far the most frequently investigated, and at

Nontuberculous Mycobacteria (NTM). DOI: https://doi.org/10.1016/B978-0-12-814692-7.00001-2

present the sequence of this locus is available, for every known species, in public databases.

Comparative studies between 16S rRNA sequence and DDH lead to identify a 16S similarity of 97% as the threshold corresponding to 70% DDH (Gevers et al., 2005). Despite the cutoff being subsequently revised and raised to 98.8%–99% (Stackebrandt and Ebers, 2006) it still remains unsuitable to appraise the divergence between several species within the genus *Mycobacterium*.

The genus *Mycobacterium* is characterized, at phenotypic level, by unique characteristics. In the cell wall, extremely rich in lipids, the mycolic acids play a major role. Although they are present in a few other *Actinobacteria*-related genera, only the genus *Mycobacterium* has chains as long as 60–90 C (Brennan and Nikaido, 1995). On the basis of growth rate, the mycobacteria are conventionally divided into two groups: the slow growers require more than 1 week to develop visible colonies on solid media, while the rapid growers may require 3–7 days, thus growing slower in comparison with other cultivable bacteria (Tortoli, 2003).

Genetically, the mycobacteria differ from the large majority of bacteria for the high G + C content (ranging from 62% to 70%). The number of copies of the ribosomal operon is low: two copies in the rapid growers and only one in the slow growers; there are very few exceptions (Böddinghaus et al., 1990).

The variability of 16S rRNA is moderate among mycobacteria with a number of species presenting 100% identical sequence (Tortoli, 2003). The 16S rRNA is about 1500 bp long and includes mostly highly conserved regions with few interposed variable traits. The two major variable strings are located in the first third of the gene; they are known as hypervariable regions A (*E. coli* positions 130–210) and B (*E. coli* positions 430–500) with the latter including the helix 18. Minor variable regions are located in the remaining two thirds of the gene (Stahl and Urbance, 1990). The hypervariable region A hosts most of the species-specific polymorphisms. The hypervariable region B is characterized by three mayor formats; it may host, in the helix 18, a 14-nucleotide insertion, a 12-nucleotide insertion, or no insertion at all.

Since the first taxonomic analyses of the genus *Mycobacterium* based on the sequence of 16S rRNA, a number of features have emerged. Rapid and slow growers appeared clearly separated in the phylogenetic tree. All the rapidly growing species had a short helix 18 (no insertion in the hypervariable region B). The majority of slow growing species had a long helix 18: 12-nucleotide insertion in the hypervariable region B. The slow growers *M. terrae* and *M. nonchromogenicum* had a 14-nucleotide insertion and were located in a separate clade interposed between slow and rapid growers. *M. simiae* did not have insertion in the hypervariable region B, but clustered with slow growers according to its phenotype (Rogall et al., 1990; Stahl and Urbance, 1990) (Fig. 1.1).

In subsequent years, the inclusion in the analysis of further species did not produce substantial changes in the topology of the phylogenetic tree. Among the newly added species, the new rapid growers clustered with the previous ones and confirmed the short helix 18; the species which added to the clade of *M. terrae* and *M. nonchromogenicum* had the 14-nucleotide insertion; most of the slow growers confirmed the presence of a 12-nucleotide insertion. The few slow growers with short helix 18 grouped with *M. simiae* except two, *M. doricum* and *M. tusciae*, which fell among the rapid growers (Figs. 1.2 and 1.3).

On the basis of such results, the presence of signatures within the 16S rRNA was theorized:

- The short helix 18 is a marker of rapid growers.
- A small group of slow growers related to *M. simiae* (*M. simiae* complex), recognizable for sharing the same sequence in the hypervariable region B, have the short helix 18 as well.
- The long helix 18 (12-nucleotide insertion) is a marker of slow growers.

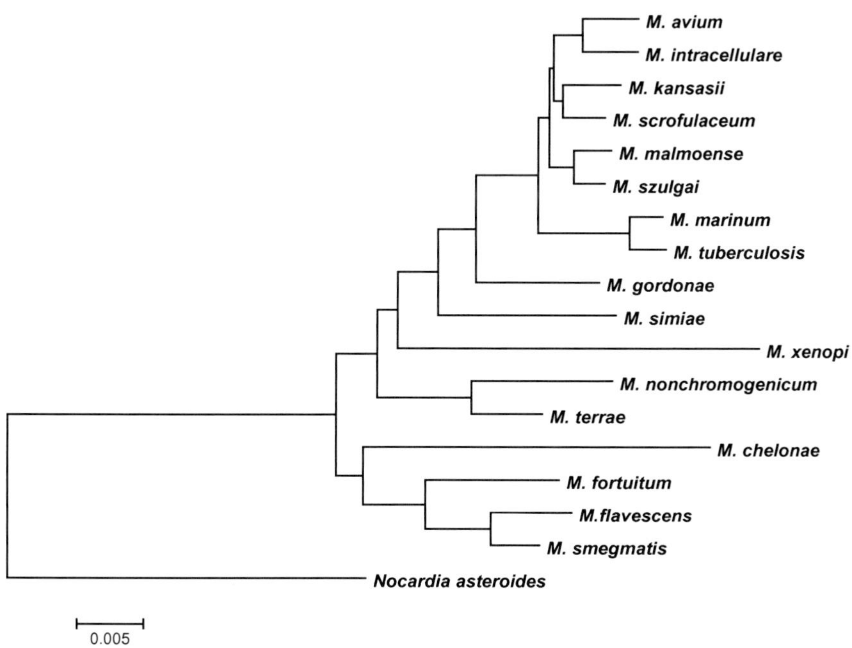

FIGURE 1.1 Phylogenetic tree inferred with the neighbor-joining algorithm from the sequence dataset investigated by Rogall et al. (1990). The tree was rooted with *Nocardia asteroids* as outgroup. Bar, 0.005 substitutions per nucleotide position.

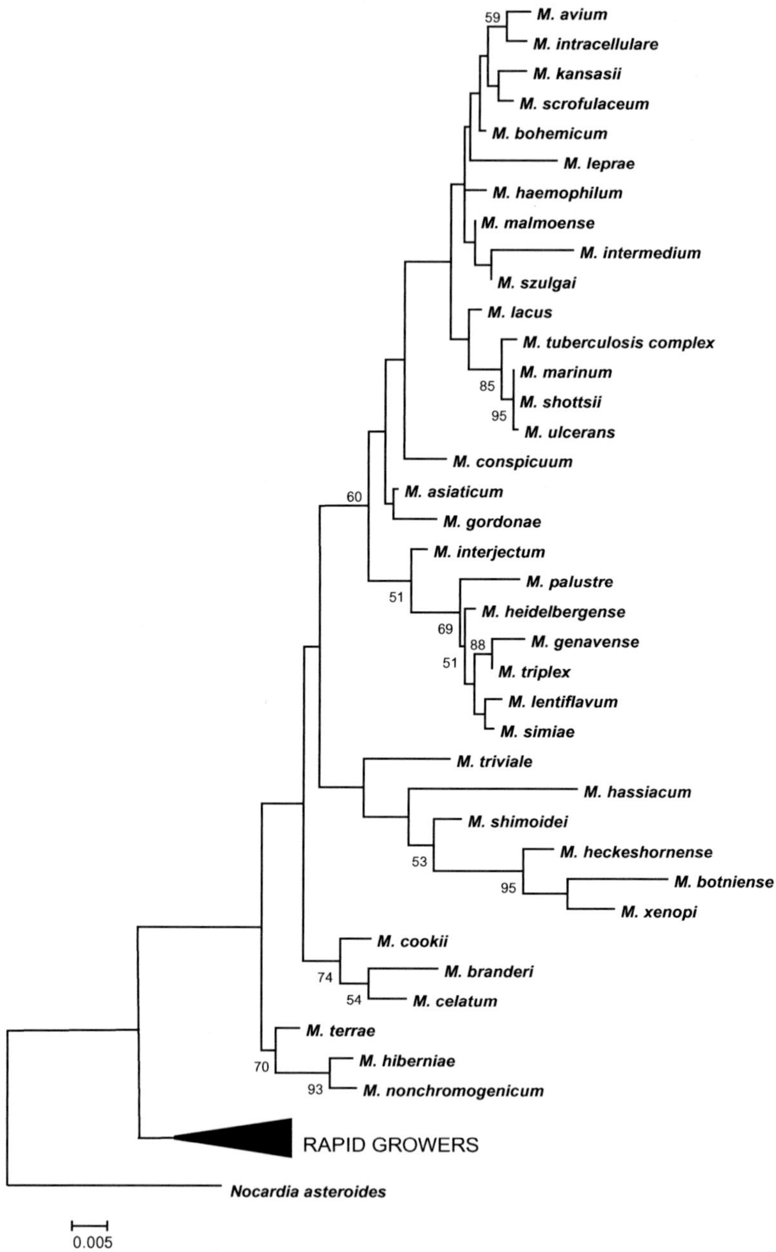

FIGURE 1.2 Phylogenetic tree of slow growers inferred with the neighbor-joining algorithm from the sequence dataset investigated by Tortoli (2012), bootstrapped 500 times. The tree was rooted with *Nocardia asteroides* as outgroup. Bar, 0.005 substitutions per nucleotide position.

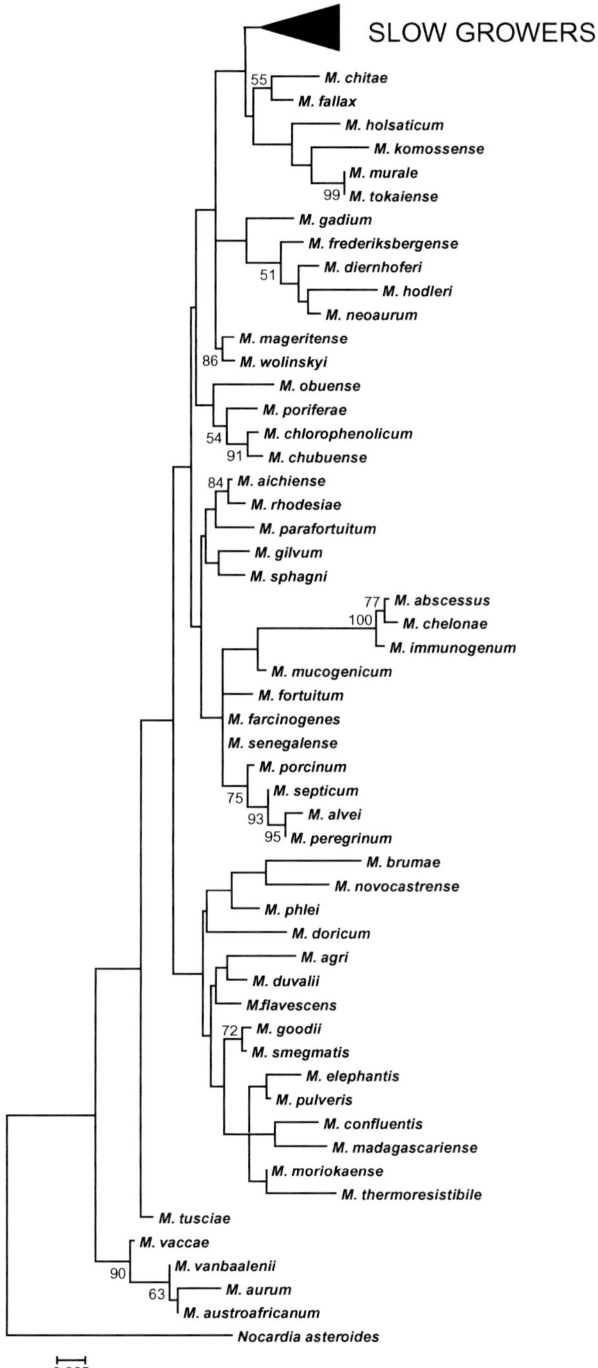

FIGURE 1.3 Phylogenetic tree of rapid growers inferred with the neighbor-joining algorithm from the sequence dataset investigated by Tortoli (2012), bootstrapped 500 times. The tree was rooted with *Nocardia asteroides* as outgroup. Bar, 0.005 substitutions per nucleotide position.

- A small group of slow growers related to *M. terrae* (*M. terrae* complex) have an even longer helix 18 (14-nucleotide insertion).

M. doricum and *M. tusciae* where regarded as the exceptions to the theorem above.

In the first decade of the 21st century, with the objective of improving the robustness of the phylogenetic trees, concatenated sequences of a few housekeeping genes were used for inference (Devulder et al., 2005; Mignard and Flandrois, 2008; Tortoli, 2012). The reduction of the weight of 16S rRNA, due to the presence of other genes in the concatenated sequence, was expected to minimize or even to sponge out the theory of the signatures. This was not the case, the 16S signatures continued to predict the topology of the phylogenetic supertree. *M. doricum* and *M. tusciae* were confirmed as the only exceptions.

The ultimate taxonomy of the genus *Mycobacterium* is being written right now. So far, two studies based on whole genomes have dealt with the phylogeny of the genus *Mycobacterium* (Fedrizzi et al., 2017; Tortoli et al., 2017). The number of genomes were different, as well as the approaches. Trees were reconstructed starting from the core genomes (concatenated sequences of the genes shared by all strains), the gene presence/absence clustering (Fedrizzi et al., 2017), and the average nucleotide identity (ANI) (Tortoli et al., 2017). The first tree was inferred from concatenated sequences using the maximum likelihood (ML) algorithm; the second was obtained, again with the ML algorithm, after converting the clusters to a binary matrix encoding the presence of each strain in the clusters; the third was constructed from the distance matrix of pairwise ANI-divergences using the ML, the neighbor-joining and the unweighted pair group method using arithmetic averages (UPGMA) algorithms. The similarity between topologies of different trees is striking.

The most ancestral branch leads to the species of the *M. abscessus-chelonae* complex. The other rapid growers emerge from a more recent branch giving rise to a major clade, accommodating the species of the *M. fortuitum-smegmatis* group, and a large number of dispersed species. Other progressively more recent branches lead to the *M. terrae* complex and to slow growers. Among slow growers several clusters of species are present. Two minor close groupings include the species related to *M. xenopi* and *M. celatum*. A large cluster includes the major pathogens (*M. tuberculosis*, *M. leprae*, *M. ulcerans*) and a number of species frequently involved in human diseases. Other clusters include the species of the *M. simiae* complex and the *M. avium* complex. Most of the 16S rRNA signatures are confirmed, only the one related to the *M. simiae* complex emerges poorly specific as part of the species presenting the signature do not cluster with *M. simiae*, but are dispersed among other slow growers not part of the complex. Interestingly, *M. doricum* and *M. tusciae* cluster again with rapid growers (Figs. 1.4 and 1.5).

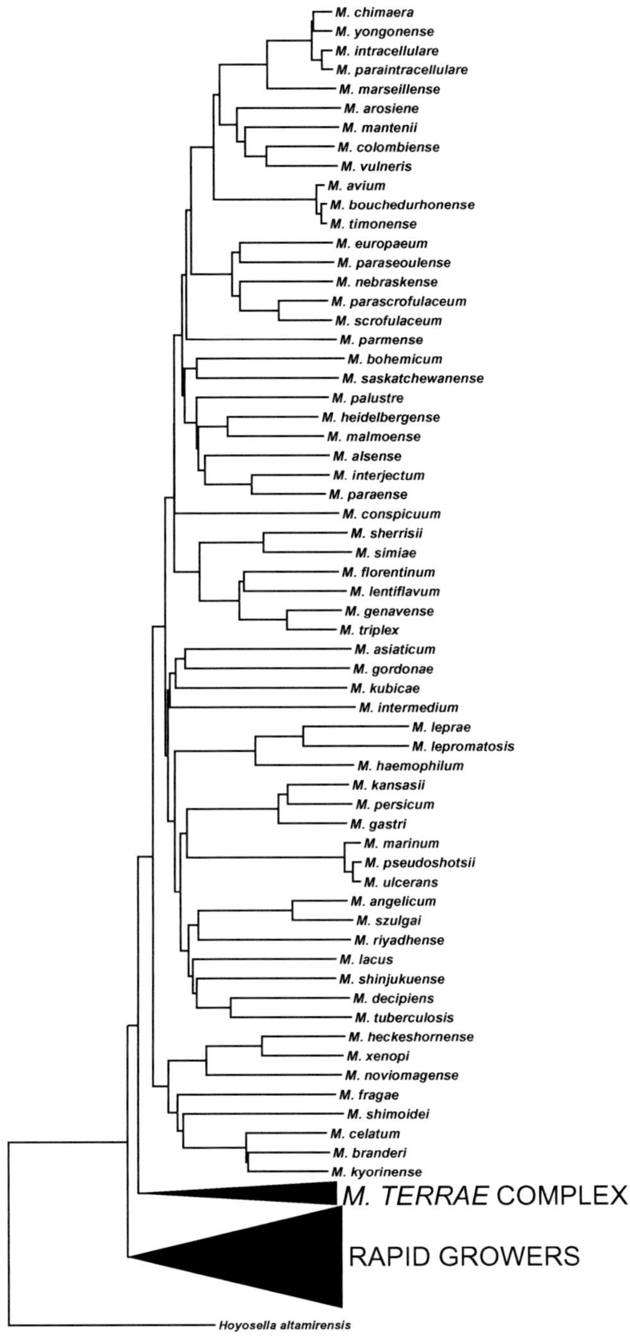

FIGURE 1.4 Phylogenetic tree of slow growers (excluding *M. terrae* complex) inferred with the maximum likelihood algorithm on 10,878 ANI-divergence scores (Tortoli, 2017), bootstrapped 500 times. The tree was rooted with *Hoyosella altamirensis* as outgroup. Bar, 2 units difference of ANI-divergence values.

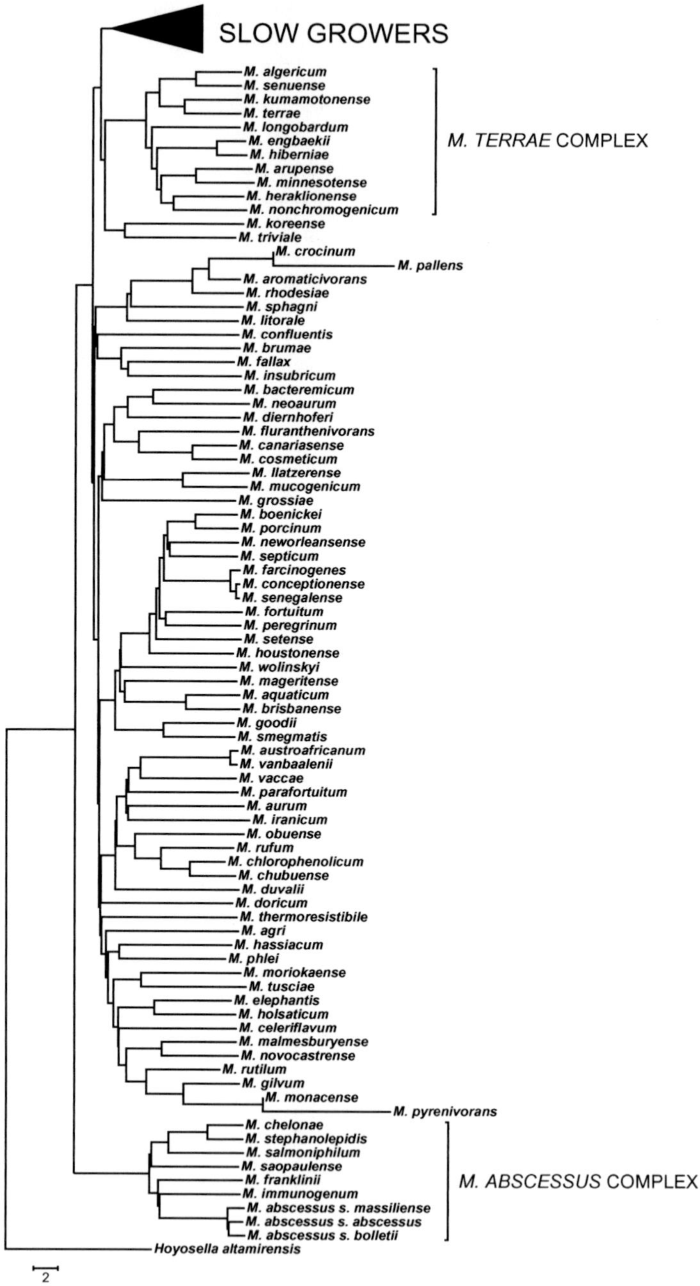

FIGURE 1.5 Phylogenetic tree of *M. terrae* complex and rapid growers inferred with the maximum likelihood algorithm on 10,878 ANI-divergence scores (Tortoli, 2017), bootstrapped 500 times. The tree was rooted with *Hoyosella altamirensis* as outgroup. Bar, 2 units difference of ANI-divergence values.

Several major traits characterize the phylogenetic scenario emerging from whole genome data. Primordial mycobacteria were rapid growers more closely related to the *M. abscessus-chelonae* complex. The evolution toward slow growers, likely associated with the acquisition of a 14-nucleotide insertion in the helix 18, leads to the species related to the present *M. terrae* complex. The deletion of two nucleotides in the helix 18 made the 12-nucleotide insertion the marker of the most successful slow growers. In contrast with previous knowledge, the lost of the whole 12-nucleotide insertion is not a specific marker of *M. simiae* complex. It is actually shared by a number of paraphyletic species.

Interestingly, the present pathogenic species and most of those frequently responsible of opportunistic diseases in humans appear to have evolved from a common ancestor. The latter hypothesis is not new, it was in fact hypothesized on the basis of the presence, in many such species, of genes related with virulence (van Ingen et al., 2012).

A recent paper (Gupta et al., 2018), published after the completion of present review, redraws in deep the classification of mycobacteria. The Authors split the genus *Mycobacterium* in five genera basing on molecular markers consisting, either in insertions/deletions of amino acids, or in proteins exclusively found in evolutionarily-related groups of species. Far from distorting the aforesaid phylogenetic reconstruction, the new genera exactly overlap the major clades of classical taxonomy: *M. abscessus* complex, other rapid growers, *M. terrae* complex and *M. triviale-related species*.

REFERENCES

Barbu, E., Lee, K.Y., Wahl, R., 1956. Contenu en bases puriqniue et pyrimidiniques des acides deoxyribonucleiques bacteriques. Ann. Inst. Pasteur. (Paris) 91, 212–224.

Böddinghaus, B., Rogall, T., Flohr, T., Blöcker, H., Böttger, E.C., 1990. Detection and identification of mycobacteria by amplification of rRNA. J. Clin. Microbiol. 28, 1751–1759.

Brennan, P.J., Nikaido, H., 1995. The envelope of mycobacteria. Annu. Rev. Biochem. 64, 29–63.

Chan, J.Z., Halachev, M.R., Loman, N.J., Constantinidou, C., Pallen, M.J., 2012. Defining bacterial species in the genomic era: insights from the genus *Acinetobacter*. BMC. Microbiol. 12, 302.

Devulder, G., Perouse, D.M., Flandrois, J.P., 2005. A multigene approach to phylogenetic analysis using the genus *Mycobacterium* as a model. Int. J. Syst. Evol. Microbiol. 55, 293–302.

Fedrizzi, T., Meehan, C.J., Grottola, A., Giacobazzi, E., Fregni Serpini, G., Tagliazucchi, S., et al., 2017. Genomic characterization of nontuberculous mycobacteria. Sci. Rep. 7, 45258.

Gevers, D., Cohan, F.M., Lawrence, J.G., Spratt, B.G., Coenye, T., Feil, E.J., et al., 2005. Opinion: re-evaluating prokaryotic species. Nat. Rev. Microbiol. 3, 733–739.

Gupta, R.S., Lo, B., Son, J., 2018. Phylogenomics and comparative genomic studies robustly support division of the genus Mycobacterium into an emended genus Mycobacterium and four novel genera. Front. Microbiol. 9, 67.

Lehmann, K.B., Neuman, R., 1896. Atlas und Grundriss der Bakteriologie und Lehrbuch der speziellen bakteriologischen Disagnostik. Lehmann, J.F., Munich.

Linnaeus, C., 1735. Systema naturae, sive regna tria naturae systematice proposita per classes, ordines, genera et species. Apud Theodorum Haak, Leiden.

Mignard, S., Flandrois, J.P., 2008. A seven-gene, multilocus, genus-wide approach to the phylogeny of mycobacteria using supertrees. Int. J. Syst. Evol. Microbiol. 58, 1432–1441.

Rogall, T., Wolters, J., Flohr, T., Böttger, E.C., 1990. Towards a phylogeny and definition of species at the molecular level within the genus *Mycobacterium*. Int. J. Syst. Bacteriol. 40, 323–330.

Sanger, F., Nicklen, S., Coulson, A.R., 1977. DNA sequencing with chain-terminating inhibitors. Proc. Natl. Acad. Sci. U.S.A. 74, 5463–5467.

Stackebrandt, E., Ebers, J., 2006. Taxonomic parameters revised: tarnished gold standards. Microbiol. Today 33, 152–155.

Stahl, D.A., Urbance, J.W., 1990. The division between fast- and slow-growing species corresponds to natural relationship among the mycobacteria. J. Bacteriol. 172, 116–124.

Tortoli, E., 2003. Impact of genotypic studies on mycobacterial taxonomy: the new mycobacteria of the 1990s. Clin. Microbiol. Rev. 16, 319–354.

Tortoli, E., 2012. Phylogeny of the genus *Mycobacterium:* many doubts, few certainties. Infect., Genet. Evol. 12, 827–831.

Tortoli, E., Fedrizzi, T., Meehan, C.J., Trovato, A., Grottola, A., Giacobazzi, E., et al., 2017. The new phylogeny of the genus *Mycobacterium:* the old and the news. Infect. Genet. Evol. 56, 19–25.

van Ingen, J., Boeree, M.J., van Soolingen, D., Iseman, M.D., Heifets, L.B., Daley, C.L., 2012. Are phylogenetic position, virulence, drug susceptibility and in vivo response to treatment in mycobacteria interrelated? Infect. Genet. Evol. 12, 832–837.

Watson, J.D., Crick, F.H., 1953. Molecular structure of nucleic acid: a structure for deoxyribose nucleic acid. Nature 171, 737–738.

Wayne, L.G., Brenner, D.J., Colwell, R.R., Grimont, P.A.D., Kandler, O., Krichevsky, M.I., et al., 1987. Report of the ad hoc committee on reconciliation of approaches to bacterial systematics. Int. J. Syst. Bacteriol. 37, 463–464.

Chapter 2

Identification of Nontuberculous *Mycobacterium*: Conventional Versus Rapid Molecular Tests

Ali Akbar Velayati, Parissa Farnia and Shima Saif
Mycobacteriology Research Centre (MRC), National Research Institute of Tuberculosis and Lung Disease (NRITLD), Shahid Beheshti University of Medical Sciences, Tehran, Iran

SPECIES CONCEPT IN *MYCOBACTERIUM*

Mycobacteria represent a very ancient genus of bacteria. They have been around for millions of years and have adapted to nearly every environment on Earth, for example, soil, water, dust, and air (Falkinham, 1996, 2009, 2010, 2011). The study of mummified ibises and cats from Egyptian tombs has provided valuable evidence of their antiquity (Daniel, 2006). The evidence of these organisms was also reported in a New World mummy from Peru, dated AD 700 (Rastogi et al., 2001). The genus *Mycobacterium* encompasses a number of medically important species that cause chronic pulmonary and nonpulmonary diseases. The chronic pulmonary diseases were recognized in Mediterranean culture during the 16th century, and initially termed phthisis by Sylvius in 1979. The generic name of *Mycobacterium* was initially designated to the way that these organisms grew in a mold-like fashion on liquid media (Lehmann and Neumann, 1896). They belong to the family Mycobacteriaceae within the order Actinomycetales (Table 2.1). For many years, only three species of *Mycobacterium*, that is, *M. tuberculosis, M. bovis*, and *M. leprae* were recognized as human pathogens Barksdale and Kim (1977). The main characteristics of mycobacteria are absence of motility, the morphology of the bacilli (slightly curved and rod shaped), and characteristic resistance to acid–alcohol following coloration with phenicated fuchsin (Ziehl–Neelsen stain) (Ziehl, 1882; Zopf, 1883; Velayati and Farnia, 2016). The morphology of mycobacterial colonies on solid media is a stable character of a given strain, though variants often arise by spontaneous

Nontuberculous Mycobacteria (NTM). DOI: https://doi.org/10.1016/B978-0-12-814692-7.00002-4

TABLE 2.1 Classification of *Mycobacterium*

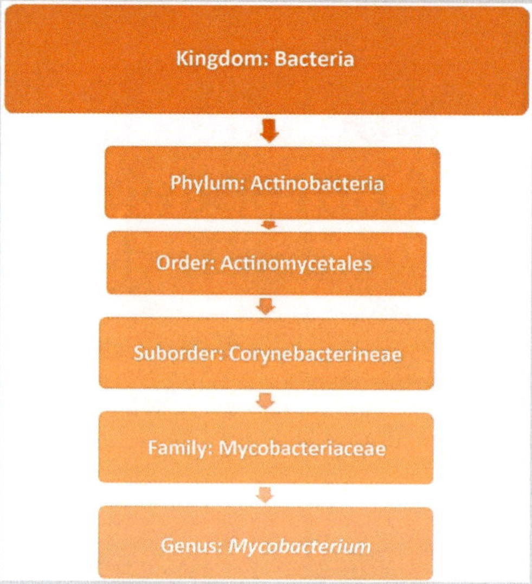

mutations (Fregnan and Smith, 1962; Vestal and Kubica, 1966). The basic types of colonies, are "rough" and "smooth"—the meaning of the description being self-evident (Vestal and Kubica, 1966; Vincent, 1994).

Classification of *Mycobacterium*

The identification of the genus *Mycobacteria* began in the late 1800s with the discovery of the leprosy bacillus (originally named *Bacillus leprae*) (Hansen, 1880) and the tubercle bacillus (originally named *Bacterium tuberculosis*) (Zopf, 1883). But their classification began in 1896 when Lehmann and Neumann placed the genus *Mycobacterium* in the Mycobacteriaceae family, Actinomycetales order, and Actinomycetes class (Lehmann and Neumann, 1896). In that early period, classification was based on one or more characters or patterns of characters which all members of the group have and which members of other groups do not have (Wayne et al., 1971; Wayne, 1981). The defined characteristics were morphology (rod-shaped, nonmotile), and acid–alcohol fastness, which led to difficulties in clearly distinguishing members of the genus *Mycobacterium* from closely related genera, that is, *Nocardia*, *Rhodococcus*, and *Corynebacterium*. Therefore, additional criteria were developed for differentiating these genera; including gram staining, growth rate, and mycolic acid profiles (Wayne et al., 1971; Wayne, 1981). In 1935, Pinner was among the first scientists who applied the term “atypical acid-fast microorganisms” to strains that caused human diseases but could not be differentiated from *M. tuberculosis* (Pinner, 1935). In 1949 Cuttino and McCabe described a fatal disseminated infection in a child caused by a newly recognized microorganism that they named “*Nocardia intracellularis*,” later on renamed as *M. intracellulare* (Cuttino and McCabe, 1949; Runyon, 1959; Runyon, 1965). In another study, Norden and Linel (1951) reported granuloma caused by “*M. balne*,” which was later on recognized as synonymous with the fish pathogen described by Aronson in 1926 (Norden and Linell, 1951; Aronson, 1926). In 1956, Masson and Prissick described a pigmented *Mycobacterium* that caused cervical adenitis in children which was named *M. scrofulaceum* (Prissick and Masson, 1956). Then, *M. kansasi* (1953) was reported as human pulmonary disease by Buhler and Pollak. Based on a number of reported cases that caused mycobacterial human pathogens (Buhler and Pollak, 1953), Timpe and Runyon used the term “atypical” that Pinner had used to described clinical isolates that did not appear to be the human tubercle bacilli (Runyon, 1959, 1965; Leao et al., 2004) (Table 2.2).

Furthermore, in 1957 Runyon EH, classified members of *Mycobacterium* based on their growth, pigmentation, and clinical significance for groups I, II, and III including only slowly growing mycobacteria, that is, organisms which require more than 1 week to grow, whereas group IV included rapidly growing mycobacteria which require 1 week or less for culture. Runyon group I includes photochromogenic species, in which colonies acquire pigmentation in the presence of light only (Velayati and Farnia, 2016; Runyon, 1959, 1965). Group II includes scotochromogens species that are pigmented in the presence or absence of light. Group III comprises nonchromogenic species. This classification system, however, has limits. For example, members of the *Mycobacterium avium–intracellulare* complex are nonphotochromogens in Runyon’s scheme, but some isolates produce slightly pigmented colonies, potentially causing incorrect classification as scotochromogen (Daniel, 2006; Rastogi et al., 2001; Runyon, 1959, 1965).

TABLE 2.2 Kubica’s Taxonomic Scheme of Mycobacteria

Group I	*M. kansasii, M. marinum*
Group II	*M. scrofulaceum, M. gordonae, M. flavescens*
Group III	*M. xenopi, M. avium, M. intracellulare, M. gastri, M. terrae, M. trivale*
Group IV	*M. fortuitum, M. vaccae, M. abscessus, M. smegmatis, M. phlei*
Principal human pathogens	*M. tuberculosis, M. bovis, M. africanum*

Sneath (1968, 1970) approved that many cultures of *Mycobacterium* have been mislabeled, so great care must be taken to check them against their original description (Sneath, 1968, 1970). He showed by comparing strains that it is difficult to compensate for differential growth rate. It has been repeatedly observed that particular slow-growing strains are scored as giving a negative response, but give positive responses upon longer incubation. In a comparison of slow and fast-growing strains by numerical analyses, the slow-growing strains are usually clustered together, based upon their negative responses (Sneath, 1968, 1970). Then he proposed the terms *vigor* and *pattern* in order to compensate the growth variation (Sneath, 1968, 1970). The separation of the genus into two major groups on the basis of the growth rate of the individual species forms the basis of mycobacterial taxonomy.

Rapidly growing mycobacteria are common saprophytes in natural habitats, but have received much less attention than the clinically more relevant slow-growing species (Hartmans and de Bont, 1992). In this regard, Tsukamura (1984) highlights the term "nonpathogens" that apply to rapid grower *Mycobacterium* that may not really exist. Wayne et al. (1971) carried out a cooperative numerical taxonomic analysis of slowly growing scotochromogenic mycobacteria of Runyon's group II (Wayne et al., 1971; Wayne, 1981). These workers defined four clusters; three of the clusters corresponded to the named species (*M. flavescens, M. gordonae*, and *M. scrofulaceum*). The fourth cluster may represent a new species (Runyon, 1959, 1965). Cross and Goodfellow (1973) proposed a similar, but more extensive grouping; they separated *M. leprae* from *Mycobacterium* genus. Additionally, based on pigment production, he subdivided the rapid grower *Mycobacterium* into nonpigmented mycobacteria and scotochromogenic mycobacteria. This division had no formal taxonomic standing, but was useful clinically and in identification schemes (Table 2.3).

Numerical classifications, which involve screening many strains for a large number of characteristics, have been most effective in determining relationships at the subgeneric level. Beginning with the work of Bojalil et al. (1962), much has been done to construct the definitive numerical taxonomies of mycobacteria. In this system, each property or characterization of stain is assigned an equal weight and strains are grouped according to the extent of shared properties or characteristics of that strain (Sneath, 1968; Bojalil et al., 1962).

As expected, the discriminatory power of the approach, as well as the reliability and significance of the clustering of strains, increases as the number of independent characteristics measured increases. In general, it provided a thorough description of each species and facilitated the discovery of tests that could reliably distinguish the various species (Sykes and Skinner, 1973; Shinnick and Good, 1994). A limitation of numerical taxonomy is that it measures phenotypic similarities which do not necessarily reflect genetic relatedness (Sykes and Skinner, 1973; Shinnick and Good, 1994). Kubica (1973) has formulated a simplified taxonomic scheme that facilitates

TABLE 2.3 Cross and Goodfellow System of Classification (1973)

Group	Name of Species
M. Leprae and other noncultivable species (e.g., *M. lepraemurium*)	
M. tuberculosis (also include *M. bovis, M. microti*)	
Photochromogenic mycobacteria	*M. kansasii, M. marinum*
Slowly growing scotochromogenic mycobacteria	*M. scrofulaceum, M. marianum, M. gordonae, M. flavescens*
Nonchromogenic slow-growing niacin-negative mycobacteria	*M. avium, M. intracellulare, M. terrae, M. gastri, M. ulcerans*
Rapidly growing nonpigmented mycobacteria	*M. fortuitum, M. abscessus, M. smegmatis, M. diernhoferi, M. chitae*
Rapidly growing scotochromogenic mycobacteria	*M. gilvum, M. obuense, M. thermoresistibile, M. vaccae, M. duvalii,M. flavescens, M. phlei, M. rhodesiae*

identification of most clinically important mycobacteria recovered from humans (Kubica, 1973). The scheme is based on growth rate and 11 other simple tests selected from 82 taxonomic characters (Bradley and Bond, 1974; Kubica, 1973; Bradley and Bond, 1974) (Table 2.2). The selected tests for the abbreviated scheme were: niacin production; nitrate reduction; catalase production; photochromogenicity; tween hydrolysis; tellurite reduction; NaCl tolerance; arylsulfatase activity, and growth on MacConkey medium (Kubica, 1973; Bradley and Bond, 1974; Runyon et al., 1959; Lutz, 1992; Kubica and Vestal, 1961; Kubica et al., 1972; Kent and Kubica, 1985).

Although, as the numbers of strains, species, and properties studies increased, the assimilation of a large amount of phonetic information into a coherent picture of relatedness became more difficult (Kent and Kubica, 1985). To facilitate the analysis of relatedness of strains and species, an International Working Group on Mycobacterial Taxonomy (IWGMT) was established in 1967 (Wayne et al., 1989, 1991). Through the concerted efforts of IWGMT and individual investigators, the *Mycobacterium* genus has become one of the best-described genera. For example, it become clear that numerical taxonomy does not necessarily reflect genetic relatedness of species or subspecies. And more reproducible tests like DNA–DNA hybridization should be adapted (Wayne and Kubica, 1986; Hill et al., 1972). Where investigated, clusters identified by numerical taxonomy agreed reasonably well with species defined by DNA–DNA hybridization, but the relationships of clusters to one another predicted by numerical taxonomy did not agree well with the results of the DNA–DNA hybridization analysis in all instances (Shinnick and Good, 1994; Goodfellow and Wayne, 1982). Bradley and Bond (1974) review some of the basic principles of measuring DNA homology and its significance and application in mycobacterial classification. Hill et al. (1972) and Baess (1974) have detected carbohydrates in DNA preparations and suggested that the presence of a DNA–polysaccharide complex which interfered with adsorption at 260 nm in the spectrophotometer. In 1980, the international code of nomenclature of the bacteria has published the names of 41 *Mycobacterium* species. Goodfellow and Wayne (1982) have updated the description of the genus in their review: "Mycobacteria are aerobic, acid–%alcohol fast actinomycetes that usually produce slightly curved or straight rods and coccoid elements; aerial hyphae are usually absent (Tables 2.2 and 2.3). Strains have meso-diamino-pimelic acid, arabinose and giTTF lose in the peptidoglycan, produce straight-chain unsaturated fatty acids, contain complex mixtures of mycolic acids (C-60 to C-90) and dihydrogenated menaquinones with nine isoprene units, and the G + C content of their DNA ranges from 62% to 70%" (Goodfellow and Wayne, 1982).

Kazda (1983) classified the mycobacteria into four groups on the basis of ecologically relevant properties (Table 2.4). Kubica and Wayne, in 1984, included three new species of *Mycobacterium* (*M. chelonae* subsp. chelonae; *M. chelonae* subsp. *abscessus*, *M. sphagni*) which bring the number of species to 44 (Kubica and Wayne, 1984). By the end of 1984, there were 52 described species (Table 2.5). In the early 1999s, several studies were published which covered the phylogenetic structure of the fast- and slow-growing mycobacteria (Smida et al., 1988; Rogall et al., 1990; Stahl and Urbance, 1990; Wayne and Diaz, 1985). These data supported the clustering of Mycobacteria according to their growth behavior and indicated that the mostly clinical, slow-growing strains evolved from their fast-growing relatives (Pitulle et al., 1992; Wayne and Diaz, 1985).

More recently, the analysis of large information-bearing molecules such as nucleic acids has been used to probe the genetic relatedness of bacteria. For example, immunologic cross-reactivity of catalases has been used as a measure of genetic relatedness among mycobacteria (Wayne and Lin, 1982; Wayne and Diaz, 1985), and nucleotide sequences of ribosomal RNA genes have been used extensively to construct phylogenetic trees (Stackebrandt et al., 1987). The basic idea that makes ribosomal RNA genes particularly useful as "molecular evolutionary clocks" are: (1) rRNA genes are ubiquitous among the bacteria and eukaryotes; (2) rRNA functions have been highly conserved over extremely long evolutionary times; (3) rRNA sequences are sufficiently long to contain a large amount of phylogenetic information; and (4) rRNA molecules display extensive secondary structure which helps align the nucleotide sequences and identify differences (Shinnick and Good, 1994). Phylogenetic trees of the genus *Mycobacterium*

TABLE 2.4 Kazda Classification Based on Ecologically Relevant Properties

Classification	Species
Obligate pathogens (unable to multiply outside living beings)	*M. tuberculosis* *M. bovis* *M. africanum* *M. asiaticum* *M. malmoense* *M. microti* *M. simiae* *M. szulgai* *M. haemophilum*
Facultative pathogens	*M. leprae* *M. paratuberculosis* *M. ulcerans*
Potential pathogens	*M. avium* *M. chelonae* *M. fortuitum* *M. intracellulare* *M. kansasii* *M. marinum* *M. senegalense* *M. scrofulaceum* *M. xenopi*
Saprophytic species	*M. gordonae* *M. nonchromogenicum* *M. triviale* *M. terrae* *M. gastri* Most of rapid grower species

have been constructed using the sequence of 16S rRNA genes (Rogall et al., 1990; Pitulle et al., 1992) and, as expected, all species of *Mycobacterium* are closely related to each other and distantly related to members of other genera (Shinnick and Good, 1994). At present, more than 150 recognized or proposed species meets the standard for inclusion into the genus of *Mycobacterium* (Table 2.6). These species are usually grouped into two major divisions, "rapidly growing" and "slowly growing," based on the time required for visible colonies to appear on a solid medium after plating sufficiently dilute suspensions to give well-separated colonies (i.e., presumably colonies arising from single cells). The appearance of colonies requires <7 days for the rapidly growing species and >7 days for the slowly growing species (Tables 2.6 and 2.7) (Daniel, 2006; Rastogi et al., 2001; Kent and Kubica, 1985; Kubica and Wayne, 1984) (Tables 2.6 and 2.7).

TABLE 2.5 The List of *Mycobacterium* Published by Kubica and Wayne (1984)

Taxon	Growth Rate	Taxon	Growth Rate
M. africanum	Rapid	*M. malmoense*	Slow
M. asiaticum	Slow	*M. marinum*	Slow
M. aurum	Rapid	*M. microti*	Slow
M. avium	Slow	*M. nonchromogenicum*	Slow
M. bovis	Slow	*M. neoaurum*	Rapid
M. chelonae	Rapid	*M. parafortuitum*	Rapid
M. chitae	Rapid	*M. paratuberculosis*	Slow
M. duvalii	Rapid	*M. phlei*	Rapid
M. farcinogenes	Rapid	*M. scrofulaceum*	Slow
M. flavescens	Rapid	*M. senegalense*	Rapid
M. fortuitum	Rapid	*M. simiae*	Slow
M. gadium	Rapid	*M. smegmatis*	Rapid
M. gastri	Slow	*M. szulgai*	Slow
M. gilvum	Rapid	*M. terrae*	Slow
M. gordonae	Slow	*M. thermoresistibile*	Rapid
M. haemophilum	Slow	*M. triviale*	Slow
M. intracellulare	Slow	*M. tuberculosis*	Slow
M. kansasii	Slow	*M. ulcerans*	Slow
M. komossense	Rapid	*M. vaccae*	Rapid
M. leprae	Slow	*M. xenopi*	Slow
M. lepraemurium	Slow		

LABORATORY PROCEDURES FOR *MYCOBACTERIUM* IDENTIFICATION

The Runyon classification of nontuberculous mycobacteria (NTM) can be used as a guideline when choosing which appropriate identification tests have to be performed (WHO, 1998; Pitulle et al., 1992). Based on the rate of growth and production of pigment in the dark or light they are classified. Generally, it is recommended to perform the tests with active cultures over 2 weeks for the rapid growers and 3–4 weeks for the

TABLE 2.6 More Than 170 Recognized or Proposed Species Meet the Standard for Inclusion in the Genes of *Mycobacterium*

Growth at different temperatures (25°C, 30°C, 37°C, 42°C, 45°C)	Growth at different temperatures (25°C, 30°C, 37°C, 42°C, 45°C)
Pigmentation	Pigmentation
Biochemical tests	Biochemical tests
Heat-resistant catalase	Niacin production
Semiquantitative catalase	Nitrate reductase
Tween 80 hydrolysis	Iron uptake
Urease	Acid phosphatase
Niacin production	3-day arylsulfatase
Nitrate reductase	10-day arylsulfatase
Acid phosphatase	Pyrazinamidase
3-day arylsulfatase	α-Esterase
10-day arylsulfatase	β-Esterase
Pyrazinamidase	β-Galactosidase
α-Esterase	Degradation of *p*-aminosalicylate
β-Esterase	Acetamidase
β-Galactosidase	Isonicotinamidase
Mycolic acid profile	Allantoinamidase
Mycolic acid pyrolysis esters	Succinamidase
DNA–DNA hybridization analysis	Production of acid from L-arabinose, xylose, or dulcitol
16S rRNA sequence	Utilization of xylose, oxalate, citrate, sorbitol, mannitol, or trehalose
	Mycolic acid profile
	Mycolic acid pyrolysis esters
	DNA–DNA hybridization analysis
	16S rRNA sequence

slow growers (Lutz, 1992; Kent and Kubica, 1985; Pitulle et al., 1992; World Health Organization, 1998). Cultures over more than 5 weeks of growth is inappropriate. The minimum number of colonies to perform an identification test is 20–25; if it is low in number, it is advisable to make a subculture, and then perform the tests. The primary steps for identification

TABLE 2.7 Clinically Important Slow-Grower NTM; Nonphotochromogens Group

	Temp. Range
M. avium complexes	22–45
M. xenopi	35–45
M. shimoidei	30–45
M. gastri	25–40
M. terrae complex	22–37
M. malmoense	22–37
M. triviale	22–37
M. genavense	24–45

is to confirm acid-fast bacilli in the appeared colonies, which can be performed by Ziehl–Neelsen or Kinyoun staining (Tsukamura, 1984; Metchock et al., 1999). Positive and negative controls should always be used when performing identification tests.

Primary Steps in Identification of *Mycobacterium* (Table 2.7)

1. *Growth rate*: The *Mycobacterium* grouped with slow-growing if it takes more than 7 days of incubation for mature colonies to be visible to the unaided eye. Rapid growers usually mature within 7 days (Lutz, 1992; Kent and Kubica, 1985; Pitulle et al., 1992; World Health Organization, 1998).
2. *Photoinduction*: Based on pigment production, the mycobacteria are divided into three main groups: (a) photochromogens: cultures whose colonies are nonpigmented when grown in the dark, but become pigmented after they are exposed to light for one or more hours; (b) scotochromogens: cultures whose colonies produce pigment when grown either in the dark or the light; and (c) nonphotochromogens: cultures whose colonies are nonpigmented or possess only pale shades of pigment and whose color is unaffected by exposure to light (Fig. 2.1) (Pitulle et al., 1992; World Health Organization, 1998).
3. *Growth at Different Temperatures*: There are three main temperatures for *Mycobacterium* growth; at 25°C, 37°C, and at 45°C (Daniel, 2006; Rastogi et al., 2001; Pitulle et al., 1992; World Health Organization, 1998).

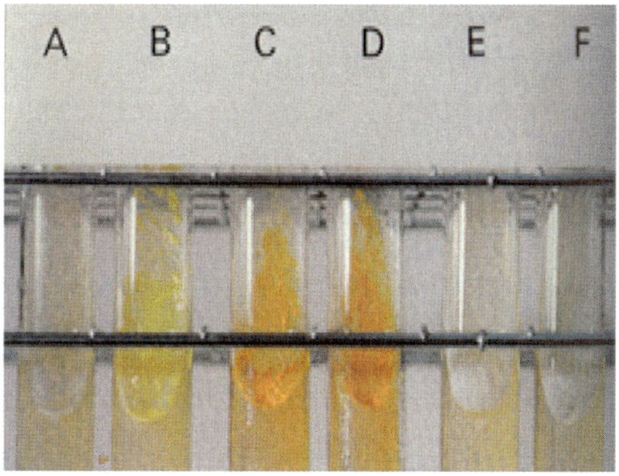

FIGURE 2.1 Photoinduction test.

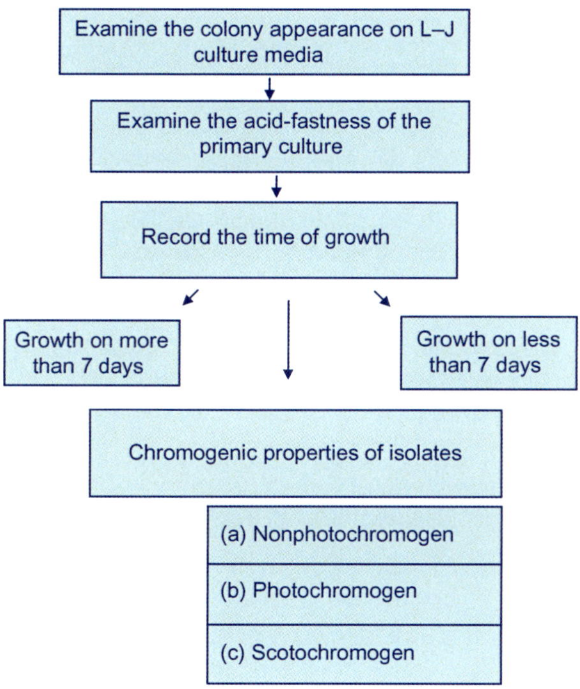

Flowchart (1) of Primary steps in phenotypic identification of *Mycobacterium*.

Photo Induction Procedure

1. First inoculate three culture tubes of agar- or egg-based medium with a suspension of bacteria that will yield isolated colonies when grown. Shield two tubes from light by wrapping with aluminum foil. Incubate all

tubes at 35–37°C. After the growth is observed on the surface of unshielded culture tubes, remove the shield from one of the culture tubes and record any pigmentation. Expose it to light from a 60 W lamp at a distance of 20–25 cm for 1 h (be certain the cap of the tube is loosened during light exposure), replace shield (with cap still loose), and reincubate culture for 6–12 h, or overnight (Daniel, 2006; Rastogi et al., 2001; Pitulle et al., 1992; World Health Organization, 1998). Take all culture tubes (in light/in dark/and in dark vs light) and record the pigment production in them (Fig. 2.1).

Reading. Growth in tubes A are not pigmented, whereas under a 60 W lamp pigmentation appeared for 24, 48, and 72 h in tube B. These cultures are photochromogenic (Leao et al., 2004; Pitulle et al., 1992; World Health Organization, 1998). Growth is pigmented in tubes C and D in light and dark; these cultures are scotochromogenic. Growth nonpigmented in tubes E and F in light and dark are nonchromogenic (Fig. 2.1).

Niacin Test

Niacin or nicotinic acid plays an important role in the oxidation–reduction reaction that occurs during the mycobacterial mechanism. The accumulation of niacin in the culture medium which these organisms are growing occurred. A positive niacin test is preliminary evidence that a buff-colored, slow-growing, rough colony may be *M. tuberculosis*. The test can be performed manually or by using commercial strips (BBL_Difco) (WHO, 1998; Leao et al., 2004; Hartmans and de Bont, 1992; Shinnick and Good, 1994; Kubica, 1973; Bradley and Bond, 1974; Runyon et al., 1959; Pitulle et al., 1992).

Materials

4% aniline, 10% cyanogens bromide, distill water (DW) or 0.85% saline, 95% ethanol.

Preparation

4% aniline solution

To 96.0 mL of 95% ethanol, add 4 mL of colorless aniline. Store it in a brown bottle in the refrigerator. If the solution turns yellow, discard it.

10% cyanogen bromide

Dissolve 5 g cyanogen bromide in 50 mL DW. Store in a tightly capped brown bottle in the refrigerator.

0.85% saline

Dissolve 0.85 g sodium chloride in 100 mL DW. Sterilize by autoclaving.

Classical Procedure

1. To the culture slant, add 1.0 mL of DW or saline.
2. Place the tube in a slant position so that the fluid covers the entire surface of the medium. Allow the extraction to proceed for at least 15 min. Remove 0.5 mL of the extracted fluid and transfer to a fresh screwcap tubes. Add 0.5 mL of the aniline solution and then add 0.5 mL of the cyanogen bromide solution. Observe for the formation of a yellow color, which should develop immediately (Runyon et al., 1959; Lutz, 1992; Pitulle et al., 1992).

Reading

Observe the color of the liquid in the bottom of the tube against a white background (Fig. 2.2).

Paper Strip Procedure

Into 0.6 mL of the culture extract, insert the strip with identification end-up. Allow the test system to react at room temperature for 15–20 min. Observe the color of the liquid in the bottom of the tube (Pitulle et al., 1992).

Reading

Positive; observe the yellow color in the bottom of the tube against a white background. Please take a note that a yellow color only on strip is not considered positive, but may be due to oxidation of chemicals. Especially at the top of the strip (Fig. 2.3) (Kent and Kubica, 1985; Pitulle et al., 1992). Negative; no color change.

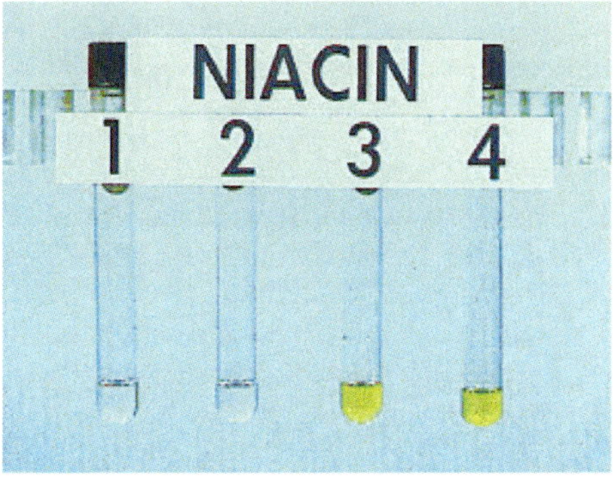

FIGURE 2.2 Niacin positive test is yellow in color.

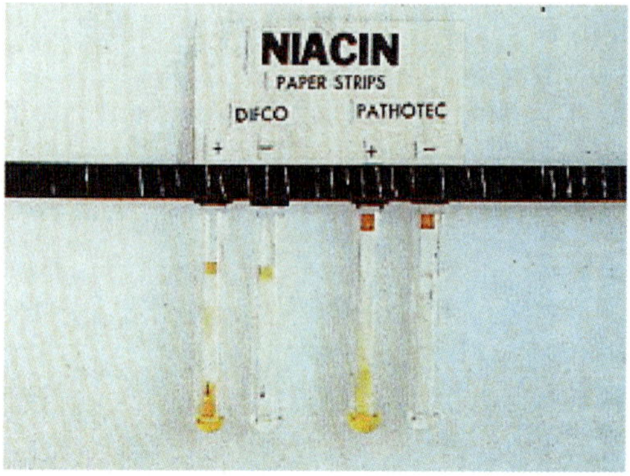

FIGURE 2.3 Niacin test using commercial strips.

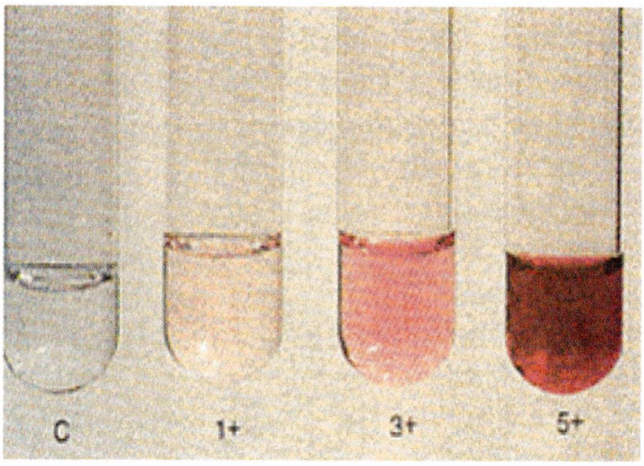

FIGURE 2.4 Nitrate positive test can vary from light pink to dark red.

Nitrate Reduction Test

All mycobacteria should be tested for nitrate-reducing ability about 3–4 weeks after colonies are visible (Fig. 2.4).

Materials

Screwcap test tubes (16 × 125 mm^2), water bath at 37°C, loop or applicator sticks, $NaNO_3$, KH_2PO_4, $Na_2HPO_4 \cdot 12H_2O$, HCL, sulfanilamide, *N*-naphthylethylenediamine dihydrochloride, distilled water (DW), zinc powder.

Preparation

1. Substrate 0.01 M sodium nitrate in 0.022 M phosphate buffer; PH = 7.0. Dissolve the following chemicals {$NaNO_3$(0.085 g) + KH_2PO_4(0.117 g) + $Na_2HPO_4 \cdot 12H_2O$(0.485 g)}, in 100 mL distilled water and sterilize

by autoclaving. Carefully add 50.0 mL concentrated HCL to 50.0 mL distilled water (Leao et al., 2004; Kubica, 1973; Pitulle et al., 1992; World Health Organization, 1998).

2. Dissolve 0.2 g sulfanilamide in 100.0 mL distilled water.
3. Dissolve 0.1 g *N*-naphthylethylenediamine dihydrochloride in 100.0 mL DW. Store the substrate and reagents in the dark under refrigeration.

Procedure

By using sterile loop or applicator stick, emulsify two spadesful of growth into the water. Add 2.0 mL of $NaNO_3$ substrate solution. Shake by hand and incubate in upright position in 37°C water bath for 2 h. Remove from the temperature bath. Add one drop of HCL reagent, two drops of Sulfanilamide reagent and add two drops of *N*-naphthylethylenediamine dihydrochloride reagent (Leao et al., 2004; Hartmans and de Bont, 1992; Pitulle et al., 1992).

Results

Positive: Red color development, which may range from positive pale pink (±) to deep red (5+).
Negative: No color development. Confirm true negative by adding a small amount of zinc powder to all the negatives tubes. The zinc catalytically reduces nitrate to nitrite, and therefore if there is $NaNO_3$ in the suspension, when adding zinc will immediately turn it to pink and proves it is a true negative.

Nitrite Test Strips for Detection of Nitrate Reduction

Remove two spadesful of growth from culture and add to the saline. Carefully insert a nitrite test strip so the bottom of the strip is inserted first. Hold the tube in a vertical position and incubate at 37°C for 2 h. After 2 h, carefully, be sure that liquid covers the strip.

Reading

Positive: Top portion of strip changing to light blue or dark blue (difco); central portion of strip changing to pink (General diagnostics).
Negative: No color change.

Catalase Test

Catalase is an intracellular, soluble enzyme capable of splitting hydrogen peroxide into water and oxygen, $2H_2O_2 \approx 2\ H_2O + O_2$. The presence of oxygen bubbles in the reaction mixture indicates catalase activity. Catalase is measured in two ways (Pitulle et al., 1992; World Health Organization, 1998; David, 1974):

1. By the ability of the catalase enzyme to remain active after heating, a measure of the heat stability of the enzyme (heat-stable). The quantitative differences in catalase activity and differences in heat stability detected at 68°C have proven to be useful taxonomic tools (Fig. 2.5).
2. By the relative activity of the enzyme, as determined by the height of a column of bubbles of oxygen formed by the action of untreated enzyme produced by mycobacteria (semiquantitative (SQ) catalase test) (Fig. 2.6).

Materials

Water bath at 68°C, loop or inoculating spade, 30% hydrogen peroxide, 10% tween 80, 0.067 M phosphate buffer, PH = 7.0.

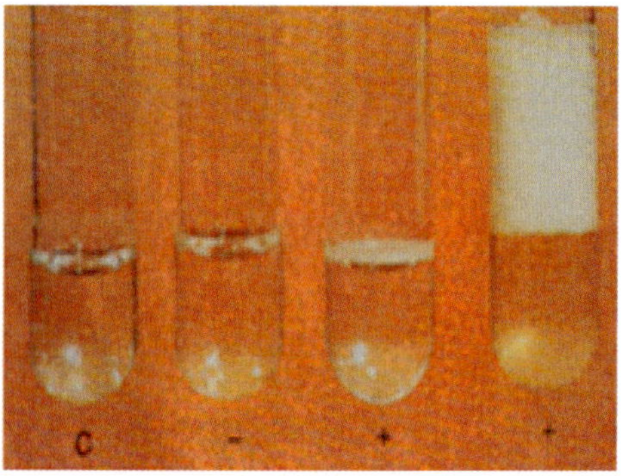

FIGURE 2.5 Heat-stable catalase (HS) test.

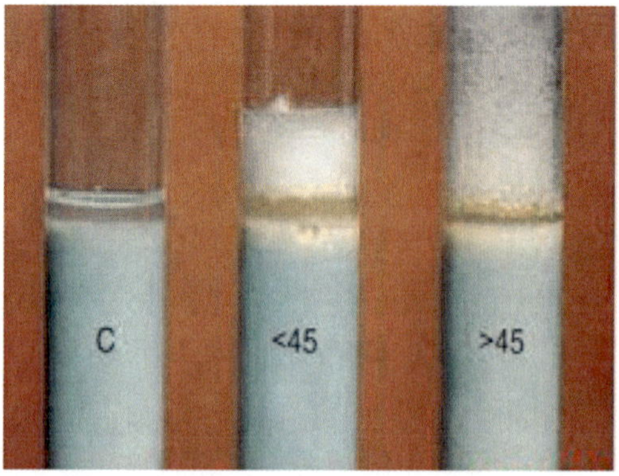

FIGURE 2.6 Semiquantitative catalase test.

Preparation

10% tween 80

Mix 10.0 mL of tween 80 with 90.0 mL DW. Autoclave for 10 min at 121°C. Immediately before performing the test, mix equal parts of superoxide and the tween 80 solution.

0.067 M phosphate buffer, PH 7.0

{Na_2HPO_4 (9.47 g/1000 mL DW); KH_2PO_4 (9.07/1000 mL DW)}.

Mix 61.1 mL of disodium phosphate with 38.9 mL of monopotassium phosphate. Check the PH.

Procedure 1 (Heat-Stable Catalase)

With a sterile pipette, dispense 0.5 mL of the phosphate buffer (PH = 7.0) into sterile screwcap test tubes. With a sterile loop or applicator stick, emulsify one spadeful of growth from the culture into a phosphate buffer. Place the tubes in 68°C for 20 min. Time and temperature is critical. Remove from heat and cool to room temperature. Add 0.5 mL of the tween-peroxide mixture. Recap tubes loosely.

Results

Positive: Formation of bubbles.
Negative: No bubbles.

Semiquantitative Catalase Test

Examine the 14-day-old Löwenstein–Jensen (LJ) culture media to ascertain that growth has occurred. (Instead of slant position, LJ culture media should be made in upright position to provide a butt or deep of solid medium.) Add 1 mL of the freshly prepared tween-peroxide mixture (as mentioned earlier), replace caps loosely and allow to stand at room temperature for 5 min. Measure the height of the foam column (Pitulle et al., 1992; David, 1974).

Results

Low catalase activity: less than 45 mm of foam.
High catalase activity: more than 45 mm of foam.

THIOPHEN-2-CARBOXYLIC ACID HYDRAZIDE

Susceptibility Test

This test is valuable for distinguishing *M. bovis* from *M. tuberculosis*. *M. bovis* is susceptible to low-concentration of thiophen-2-carboxylic acid hydrazide (TCH) (Leao et al., 2004; Kent and Kubica, 1985; Pitulle et al., 1992; World Health Organization, 1998).

Materials

LJ culture media or 7H-10 Middle brook.
Thiophen-2-carboxylic acid hydrazide.

Preparation

Prepare two batches of LJ culture media. One batch is drug-free control medium and the other one contained 10 μg/mL of TCH.

Procedure

Dilute a 7-day-old culture to 10^{-3} and 10^{-5} in sterile saline or water. Inoculate one control and one drug-containing medium with 100 μL of each dilution. Incubate for 3 weeks at 35°C with 10% carbon dioxide.

Reading

Resistant to TCH; growth in both media (with and without TCH).
Sensitive to TCH; growth only in the medium without TCH.

Tween Hydrolysis Test

The test medium contains tween 80 and the indicator is neutral red in a buffered neutral solution. Normally, neutral red is red at PH = 7.0, but when bound by lipids or by tween 80, the indicator dye takes on the amber or straw color which ultimately occurred at alkaline PH. When the bacterial esterases split the tween 80, it will no longer act to bind the indicator dye and the neutral red then reverts to its usual color at PH 7.0, which is red, and the test is called positive (Leao et al., 2004; Kent and Kubica, 1985; Pitulle et al., 1992).

Materials

Tween 80

0.067 M phosphate buffer, PH = 7. Aqueous neutral red, 0.1%, screwcap tubes, 16 × 125 mm.

Preparation

Phosphate buffer, pH = 7.0{Na_2HPO_4(0.067 M), 9.47 g/1000 mL DW (solution a), KH_2PO_4 (0.067 M), 9.07/1000 mL DW (solution b)}. To prepare 100 mL of PH = 7.0 buffer, mix 61.1 mL of solution "a" with 38.9 mL of solution "b." Check the PH of the final solution.

Substrate Medium

Add the following in order, to 100 mL of PH = 7.0 phosphate buffer; 0.5 mL of tween 80 and 2 mL of a 0.1% aqueous solution of neutral red. Dispense the substrate in 2 mL amounts into 16 × 125 mm^2 screwcap tubes. Autoclave at 121°C for 10 min. The substrate should be straw or amber color after autoclaving. Store in the dark at 4°C for no more than 2 weeks.

Procedure

Inoculate the substrate with a spadeful of test organisms from an actively growing young culture slant. Incubate at 35°C. Examine after 1, 5, and 10 days. Do not shake tubes while reading them.

Reading

Positive: Observe after 5 and 10 days. Change of color from amber to pink or red. Take note that it is necessary that the medium also changes in color. Negative: If colonies are red, but the medium remains amber, the test is reported as negative.

Control

M. kansasii server as a positive control (Fig. 2.7).

Urease Test

The ability of a culture to hydrolyze urea (releasing ammonia) is useful in identifying both scotochromogens and nonphotochromogens (Fig. 2.8)

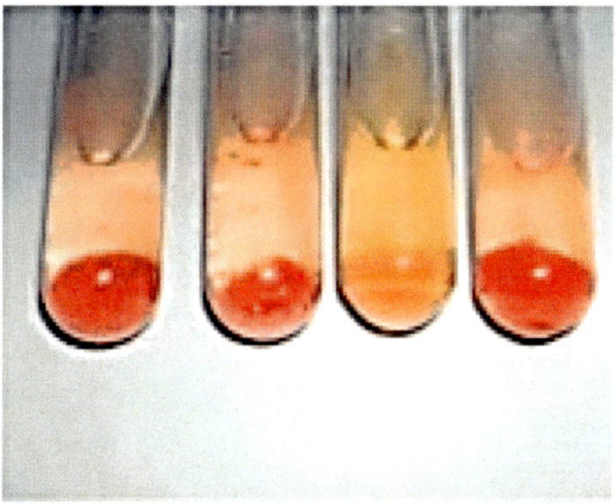

FIGURE 2.7 Tween hydrolysis tests.

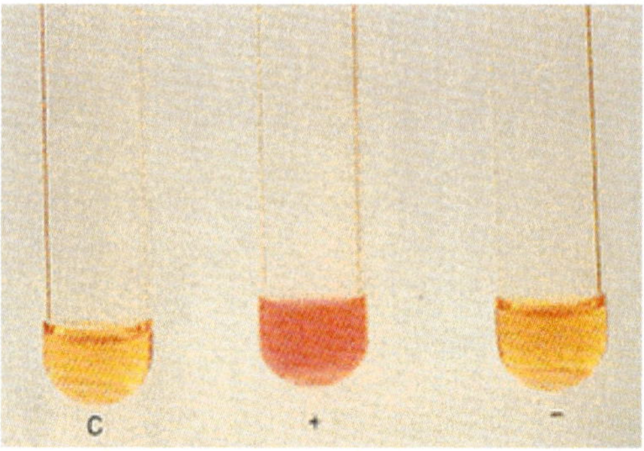

FIGURE 2.8 Urease test, the tube in the left is negative and in the center is positive.

(Leao et al., 2004; Kent and Kubica, 1985; Smida et al., 1988; World Health Organization, 1998).

Materials

Difco–bacto concentrate urea, distilled water, screwcap tubes (13 × 100 mm^2).

Preparation

Aseptically mix 1 part of Difco–bacto urea concentrate with 9 parts of distilled water. Dispense 3 mL amounts into sterile screwcap tubes.

Procedure

Inoculate the liquid substrate with a spadeful of young, actively growing culture slant. Incubate at 35°C. Observe for a color change.

Results

Positive: Color change from amber to pink or red.
Negative: No color change.

Controls

M. scrofulaceum or *M. bovis* as a positive control.

THREE- AND FOURTEEN-DAY ARYLSULFATASE TEST

Arylsulfatase is an enzyme capable of hydrolyzing the bond between the sulfate group and the aromatic ring structure in compounds with the general

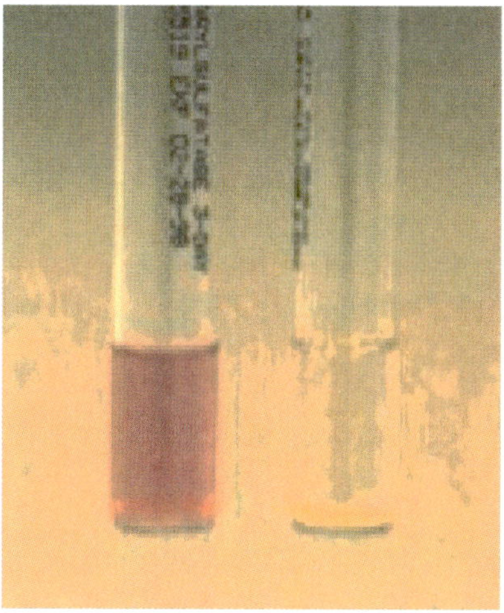

FIGURE 2.9 Three-day arylsulfatase test.

formula R-OSO_3H. When the substrate potassium phenolphthalein disulfate is incorporated into a growth medium, the presence of free phenolphthalein liberated by enzymatic hydrolysis is indicated by a red color that develops when alkali is added (Fig. 2.9) (Hartmans and de Bont, 1992; Kent and Kubica, 1985; Smida et al., 1988; Rogall et al., 1990; Stahl and Urbance, 1990; Wayne and Diaz, 1985; Stackebrandt et al., 1987; Pitulle et al., 1992; World Health Organization, 1998).

Materials

Phenolphthalein disulfate (tripotassium salt), Middle brook $7H_9$ broth or Dubos tween-albumin, albumin dextrose catalase enrichment, sodium carbonate, DW.

Preparation

Stock substrate solution

Dissolve 2.6 g of phenolphthalein disulfate (tripotassium salt), in 50 mL of DW to make a 0.08 M of stock solution. Sterilize by filtration through 0.22 μm pore size membrane filter and store at 5°C.

2 N sodium carbonate

Dissolve 10.6 g of an anhydrous, Na_2CO_3 in 100 mL of DW. Sterilize by filtration through 0.22 μm pore size membrane filter and store at room temperature.

Liquid medium

Use either Middle brook $7H_9$ or Dubos broth. Prepare two flasks, each with 200 mL of the sterile liquid medium. For the 3-day test medium: ascetically add 2.5 mL of the 0.08 M stock substrate to one flask of 200 mL of complete liquid medium to provide a 0.001 M substrate. For the 14-day test medium; aseptically add 7.5 mL of the 0.08 M stock substrate to the other flask of 200 mL of complete liquid medium to provide a 0.003 M substrate. Dispense each working solution in 2 mL amounts into screwcap test tubes and store at room temperature or in a refrigerator (Leao et al., 2004; Kent and Kubica, 1985).

Reading

Positive: Immediate color change to pink or red after addition of carbonate solution (compare results with a set of color standards and record intensity of color (− to 5+).
Negative: No color change.

Control

Positive: A tube of uninoculated substrate inoculated with *M. fortuitum.*

OXYGEN PREFERENCE TEST

The medium is middle brook $7H_9$ or Kirchner broth rendered semisolid by the addition of 0.1% pure agar and dispensed in 10 mL amounts in screw-capped bottles. Pipette 0.2 mL of the bacterial suspension about 1 cm below the surface of the medium and mix carefully, avoiding bubbles and aeration. Incubate at 37°C for 18 days. Aerobic strains grow at or near the surface, while microaerophilic strains grow as a band 10–20 mm below the surface, sometimes extending upward (Leao et al., 2004; Pitulle et al., 1992; Zhang et al., 1992).

Control

Microaerophilic: *M. bovis*/*M. africanum.*
Aerobic: *M. bovis.* B.C.G/*M. tuberculosis.*

Detection of Pyrazinamidase Activity

Middle brook $7H_9$ broth (1000 mL), pyrazinamide (100 mg), sodium pyruvate (2 g), agar (15 g). Dissolve by steaming. Dispense in 5 mL amounts in screw-cap bottles and sterilize by autoclaving at 115°C min. Allow to cool in the upright position so that butts rather than slopes are formed (Hartmans and de Bont, 1992; Pitulle et al., 1992; World Health Organization, 1998).

Procedure

Inoculate a heavy inoculum from afresh culture slant onto the surface of each of two tubes of medium. So that the inoculum is visible to the naked eye. Incubate at 37°C for 4 days. Remove one tube from incubator and add 1 mL of freshly prepared ferrous ammonium sulfate ($Fe(NH_4)_2(SO_4)_2$, $6H_2O$) 1% to the tube. Place the tubes in a refrigerator to prevent growth of contaminants. After 4 h, examine tubes for a pink band in the agar. To read hold the tubes against a white background and use bright room light. After 7-day incubation, remove the second tube from incubator and perform as with the first tube (if 4-day test is positive, it is not necessary to hold the second tube for 7-day test).

Reading

Positive: A pink band in the agar.
Negative: No pink band in the agar.

Control

Positive control = *M. tuberculosis*, *M. avium*, *M. marinum*.

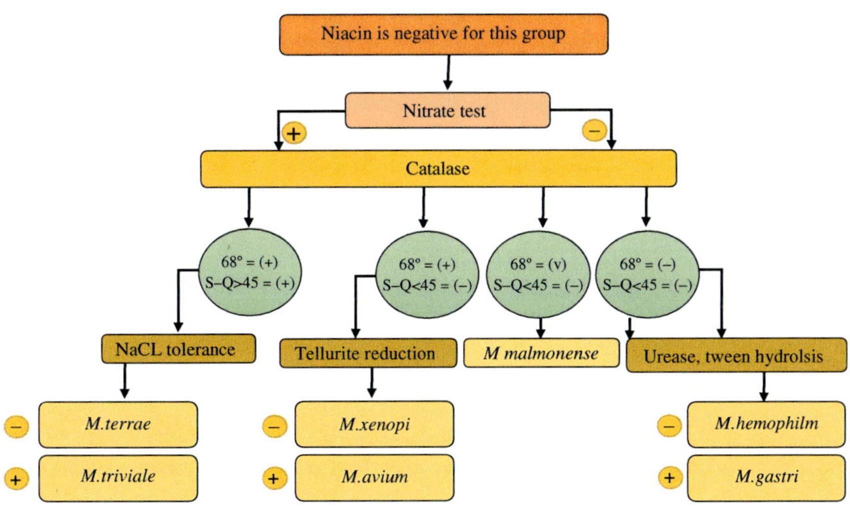

Flowchart (2) of identification of clinically important slow-grower NTM; nonphotochromogens group.

	Temp. Range
M. Kansasii	**25-40**
M. Marinum	**25-35**
M. simiae	**22-37**
M. asiaticum	**33-37**

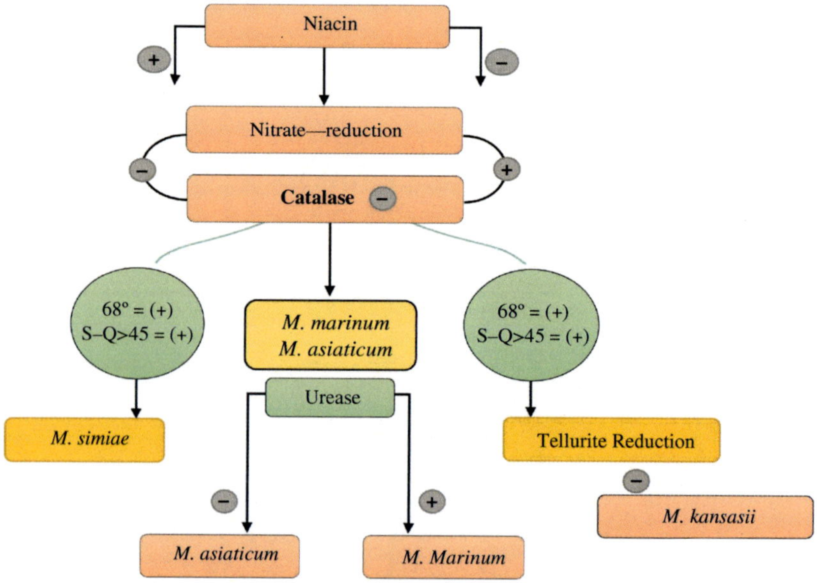

Flowchart (3) of identification of clinically important slow-grower NTM photochromogens group.

	Temp. Range
M.Scrofulaceum	**22-37**
M. szulgai	**22-37**
M.gordonae	**22-37**
M.flavescens	**25-42**

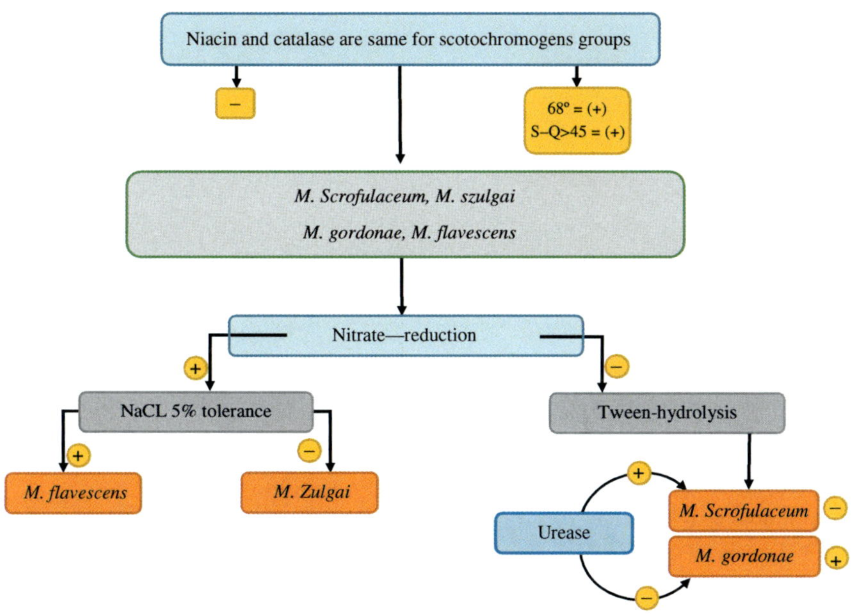

Flowchart (4) of identification of clinically important NTN; scotochromogens group.

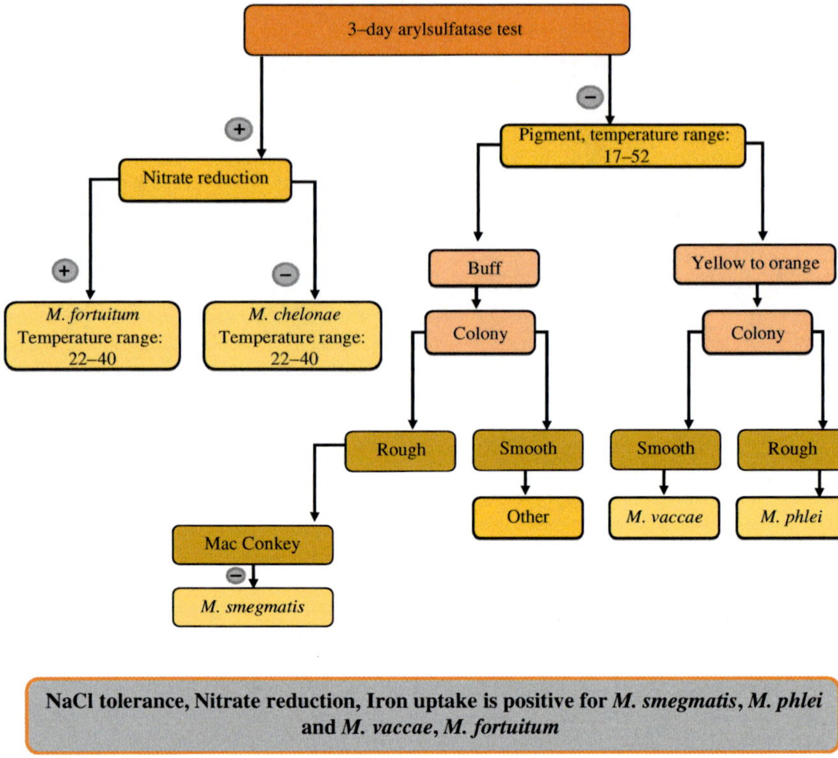

Flowchart (5) of identification for clinically important rapid grower NTM.

Disinfection of Nontuberculous Mycobacteria in Laboratory

Laboratory workers must know NTM are much more resistant to disinfectants than other bacteria and viruses. They are more resistant to chlorine, benzalkonium chloride, cetylpyridinium chloride, quaternary ammonium compounds, and phenolic- or glutaraldehyde-based disinfectants (Cole et al., 1990; Tiwari et al., 2003). The most important disinfectants for NTM are 70% alcohol and 5% phenol. Although, the effect depends on three main factors: (1) the concentration of the disinfectant; (2) the duration of disinfection; and (3) the concentration of mycobacteria in the solution or on the surface (World Health Organization, 1998). It is important to keep in mind that NTM, in particular rapid growing Mycobacteria such as *M. chelonae* or *M. fortuitum*, are more resistant to disinfection. Generally, any solution can be free of environmental mycobacteria by autoclaving. If the solution cannot be autoclaved, sterile water must be used to prepare the solution, and the final products must be sterilized by filtration (Lutz, 1992). Overall, as NTM is present everywhere in the environment, laboratory staff should be extremely careful and take all necessary precautions in order to avoid false positive cultures. For this reason, a high level of cooperation should exist between laboratory and clinicians (Lutz, 1992). A standard lab should have regular statistical analysis of test results with regular quality control tests.

Disinfectants

Discard jars; phenolic compounds (minimal concentration is 2% or alkaline glutaraldehyde minimal concentration is 2%), followed by autoclaving (World Health Organization, 1998).

Cleaning Bench Surface

Phenolic compounds (minimal concentration is 2%), alkaline glutaraldehyde minimal concentration is 2%), 70% alcohol, decontamination of BSC filters by formaldehyde gas.

Molecular Procedures for Identification of *Mycobacterium*

For a rapid diagnosis of *Mycobacterium*, molecular biology assay have been developed and applied since the 1980s (Daniel, 2006; Rastogi et al., 2001). Among these, polymerase chain reaction restriction fragment length polymorphisms (PCR-RFLP) and deoxyribonucleic acid sequencing based on 16sRNA had the most important applications (Rastogi et al., 2001). Using PCR-RFLP of *hsp65*, the internal transcribed spacer 16S–23S (ITS region) or 16rDNA involves amplification of specific base pair, followed by digestion with *Bst*EIII and *Hae*III (Devallois et al., 1997). The different resulting pattern permits proper identification (Devallois et al., 1997; Embley and Stackebrandt, 1994; Brunello et al., 2001; Hafner et al., 2004). The ITS 16S–23S region is more variable than other methods and allows further discrimination among organisms at the subspecies level (Boddinghaus et al., 1990; Cloud et al., 2002; Cook et al., 2003; Clarridge, 2004). The methodology used are from research work performed and standardized by Rastogi et al. (2001), Rogall et al. (1990), Woods (2002), Lutz (1992), World Health Organization (1998), Clarridge (2004), Thierry et al. (1990), Höfler (1994), Levy-Frebault and Portaels (1992), Telenti et al. (1993), Roth et al. (1998), Ringuet et al. (1999), Taylor et al. (1997), Sola et al. (2001), Tortoli et al. (2001), Tortoli (2006), Turenne et al. (2001), Van Soolingen et al. (1995).

Stock Solution and Buffers

Lysozyme (10 mg/mL)

Dissolve 10 mg lysozyme in mL distilled water. Store in small aliquots at −20°C for no longer than 1 year.

10% sodium dodecyl sulfate (SDS)

10 g SDS/100 mL DW; dissolve by heating at 65°C for 20 min. Do not autoclave. Store at room temp for no longer than 1 month.

1 × TE buffer

First prepare the following stock buffers.

1 M Tris—1 L

Dissolve 121.14 g tris(hydroxymethyl)aminomethane (tris; MW = 121.14 in 800 mL H_2O). Adjust PH to the desired value by adding concentrated HCL as given as follows:

1 M tris with PH 7.4 need approximately 70 mL of concentrated HCL.

1 M tris with PH 7.6 need approximately 60 mL of concentrated HCL.

1 M tris with PH 8.0 need approximately 42 mL of concentrated HCL.

Adjust volume to 1 L with H_2O. Sterilize by autoclaving and store at room temperature

0.5 M Na_2 EDTA (PH 8.0)—1 L

Dissolve 186.12 g disodium ethylenediaminetetraacetate-2 H_2O (Na_2 EDTA-2H_2O, MW = 372.24) in 800 mL H_2O; stir vigorously on a magnetic stirrer. Adjust to PH 8.0 with NaOH (± 20 g Na OH pellets) and adjust volume to 1 L with H_2O. Divide into aliquots and sterilize by autoclaving. Store at room temperature (remember the disodium salt of EDTA will only dissolve when the PH of the solution is adjusted to 8.0 by the addition of NaOH).

1 × TE buffer

Add 10 mL 1 M tris (PH 8.0, 7.6 or 7.4) and 2 mL 0.5 M Na_2 EDTA (PH 8.0) to 800 mL H_2O. Mix and adjust volume to 1 L with H_2O. Sterilize by autoclaving. Store at room temperature.

Proteinase K

Dissolve 10 mg of proteinase in one mL of distilled water. Store in small aliquots at −20°C for no longer than 1 year.

SDS/proteinase K mix

Mix 5 μL proteinase K (10 mg/mL) with 70 μL of 10% SDS for each sample. Dissolve 29.2 g NaCL in 100 mL distilled water. Autoclave. Store at room temp for no longer than 1 year.

CTAB/NaCL solution

Dissolve 4.1 g NaCL in 80 mL distilled water. While stirring, add 10 g CTAB (*n*-cetyl-*N*,*N*,*N*-trimethyl ammonium bromide). If necessary, heat solution to 650°C. Adjust the volume to 100 mL distilled water. Store at room temp for no longer than 6 months.

Chloroform/isoamyl alcohol (24:1)

Mix 24 volume of chloroform with 1 volume of isoamyl alcohol. Store at room temp for no longer than 1 year.

Isopropanol

Store at room temp for no longer than 1 year.

70% ethanol

Store at room temp for no longer than 1 year.

5% EDTA as anticoagulation

Dissolve 0.5 g of EDTA in 10 mL of distal water.

PBS buffer

Dissolve the following in 800 mL of DW; 80 g of NaCl, 2 g of KCL, 26.8 g of $Na_2HPO_4 \cdot 7H_2O$ and 2.4 g of KH_2PO_4 and make the PH to 7.4. Bring the volume to 1000 with DW.

SE buffer [(EDTA (0.5 M) + NaCL (3 M)]

3 M NaCL (MW = 58.44); dissolve 175.32 g of NaCL in 1000 mL DW.

1 M EDTA (MW = 372.54); dissolve 372.54 g of EDTA in 800 mL DW and make the PH to 8.2 and bring the volume to 1000.

0.5 M EDTA; take 250 mL from 1 M solution and bring it to 500 mL with DW.

SE buffer; one volume of 3 M NaCL + one volume of 0.5 M EDTA

Sodium acetate ($C_2H_3O_2Na$, MW = 82.03)

Dissolve 246.09 g of sodium acetate in 1000 DW and bring PH to 5.

Digestion buffer

500 mL of 1 M Tris-HCL, 20 mL of 1 M EDTA, 10 mL of NaCL, 100 mL of SDS, bring the total volume to 1000 mL.

GENOMIC DNA EXTRACTION FROM MYCOBACTERIAL CULTURE

Transfer at least one loop of mycobacterial cells into micro centrifuge tube containing 400 μL of 1× TE. Heat 20 min at 80°C to kill the cells, and cool at room temperature. Add 50–100 μL of 10 mg/mL lysozyme. Vortex and incubate for 24 h at 37°C. Mix proteinase K and SDS 10% (SDS = 70 μL + proteinase K = 2.5 μL). Add 72.5 μL of 10% SDS/proteinase K mix, vortex shortly and incubate 24 h at 65°C. Add 100 μL of 5 M NaCL, add 100 μL of CTAB/NaCL solution, which is prewarmed at 65°C for 5 min. Vortex until the liquid content becomes white (milky) (Sola et al., 2001; Van Soolingen et al., 1995).

Incubate for 10 min at 65°C. Add a volume of about 750 μL of chloroform/isoamyl alcohol. Vortex for at least 10 s, and centrifuge at room temp for 20 min at 12,000 rpm. Becomes white (milky).

Incubate for 10 min at 65°C. Transfer all of the aqueous supernatant in 180 μL amounts to a fresh micro centrifuge tube. Carefully add 0.6 volume (450 μL) isopropanol to precipitate the nucleic acids. Shake carefully by hand. Place overnight at −20°C.

Spin for 20 min at 12,000 rpm. Remove most of the supernatant; leave about 20 μL above the pellet. Add 1 mL of 70% cold ethanol and centrifuge for 1 min at 12,000 rpm. Discard the supernatant and keep the micro-tube reversely on a filter paper for 10 min.

Add 10–30 μL 1× TE and keep at 37°C for 3–4 h. Transfer and keep at 4°C in refrigerator (Sola et al., 2001; Van Soolingen et al., 1995). The extracted DNA gives a very sharp band which is visible on naked eyes on 0,.8% agarose gel (Fig. 2.10).

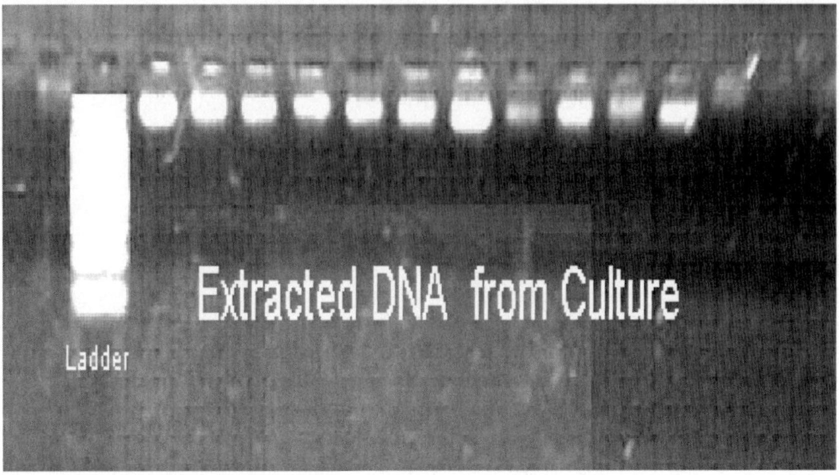

FIGURE 2.10 Electrophoresis of DNA on 0.8% agarose gel.

Extraction of Mycobacterial DNA From Paraffin-Embedded Tissue

Add 150 μL of a 5% Chelex suspension to 14 μm paraffin-embedded tissue section. Vortex thoroughly (5–10 min). The section should be completely covered with the Chelex suspension. Heat the mixture at 100°C for 30 min. The paraffin then appears floating on the surface of the solution. Paraffin should be removed by sterile needle. The tubes are centrifuged at 14,000 rpm for 10 min. Transfer the solution (beneath the paraffin), to a clean micro centrifuge tube. This solution contains chromosomal DNA and can be used for further PCR (Höfler, 1994; Sola et al., 2001).

Extraction of Mycobacterial DNA From Clinical Specimens (Sputum and Others)

Following digestion and decontamination, the sediments are removed and vertex thoroughly. To each 500 μL of sample sediments, 1000 μL of digestion buffer added and they were incubated at 80°C for 20 min. Centrifuge the sample at 14,000 rpm for 15 min. Discard the supernatant completely. To the sediment add 250 μL of digestion buffer and 140 μL of phenol. Centrifuged the samples at 14,000 g for 15 min. Transfer the supernatant to a new micro-tube, and add equal amount of chloroform-isoamyl alcohol. Centrifuged the samples at 14,800 rpm for 15 min. Transfer the supernatant into a new micro-tube. To 0.1 volume of sample add sodium acetate (3 M) and to 0.7 volume of sample add cold isopropanol. The tubes were incubated at −20°C overnight. Spin 20 min at 14,000 rpm. Remove most of the supernatant; leave about 20 μL above the pellet. Add 1 mL of 70% cold ethanol and centrifuge for 1 min at 14,000 rpm. Discard the supernatant and keep the micro-tube reversely on a filter paper for 10 min. Spin 20 min at 14,000 rpm. Remove most of the supernatant; leave about 20 μL above the pellet. Add 1 mL of 70% cold ethanol and centrifuge for 1 min at 14,000 rpm. Discard the supernatant and keep the micro-tube reversely on a filter paper for 10 min redissolve the pellet in the amount of 1× TE (10–30 μL) (Cloud et al., 2002; Sola et al., 2001; Van Soolingen et al., 1995).

Primer Preparation

When you order primer from a commercial company the following Information should be clear:

Sequence (5′–3′) of the primer, number of nucleotides.
The unit size (e.g., 100 nmol).
Purification standard (e.g., standard desalting or gel purified).

The primer will then be synthesized and sent to you in a lyophilized (freeze-dried) form, together with the following data.

Molecular weight
GC content
Tm (50 mM NaCl)
Amount of oligo (lyophilized). This could be expressed as OD260, nM and μg

Example of Primer Stock Solution Preparation

OD260 = 7.2
Weight = 197.5 μg; 32.4 nmol
Conc: 100 μM

Take, for example, a primer data sheet with the above information:

To make 100 μM/μL of stock solution: Check the weight of primer (here, it was 32.4 nmol). Then, based on this weight add the 324 μL DW to make a stock of 100 μM/μL.

To make 10 μM of working solution: Take 10 μL of stock solution and bring the volume into 100 μL.

Reaction Mixture: Adding 1 μL (of the 10 μM of working solution), to a 20 μL PCR reaction will result in a final primer concentration of 10 pmol of primer in 20 μL volume.

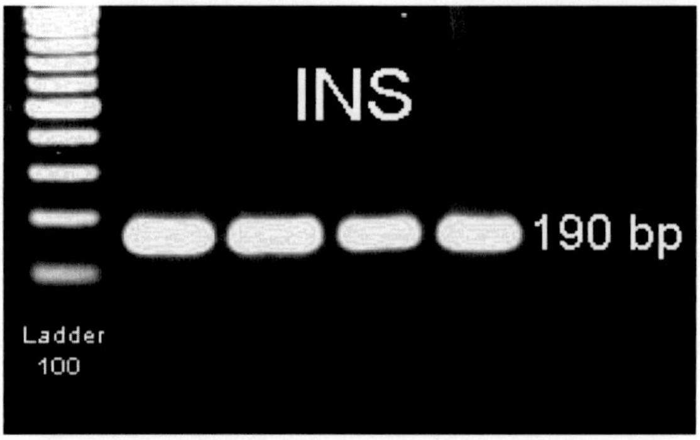

FIGURE 2.11 PCR product gives 190 bp. If it is positive, it is TB-Complex; otherwise we have to proceed for second steps.

Rapid Identification Procedures for Nontuberculous Mycobacteria Identification

(*Step 1*) Isolated DNA is always first amplified using IS6110 primer (to be sure it is not TB-Complex). (*Step 2*) Nested-PCR using TB_{11} and TB_{12} primer for *hsp65* gene. The PCR products digested with *Bst*EII, *Hae*III restriction enzyme. (*Step 3*) Most of the NTM are identified by *hsp65*, if not, then primer for 16S–23S rRNA gene were used and the PCR-products were digested with *Hae*III, *Cfo*I, *Taq*I, *MspI*, *DdeI*, *Ava*II, *Hin*fI restriction enzyme (Fig. 2.11).

(*Step 1*) Isolated DNA is always first amplified using IS6110 primer (to be sure it is not TB-Complex).

Primer name: *IS6110 (TB_1 and TB_2)*

Sequence:

A:5′ - ATC CTG CGA GCG TAG GCG TCG G - 3′

B: 5′ - CAG GAC CAC GAT CGC TGA TCC GG - 3′

Primer name and information: *TB_1*

Oligo sequence: TB_1: 5′-ATC CTG CGA GCG TAG GCG TCG G-3′

Delivered: 6.4 OD, 185.4 μg, 27.3 nmol, conc: 100.0 μM, MW: 6792, Tm: 67.7, synthesis scale: 0.04 μmol, no. bases: 22, GC%: 68.2, dissolve in: 273 μL DW.

Primer name and information: *TB_2*

Oligo sequence: TB_2: 5′-CAG GAC CAC GAT CGC TGA TCC GG-3′

Delivered: 6.0 OD, 171.5 μg, 24.4 nmol, conc: 100.0 μM, MW: 7035, Tm: 67.8, synthesis scale: 0.04 μmol, no. bases: 23, GC%: 65.2, dissolve in: 244 μL DW.

PCR mixture for 50 μL reaction

dNTP: (0.2 mM) 1 μL

PCR buffer: 5 μL

MgCl: (1.5 mM) 1.5 μL

Primer A: (8 pmol) 0.8 μL

Primer B: (8 pmol) 0.8 μL
Enzyme (Taq): (1 U) 0.25 μL
DMSO: (1%) 2 μL
DNA template: 2 μL
PCR cycle

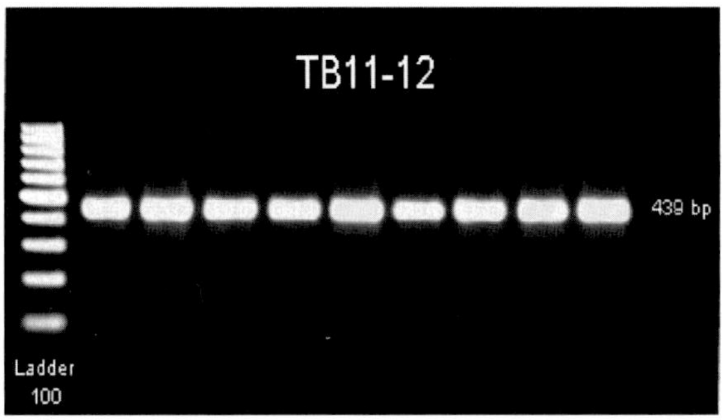

FIGURE 2.12 PCR−product using TB_{11-12} primer.

Mycobacterium simiae		
Slowly growing	*HSP*65 gene	
Photo-chromogenic	*Bste*II pattern	*Hae*III pattern
Mycobacterium simiae type 1	234-211	180-130
Mycobacterium simiae type 3	235 / 130 / 85	145 / 130 / 0 /
Mycobacterium simiae type 5	440 / 0 / 0	145 / 130 / 0 / 0

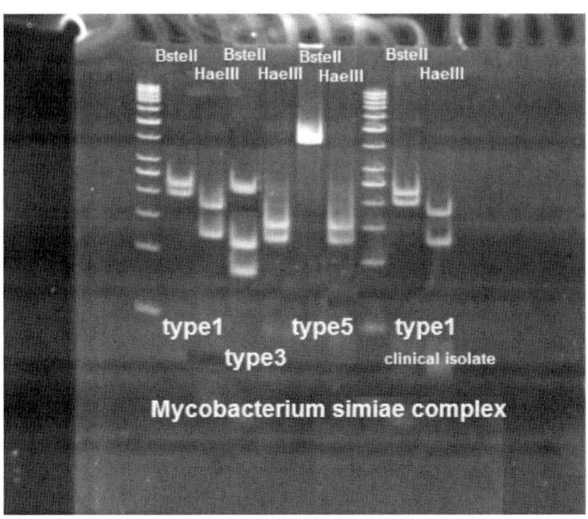

FIGURE 2.13 *Mycobacterium simiae* amplified by *hsp65* and digested with restriction enzyme.

Initial denaturation 10 min 95°C			
30 cycles	**Step1**	**20 s**	**93°C**
	Step 2	**1 min**	**65°C**
	Step 3	**20 s**	**72°C**
Final extension 10 min 72°C			

(Step 2) Nested-PCR using TB_{11} and TB_{12} primer for *hsp65* gene
Step 2.1:
Sequence:
A: 5′ - CGT AYG ACG AAG AGG CCC GT - 3′
B: 5′ - WAS GGR TCC TCS AGG ACS GC - 3′
Primer name and information: TB_{15}
Oligo sequence: TB_{15}: 5′-CGT AYG ACGAAG AGG CCC GT-3′

Mycobacterium.intracellulare		
Slowly growing	*HSP65* gene	
Non pigmented	*Bste*II pattern	*Hae*III pattern
Mycobacterium intracellulare type 1	235 / 120 / 100	145 / 130 / 60 / 0
Mycobacterium intracellulare type 5	235 / 120 / 100	130 / 100 / 60 / 55

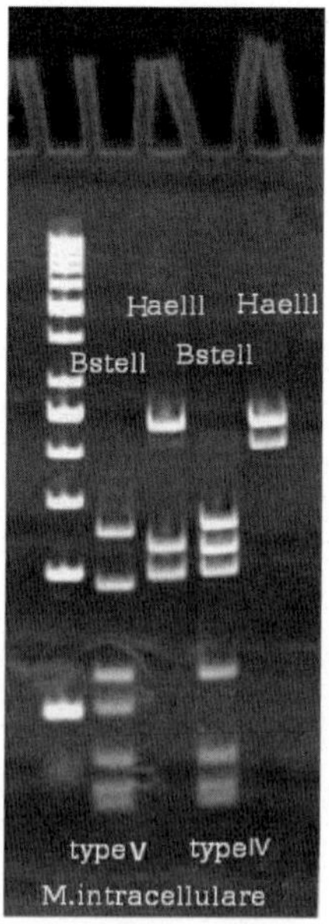

FIGURE 2.14 *Mycobacterium intracellulare* amplified by *hsp65* and digested with restriction enzyme.

Delivered: 30.8 OD (849.8 μg, 138.0 nmol), conc: 100.0 μM, MW: 6160, Tm: 62.4, synthesis scale: 0.04 μmol, no. bases: 20, GC%: 62.5, dissolve in: 1380 μL DW.

Primer name and information: TB_{17}

Oligo sequence: TB_{17}: 5′-WAS GGR TCC TCS AGG ACS GC-3′

Delivered: 29.5 OD (842.5 μg, 137.3 nmol), conc.: 100.0 μM, MW: 6136, Tm: 64.5, synthesis scale: 0.04 μmol, no. bases: 20, GC%: 67.5, dissolve in: 1373 μL DW

PCR Mixture for 50 μL reaction

dNTP: (0.2 mM) 1 μL

PCR buffer: 5 μL

Mycobacterium kansasii		
Slowly growing	*HSP*65 gene	
Photo-chromogen	*Bste*ll pattern	*Hae*lll pattern
Mycobacterium kansasii type 1	235 / 210 / 0	130 / 105 / 80 / 0
Mycobacterium kansasii type 4	235 / 120 / 85	130 / 115 / 75 / 60
Mycobacterium kansasii type 5	320 / 115 / 0	130 / 95 / 75 / 60

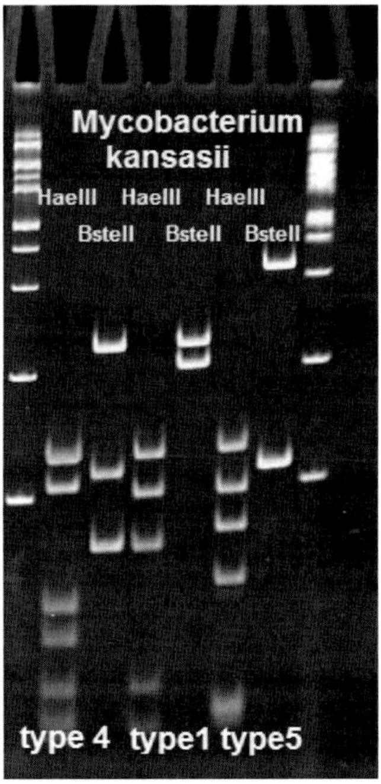

FIGURE 2.15 *Mycobacterium kansasii* amplified by *hsp65* and digested with restriction enzyme.

MgCl: (1.5 mM) 1.5 μL
Primer A: (4 pmol) 0.2 μL
Primer B: (4 pmol) 0.2 μL
Enzyme (tag): (1 U) 0.25 μL
DMSO: (1%) 2.5 μL
DNA template: 3–5 μL
PCR cycle:

Initial denaturation 5 min 95°C			
30 cycles	Step 1	20 s	95°C
	Step 2	1 min	60°C
	Step 3	40 s	72°C
Final extension 5 min 72°C			

PCR product:

Mycobacterium gordonae		
Slowly growing	*HSP65* gene	
Scotochromogen	*Bste*II pattern	*Hae*III pattern
Mycobacterium gordonae type 1	235 / 210 / 0	160 / 115 / 60 / 0
Mycobacterium gordonae type 3	235 / 120 / 100	130 / 115 / 0 / 0
Mycobacterium gordonae type 4	320 / 115 / 0	130 / 115 / 60 / 0
Mycobacterium gordonae type 8	235 / 210 / 0	130 / 115 / 0 / 0

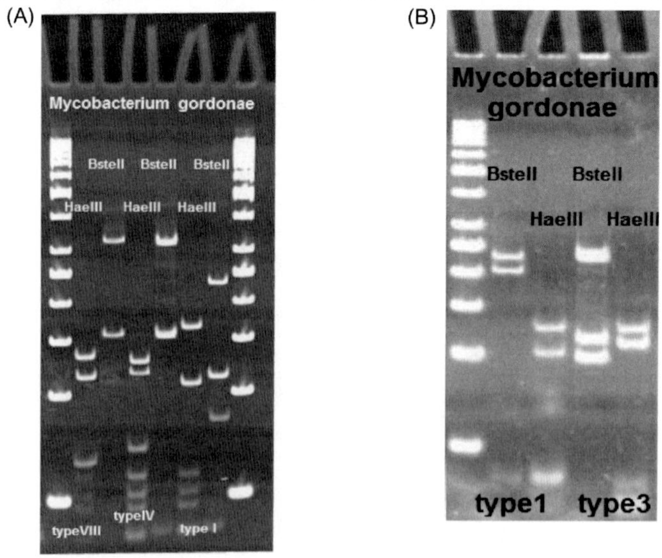

FIGURE 2.16 (A, B) *Mycobacterium gordonae* amplified by *hsp65* and digested with restriction enzyme.

Primer TB_{15} (5′-CGT AYG ACG AAG AGG CCC GT-3′) and TB_{17} (5′-WAS GGR TCC TCS AGG ACS GC-3′) amplified a 470 bp fragment.

Step 2.2 (hsp65 (TB_{11} and TB_{12}) primer)

Primer name and information TB_{11}

Oligo sequence: TB_{11}: A: 5′-ACC AAC GAT GGT GTG TCC AT-3′

Delivered: 7.8 OD (218.4 μg, 35.7 nmol), conc: 100.0 μM, MW: 6117, Tm: 57.3, synthesis scale: 0.04 μmol no. bases: 20, GC%: 50.0, dissolve in: 357 μL W.

Primer name and information: *TB_{12}*

Oligo sequence: TB_{12}: B: 5′-CTT GTC GAA CCG CAT ACC CT-3′

Mycobacterium peregrinum		
Rapidly growing	*HSP65* gene	
Non pigmented	*Bste*II pattern	*Hae*III pattern
Mycobacterium peregrinum type 1	235 / 210 / 0	145 / 140 / 100 / 50
Mycobacterium peregrinum type 2	235 / 210 / 0	140 / 125 / 100 / 50
Mycobacterium peregrinum type 3	235 / 130 / 85	145 / 140 / 100 / 60

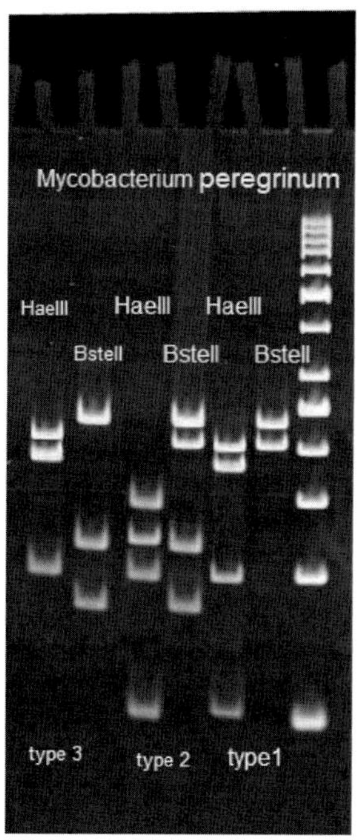

FIGURE 2.17 *Mycobacterium peregrinum* amplified by *hsp65* and digested with restriction enzyme.

Delivered: 7.6 OD (227.5 μg, 37.8 nmol), conc: 100.0 μM, MW: 6013, Tm: 59.4, synthesis scale: 0.04 μmol, no. bases: 20, GC%: 55.0, dissolve in: 378 μL DW.

Polymerase Chain Reaction Mixture for 50 μL Reaction

PCR buffer: 5 μL

- Primer A: (8 pmol) 0.8 μL
- Primer B: (8 pmol) 0.8 μL
- DNA template: 2 μL
- dNTP: (0.2 mM) 1 μL
- MgCl: (1.5 mM) 1.5 μL
- Enzyme (Taq): (1 U) 0.25 μL
- DMSO: (1%) 2 μL
- PCR cycle:

Mycobacterium terrae		
Slowly growing	*HSP65* gene	
Non pigmented	*Bste*II pattern	*Hae*III pattern
Mycobacterium terrae type 1	320 / 115 / 0	180 / 130 / 0 / 0
Mycobacterium terrae type 3	320 / 115 / 0	140 / 60 / 50 / 0
Mycobacterium terrae type 4	235 / 210 / 0	140 / 115 / 70 / 0

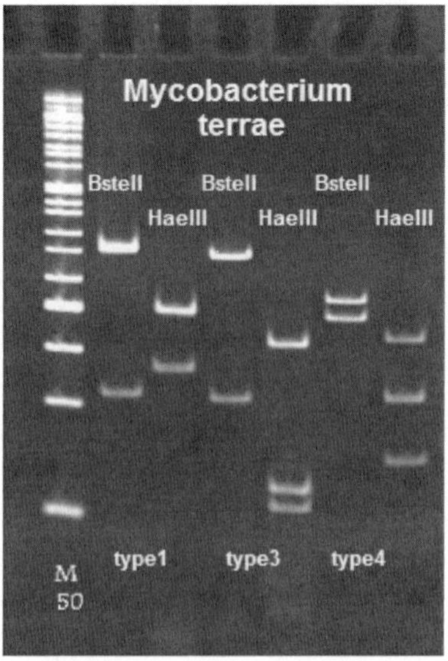

FIGURE 2.18 *Mycobacterium terrae* complex amplified by *hsp65* and digested with restriction enzyme.

Initial denaturation 4 min 95°C			
30 cycles	Step 1	30 s	94°C
	Step 2	1 min	56°C
	Step 3	40 s	72°C
Final extension 10 min 72°C			

Mycobacterium avium		
Slowly growing	*HSP*65 gene	
Non pigmented	*Bste*ll pattern	*Hae*lll pattern
Mycobacterium avium s. avium type 1	235 / 210 / 0	130 / 105 / 0 / 0

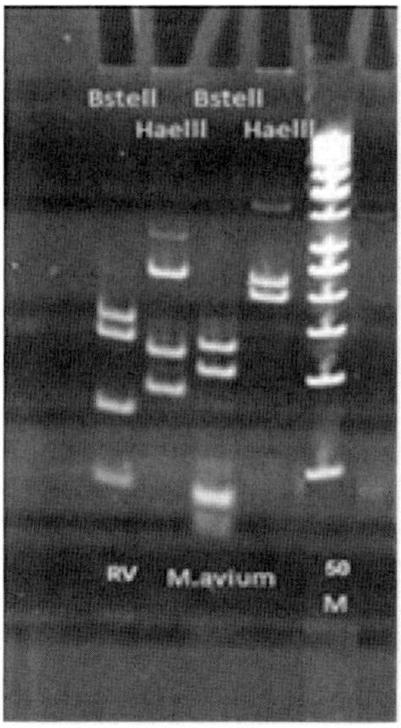

FIGURE 2.19 *Mycobacterium avium* complex amplified by *hsp65* and digested with restriction enzyme.

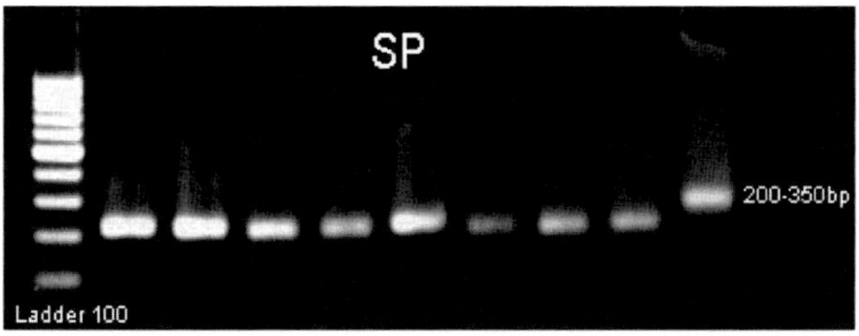

FIGURE 2.20 *PCR product*: Primer $SP_1$and SP_2 amplified a 200–350 bp fragment.

Polymerase Chain Reaction Product

Primer Tbll (5'-ACC AAC GAT GGT GTG TCC AT) and TB_{12} (5'-CTT GTC GAA CCG CAT ACC CT) amplified a 439 bp fragment between positions 398 and 836 of the published gene. The PCR product is more than 400 bp, and easily can be detected (Fig. 2.12).

Mycobacterium fortuitum		
Rapidly growing	16S-23S rRNA gene	
Non pigmented	*Cfo*I pattern	*Hae*III pattern
Mycobacterium fortuitum type 1	160/140	110/90
Mycobacterium fortuitum type 2	260	180/120
Mycobacterium fortuitum type 3	220/170/160/140	150/120/95
Mycobacterium fortuitum type 4	300/180/160/145/	120 /90 / 0 / 0
Mycobacterium fortuitum type 5	200/160/140/0	170 / 120 / 0 / 0
Mycobacterium fortuitum type 6	199/165/133/81	170/110/87
Mycobacterium fortuitum type 7	191/172/109	130/160/115/0
Mycobacterium fortuitum type 8	191/172/109	267-187-133/0
Mycobacterium fortuitum type 9	110/86/49/33	No digest

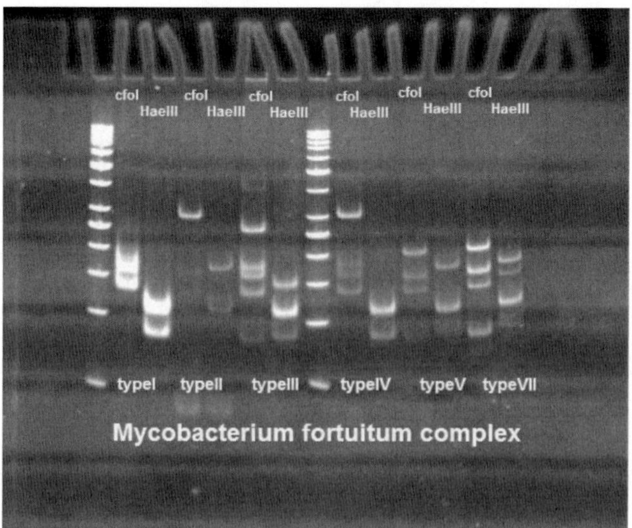

FIGURE 2.21 *Mycobacterium fortuitum complex* amplified by16S–23S rRNA and digested with restriction enzyme.

hsp65 Restriction Analysis

For B*stEII* digestion, 10 μL of PCR product is added directly to a mixture containing 1 μL of enzyme, 2 μL of restriction buffer (5 × buffer), and 18 μL of water, and the mixture is incubated for 1–16 h at 37°C.

For H*aeIII* digestion, 10 μL of PCR product is added directly to a mixture containing 1 μL of enzyme, 2 μL of restriction buffer (5 × buffer), and 18 μL of water, and the mixture is incubated for 1–16 h at 37°C.

Evaluation of Restriction Patterns

7 μL of the digested PCR product with 5 μL of loading buffer were mixed and loaded onto 8% polyacrylamide gel and visualized by UV light (Figs. 2.13–2.19). PCR can be digested or not digested by B*stEII* and H*aeIII*.

Mycobacterium parafortuitum		
Rapidly growing	16S-23S rRNA gene	
Non pigmented	*Cfol* pattern	*Hae*lll pattern
Mycobacterium parafortuitum type 1	300 / 0 / 0	120 / 50 / 0 / 0
Mycobacterium parafortuitum type 2	300 / 0 / 0	120 /100 /0 / 0

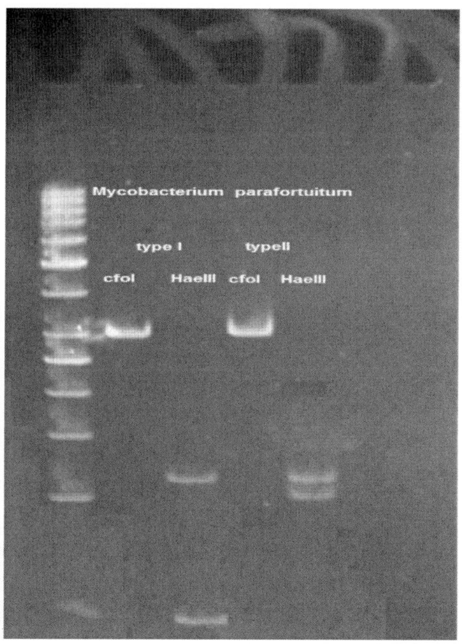

FIGURE 2.22 *Mycobacterium parafortuitum* amplified by16S–23S rRNA and digested with restriction enzyme.

Identification Using 16S–23S rRNA Primer

Primer name: *SP_1*

Oligo sequence: SP_1: 5′-ACC TCC TTT CTA AGG AGC ACC-3′

Delivered: 13.6 OD (394 μg, 62.4 nmol), conc. (volume 1 mL) 62 pmol/μL, MW: 6326 g/mol, Tm: 59.8, synthesis scale: 0.05 μmol, no. of bases: 21, GC%: 52.4, dissolve in: 624 μL DW.

Primer name: *SP_2*

Oligo sequence: SP_2: 5′-GAT GCT CGC AAC CAC TAT CCA-3′

Delivered: 9.4 OD (265 μg, 41.9 nmol), conc. (volume 1 mL) 42 pmol/μL, MW: 6335 g/molTm: 59.8, synthesis scale: 0.04 μmol, no. bases: 21, GC %: 52.4, dissolve in: 419 μL DW.

Mycobacterium chelonae		
Rapidly growing	*HSP*65 gene	
Non pigmented	*Bste*II pattern	*Hae*III pattern
Mycobacterium chelonae type 1	320 / 130 / 0	200 / 60 / 55 / 50

Rapidly growing	16S-23S rRNA gene	
Non pigmented	PCR product size	*Hae*III pattern
Mycobacterium chelonae type 1	250	No digest

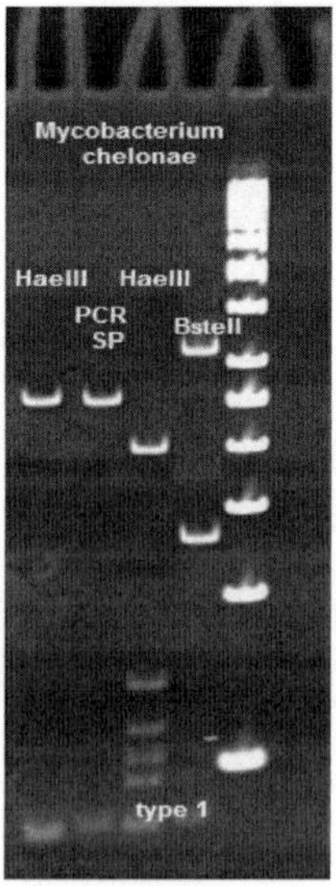

FIGURE 2.23 *Mycobacterium chelonae* amplified by *hsp65* and 16S–23S rRNA and digested with restriction enzyme.

Polymerase Chain Reaction Mixture for 50 μL Reaction

dNTP: (0.2 mM) 1 μL
PCR buffer: 5 μL
$MgCl_2$: (1.75 mM) 1.75 μL
Primer A: (8 pmol) 0.8 μL
Primer B: (8 pmol) 0.8 μL
Enzyme tag: (2.5 U) 0.25 μL
DMSO: (1%) 2 μL
Template: 1.5 μL
Water: up to 50 μL
PCR cycle (Fig. 2.20).

Mycobacterium scrofulaceum		
Slowly growing	*HSP65* gene	
Scotochromogen	*Bste*II pattern	*Hae*III pattern
Mycobacterium scrofulaceum type 1	235 / 210 / 0	145 / 130 / 95 / 0

Mycobacterium parascrofulaceum		
Slowly growing	*HSP65* gene	
Scotochromogen	*Bste*II pattern	*Hae*III pattern
Mycobacterium parascrofulaceum type 1	235 / 120 / 85	140 / 95 / 0 / 0

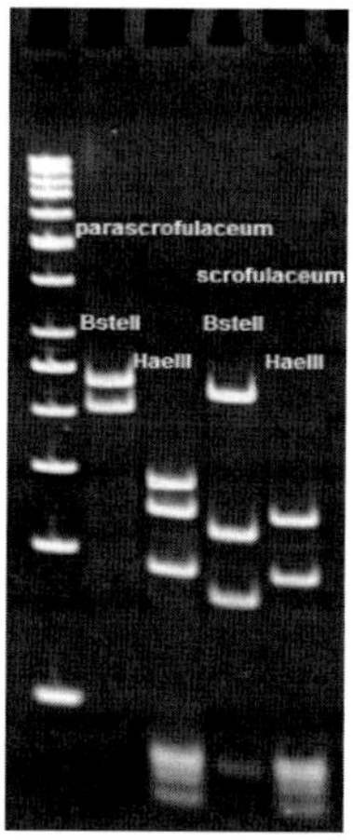

FIGURE 2.24 *Mycobacterium scrofulaceum and parascrofulaceum* amplified by *hsp65* and 16S–23S rRNA and digested with restriction enzyme.

Initial denaturation	**4 min 95°C**		
30 cycles	**Seq1**	**30 s**	**94°C**
	Seq2	**1 min**	**56°C**
	Seq3	**40 s**	**72°C**
Final extension 10 min 72°C			

Restriction Analysis

For *Hae*III digestion, 10 μL (0.1–0.5 μg) of PCR product is added directly to a mixture containing 1 μL of enzyme, 2 μL of restriction buffer (5 × buffer), and 18 μL of water, and the mixture is incubated for 16 h at 37°C. For *Cfo*I digestion, 10 μL (0.1–0.5 μg) of PCR product is added directly to a mixture containing μL of enzyme, μL of restriction buffer (5 × buffer), and 18 μL of water, and the mixture is incubated for 1 h at 37°C (Figs. 2.21–2.25).

Mycobacterium conceptionense		
Rapidly growing	*HSP65* gene	
Non pigmented	*Bste*ll pattern	*Hae*lll pattern
Mycobacterium conceptionense type 1	235 / 120 / 85	140 / 125 / 60 / 55

Rapidly growing	*16S-23S rRNA* gene	
Non pigmented	PCR product size	*Hae*lll pattern
Mycobacterium conceptionense type 1	300	300

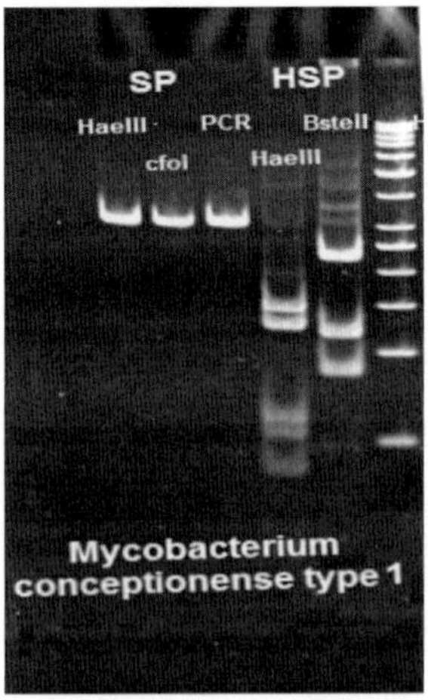

FIGURE 2.25 *Mycobacterium conceptionense* amplified by *hsp65* and 16S–23S rRNA and digested with restriction enzyme.

For *Taq*I digestion, 10 μL (0.1–0.5 μg) of PCR product is added directly to a mixture containing μL of enzyme, μL of restriction buffer (5 × buffer), and 18 μL of water, and the mixture is incubated for 16 h at 65°C.

For *Msp*I digestion, 10 μL (0.1–0.5 μg) of PCR product is added directly to a mixture containing μL of enzyme, μL of restriction buffer (5 × buffer), and 18 μL of water, and the mixture is incubated for 16 h at 37°C.

For *Dde*I digestion, 10 μL (0.1–0.5 μg) of PCR product is added directly to a mixture containing μL of enzyme, μL of restriction buffer (5 × buffer), and 18 μL of water, and the mixture is incubated for 16 h at 37°C.

For *Ava*II digestion, 10 μL (0.1–0.5 μg) of PCR product is added directly to a mixture containing μL of enzyme, μL of restriction buffer (5 × buffer), and 16 μL of water, and the mixture is incubated for 16 h at 37°C.

For *Hinf*I digestion, 10 μL (0.1–0.5 μg) of PCR product is added directly to a mixture containing μL of enzyme, μL of restriction buffer (5 × buffer), and 18 μL of water, and the mixture is incubated for 16 h at 37°C.

REFERENCES

Aronson, J.D., 1926. Spontaneous tuberculosis in salt water fish. J. Infect. Dis. 39, 314–320.

Baess, I., 1974. Isolation and purification of deoxyribonucleic acid from mycobacteria. Acta Patholog. Microb. 82 (6), 780–784.

Barksdale, L., Kim, K.-S., 1977. Mycobacterium. Bacteriol. Rev. 41 (1), 217–372.

Boddinghaus, B., Rogall, T., Flohr, T., Blocker, H., Böttger, E., 1990. Detection and identification of mycobacteria by amplification of rRNA. J. Clin. Microbiol. 28 (8), 1751–1759.

Bojalil, L., Cerbon, J., Trujillo, A., 1962. Adansonian classification of mycobacteria. J. Gen. Microbiol. 28 (2), 333–346.

Bradley, S., Bond, J., 1974. Taxonomic criteria for mycobacteria and nocardiae, Advances in Applied Microbiology, vol. 18. Elsevier, pp. 131–190.

Brunello, F., Ligozzi, M., Cristelli, E., Bonora, S., Tortoli, E., Fontana, R., 2001. Identification of 54 mycobacterial species by PCR-restriction fragment length polymorphism analysis of the hsp65Gene. J. Clin. Microbiol. 39 (8), 2799–2806.

Buhler, V.B., Pollak, A., 1953. Human infection with atypical acid-fast organisms. Report of two cases with pathologic findings. Am. J. Clin. Pathol. 23 (4), 363–374.

Clarridge, J.E., 2004. Impact of 16S rRNA gene sequence analysis for identification of bacteria on clinical microbiology and infectious diseases. Clin. Microbiol. Rev. 17 (4), 840–862.

Cloud, J., Neal, H., Rosenberry, R., Turenne, C., Jama, M., Hillyard, D., et al., 2002. Identification of *Mycobacterium* spp. by using a commercial 16S ribosomal DNA sequencing kit and additional sequencing libraries. J. Clin. Microbiol. 40 (2), 400–406.

Cole, E.C., Rutala, W.A., Nessen, L., Wannamaker, N., Weber, D., 1990. Effect of methodology, dilution, and exposure time on the tuberculocidal activity of glutaraldehyde-based disinfectants. Appl. Environ. Microbiol. 56 (6), 1813–1817.

Cook, V.J., Turenne, C.Y., Wolfe, J., Pauls, R., Kabani, A., 2003. Conventional methods versus 16S ribosomal DNA sequencing for identification of nontuberculous mycobacteria: cost analysis. J. Clin. Microbiol. 41 (3), 1010–1015.

Cross, T., Goodfellow, M., 1973. Taxonomy and classification of the actinomycetes. Soc. Appl. Bacteriol. Symp. Ser. 2, 11–112.

Cuttino, J.T., McCabe, A.M., 1949. Pure granulomatous nocardiosis: a new fungus disease distinguished by intracellular parasitism: a description of a new disease in man due to a hitherto undescribed organism, *Nocardia intracellularis*, n. sp., including a study of the biologic and pathogenic properties of this species. Am. J. Pathol. 25 (1), 1–34.

Daniel, T.M., 2006. The history of tuberculosis. Respir. Med. 100 (11), 1862–1870.

David, H.L., 1974. Carotenoid pigments of *Mycobacterium kansasii*. Appl. Microbiol. 28 (4), 696–699.

Devallois, A., Goh, K.S., Rastogi, N., 1997. Rapid identification of mycobacteria to species level by PCR-restriction fragment length polymorphism analysis of the hsp65 gene and proposition of an algorithm to differentiate 34 mycobacterial species. J. Clin. Microbiol. 35 (11), 2969–2973.

Embley, T., Stackebrandt, E., 1994. The molecular phylogency and systematics of the actinomycetes. Annu. Rev. Microbiol. 48 (1), 257–289.

Falkinham III, J., 1996. Epidemiology of infection by nontuberculous mycobacteria. Clin. Microbiol. Rev. 9 (2), 177–215.

Falkinham III, J., 2009. Surrounded by mycobacteria: nontuberculous mycobacteria in the human environment. J. Appl. Microbiol. 107 (2), 356–367.

Falkinham, J.O., 2010. Impact of human activities on the ecology of nontuberculous mycobacteria. Future. Microbiol. 5 (6), 951–960.

Falkinham III, J.O., 2011. Nontuberculous mycobacteria from household plumbing of patients with nontuberculous mycobacteria disease. Emerg. Infect. Dis. 17 (3), 419–424.

Fregnan, G., Smith, D., 1962. Description of various colony forms of mycobacteria. J. Bacteriol. 83 (4), 819–827.

Goodfellow, M., Wayne, L., 1982. Taxonomy and nomenclature. In: Ratledge, C., Stanford, J. (Eds.), The Biology of the Mycobacteria. Academic Press, London, pp. 471–521.

Hafner, B., Haag, H., Geiss, H.-K., Nolte, O., 2004. Different molecular methods for the identification of rarely isolated non-tuberculous mycobacteria and description of new hsp65 restriction fragment length polymorphism patterns. Mol. Cell. Probes 18 (1), 59–65.

Hansen, G.A., 1880. Bacillus leprae. Virchows. Arch. 79 (1), 32–42.

Hartmans, S., de Bont, J., 1992. The genus *Mycobacterium*-nonmedical. In: Balows, A., Truper, H.G., Dworking, M., et al.,The Prokaryotes: A Handbook on the Biology of Bacteria: Ecophsiology, Isolation, Identification, Applications. The Prokaryotes. 2. Springer Verlag, New York, pp. 1214–1237.

Hill, E.B., Wayne, L.G., Gross, W.M., 1972. Purification of mycobacterial deoxyribonucleic acid. J. Bacteriol. 112 (3), 1033–1039.

Höfler, G., 1994. Detection of bacterial DNA using the polymerase chain reaction (PCR). Verh. Dtsch. Ges. Pathol. 78, 104–110.

Kazda, J., 1983. The principles of the ecology of mycobacteria. In: Ratledge, C., Stanford, J. (Eds.), The Biology of the Mycobacteria, Volume 2: Immunological and Environmental Aspects, 2. Academic Press, London, pp. 323–341.

Kent, P., Kubica, G., 1985. Public Health Mycobacteriology: A Guide for Level III Lab. US Department of Health and Human Services. Public Health Services Center for Disease Control, Atlanta, GA, pp. 64–68.

Kubica, G., Wayne, L.C., 1984. The Mycobacteria—A Source Book, 42. Marcel Dekker Inc, New York Members of the Genus Mycobacterium, New York, pp. 337–343.

Kubica, G., Baess, I., Gordon, R.E., Jenkins, P., Kwapinski, J., McDurmont, C., et al., 1972. A co-operative numerical analysis of rapidly growing mycobacteria. J. Gen. Microbiol. 73 (1), 55–70.

Kubica, G.P., 1973. Differential identification of mycobacteria: VII. Key features for identification of clinically significant mycobacteria. Am. Rev. Respir. Dis. 107 (1), 9–21.

Kubica, G.P., Vestal, A.L., 1961. The arylsulfatase activity of acid-fast bacilli: I. Investigation of activity of stock cultures of acid-fast bacilli. Am. Rev. Respir. Dis. 83 (5), 728–732.

Leao, S.C., Martin, A., Mejia, G.I., Palomino, J.C., Robledo, J., Telles, M.A.S., Portaels, F., 2004. Practical Handbook for the Phenotypic and Genotypic Identification of Mycobacteria. Vanden Broele, Brugge.

Lehmann, K.B., Neumann, R., 1896. Atlas und Grundriss der Bakteriologie und Lehrbuch der speziellen bakteriologischen Diagnostik. In: Lehmann, J.F. (Ed.), Munich, Germany.

Levy-Frebault, V.V., Portaels, F., 1992. Proposed minimal standards for the genus *Mycobacterium* and for description of new slowly growing *Mycobacterium* species. Int. J. Syst. Bacteriol. 42 (2), 315–323.

Lutz, B., 1992. Identification test for mycobacteria, pp. 3.12.1–3.12.27 In: Hd, H. (Ed.), Clinical Microbiology Procedures Handbook. Section 3 Mycobacteriology. American Society for Microbiology, Washington, DC.

Metchock, B., Nolte, F., Wallace, R., 1999. Mycobacterium. In: Murray, P., Baron, E., Pfaller, M., Tenover, F., Yoken, R. (Eds.), Manual of Clinical Microbiology. American Society for Microbiology, Washington, DC, pp. 399–437.

Norden, A., Linell, F., 1951. A new type of pathogenic *Mycobacterium*. Nature 168 (4280), 826.

Pinner, M., 1935. Atypical acid-fast microorganisms. III. Chromogenie acid-fast bacilli from human beings. Am. Rev. Tuberc. Pulm. 32, 424–439.

Pitulle, C., Dorsch, M., Kazda, J., Wolters, J., Stackebrandt, E., 1992. Phylogeny of rapidly growingSneath, P. H. A. 1970 In: Prauser, H. (Ed.), The Actinomycetales. Fischer, Jena, pp. 371–377.

Prissick, F., Masson, A., 1956. Cervical lymphadenitis in children caused by chromogenic mycobacteria. Can. Med. Assoc. J. 75 (10), 798–803.

Rastogi, N., Legrand, E., Sola, C., 2001. The mycobacteria: an introduction to nomenclature and pathogenesis. Rev. Sci. Tech. Oie. 20 (1), 21–54.

Ringuet, H., Akoua-Koffi, C., Honore, S., Varnerot, A., Vincent, V., Berche, P., et al., 1999. hsp65 sequencing for identification of rapidly growing mycobacteria. J. Clin. Microbiol. 37 (3), 852–857.

Rogall, T., Flohr, T., Böttger, E.C., 1990. Differentiation of *Mycobacterium* species by direct sequencing of amplified DNA. J. Gen. Microbiol. 136 (9), 1915–1920.

Roth, A., Fischer, M., Hamid, M.E., Michalke, S., Ludwig, W., Mauch, H., 1998. Differentiation of phylogenetically related slowly growing mycobacteria based on 16S-23S rRNA gene internal transcribed spacer sequences. J. Clin. Microbiol. 36 (1), 139–147.

Runyon, E., 1965. Pathogenic mycobacteria. Adv. Tuberc. Res. 14, 235–287.

Runyon, E., Selin, M., Harris, H.W., 1959. Distinguishing mycobacteria by the niacin test. A modified procedure. American review of tuberculosis and pulmonary. Diseases 79 (5), 663–665.

Runyon, E.H., 1959. Anonymous mycobacteria in pulmonary disease. Med. Clin. North Am. 43 (1), 273–290.

Shinnick, T., Good, R., 1994. Mycobacterial taxonomy. Eur. J. Clin. Microbiol. 13 (11), 884–901.

Smida, J., Kazda, J., Stackebrandt, E., 1988. Molecular-genetic evidence for the relationship of *Mycobacterium leprae* to slow-growing pathogenic mycobacteria. Int. J. Lepr. 56, 449–454.

Sneath, P.H.A., 1970. In: Prauser, H. (Ed.), The Actinomycetales. Fischer, Jena, pp. 371–377.

Sneath, P., 1968. Vigour and pattern in taxonomy. J. Gen. Microbiol. 54 (1), 1–11.

Sola, C., Filliol, I., Gutierrez, M.C., Mokrousov, I., Vincent, V., Rastogi, N., 2001. Spoligotype database of *Mycobacterium tuberculosis*: biogeographic distribution of shared types and epidemiologic and phylogenetic perspectives. Emerg. Infect. Dis. 7 (3), 390–396.

Stackebrandt, E., Ludwig, W., Weizenegger, M., Dorn, S., McGill, T.J., Fox, G.E., et al., 1987. Comparative 16S rRNA oligonucleotide analyses and murein types of round-spore-forming bacilli and non-spore-forming relatives. J. Gen. Microbiol. 133 (9), 2523–2529.

Stahl, D.A., Urbance, J., 1990. The division between fast-and slow-growing species corresponds to natural relationships among the mycobacteria. J. Bacteriol. 172 (1), 116–124.

Sykes, G., Skinner, F.A., 1973. Actinomycetales: Characteristics and Practical Importance. Academic Press, New York.

Taylor, T.B., Patterson, C., Hale, Y., Safranek, W.W., 1997. Routine use of PCR-restriction fragment length polymorphism analysis for identification of mycobacteria growing in liquid media. J. Clin. Microbiol. 35 (1), 79–85.

Telenti, A., Marchesi, F., Balz, M., Bally, F., Böttger, E., Bodmer, T., 1993. Rapid identification of mycobacteria to the species level by polymerase chain reaction and restriction enzyme analysis. J. Clin. Microbiol. 31 (2), 175–178.

Thierry, D., Brisson-Noel, A., Vincent-Levy-Frebault, V., Nguyen, S., Guesdon, J., Gicquel, B., 1990. Characterization of a *Mycobacterium tuberculosis* insertion sequence, IS6110, and its application in diagnosis. J. Clin. Microbiol. 28 (12), 2668–2673.

Tiwari, T.S., Ray, B., Jost Jr, K.C., Rathod, M.K., Zhang, Y., et al., 2003. Forty years of disinfectant failure: outbreak of postinjection *Mycobacterium abscessus* infection caused by contamination of benzalkonium chloride. Clin. Infect. Dis. 36 (8), 954–962.

Tortoli, E., 2006. The new mycobacteria: an update. FEMS Immun. Med. Mic. 48 (2), 159–178.

Tortoli, E., Nanetti, A., Piersimoni, C., Cichero, P., Farina, C., Mucignat, G., et al., 2001. Performance assessment of new multiplex probe assay for identification of mycobacteria. J. Clin. Microbiol. 39 (3), 1079–1084.

Tsukamura, M., 1984. The "non-pathogenic" species of mycobacteria: their distribution and ecology in non-living reservoirs. Mycobacteria A Sourcebook Part B 1339–1359.

Turenne, C.Y., Tschetter, L., Wolfe, J., Kabani, A., 2001. Necessity of quality-controlled 16S rRNA gene sequence databases: identifying nontuberculous *Mycobacterium* species. J. Clin. Microbiol. 39 (10), 3637–3648.

Van Soolingen, D., Qian, L., De Haas, P., Douglas, J.T., Traore, H., Portaels, F., et al., 1995. Predominance of a single genotype of *Mycobacterium tuberculosis* in countries of East Asia. J. Clin. Microbiol. 33 (12), 3234–3238.

Velayati, A.A., Farnia, P., 2016. Atlas of *Mycobacterium tuberculosis*. Academic Press, New York.

Vestal, A., Kubica, G., 1966. Differential colonial characteristics of mycobacteria on Middlebrook and Cohn 7H11 agar-base medium. Am. Rev. Respir. Dis. 94 (2), 247–252.

Vincent, V., 1994. Les mycobacteries: bacilles de la tuberculosis. Manuel de bactériologie clinique 2, 899–922.

Wayne, L.G., 1981. Numerical taxonomy and cooperative studies: roles and limits. Rev. Infect. Dis. 3 (5), 822–888.

Wayne, L.G., Diaz, G., 1985. Identification of mycobacteria by specific precipitation of catalase with absorbed sera. J. Clin. Microbiol. 21 (5), 721–725.

Wayne, L., Lin, K.-Y., 1982. Glyoxylate metabolism and adaptation of *Mycobacterium tuberculosis* to survival under anaerobic conditions. Infect. Immun. 37 (3), 1042–1049.

Wayne, L., Dietz, T., Gernez-Rieux, C., Jenkins, P., Kappler, W., Kubica, G., et al., 1971. A co-operative numerical anaysis of scotochromogenic slowly growing mycobacteria. J. Gen. Microbiol. 66 (3), 255–271.

Wayne, L., Kubica, G., 1986. Genus *Mycobacterium* Lehmann and Neumann 1896, 363. In: Sneath, P.H.A., Mair, N.S., Sharpe, M.E., et al.,Bergey's Manual of Systematic Bacteriology, 2. Williams & Wilkins, Athens, GA, pp. 1436–1457.

Wayne, L., Good, R., Krichevsky, M., Blacklock, Z., David, H., Dawson, D., et al., 1989. Third report of the cooperative, open-ended study of slowly growing mycobacteria by the International Working Group on Mycobacterial Taxonomy. Int. J. Syst. Bacteriol. 39 (3), 267–278.

Wayne, L., Good, R., Krichevsky, M., Blacklock, Z., David, H., Dawson, D., et al., 1991. Fourth report of the cooperative, open-ended study of slowly growing mycobacteria by the International Working Group on Mycobacterial Taxonomy. Int. J. Syst. Bacteriol. 41 (4), 463–472.

Woods, G.L., 2002. The mycobacteriology laboratory and new diagnostic techniques. In: Griffith, D. (Ed.), Infect. Dis. North Am., 16. Mycobacterial Infect., pp. 127–144. (1).

World Health Organization., 1998. Laboratory Services in Tuberculosis Control. Geneva.

Zhang, Y., Heym, B., Allen, B., Young, D., Cole, S., 1992. The catalase—peroxidase gene and isoniazid resistance of *Mycobacterium tuberculosis*. Nature. 358, 591–593.

Ziehl, F., 1882. Zur Farbung des tuberkelbacillus. Deut. Med. Wochenschr. 8 (33), 451.

Zopf, W., 1883. Die Spaltpilze. Breslau: Eduard Trewendt, German.

FURTHER READING

Harada, K., 1976. The nature of mycobacterial acid-fastness. Stain Technol. 51 (5), 255–260.

Saviola, B., Bishai, W., 2006. The genus *Mycobacterium*–medical, Prokaryotes, 3. Springer, pp. 919–933.

Chapter 3

Susceptibility Testing of Nontuberculous Mycobacteria

Imran Ahmed, Rumina Hasan and Sadia Shakoor
Department of Pathology and Laboratory Medicine, Aga Khan University, Karachi, Pakistan

Nontuberculous mycobacteria (NTM) are environmental microorganisms widely distributed in nature, but mainly in aquatic niches. With the advancement in identification of microorganisms, many new species are being described with increasing reports of NTM infections from around the world.

Antimicrobial resistance is a global issue and it has not spared NTM. Many countries have reported clarithromycin resistance among various NTM species (Fig. 3.1) (Chew et al., 2017; Zheng et al., 2017; Li et al., 2017; Zhao et al., 2014; Nie et al., 2014; Kim et al., 2015; Lee et al., 2014; Koh et al., 2016; Heidarieh et al., 2016; Ramis et al., 2015; Candido et al., 2014; Coolen et al., 2015; Bryant et al., 2016; Chua et al., 2015; Boyle et al., 2016; Luo et al., 2015; Kadota et al., 2016; Broda et al., 2013). Spread of clarithromycin resistance raises a concern about dissemination of resistance strains, but also highlights the need to implement antimicrobial susceptibility testing in NTM.

Given the frequency of NTM in the environment, both, the American Thoracic Society (ATS) (Griffith et al., 2007) and the British Thoracic Society guidelines for the management of NTM pulmonary disease (Haworth et al., 2017) recommend that susceptibility testing for NTM should only be carried out for isolates where there is clinical suspicion of disease. This is to avoid unnecessary cost of testing and treatment, as well as to conserve laboratory resources. However, with the increasing frequency of nonpulmonary NTM infections, susceptibility testing will increasingly be attractive to predict success of treatment for clinically significant isolates from blood, sterile body fluids, and skin and soft tissue infections.

Nontuberculous Mycobacteria (NTM). DOI: https://doi.org/10.1016/B978-0-12-814692-7.00003-6

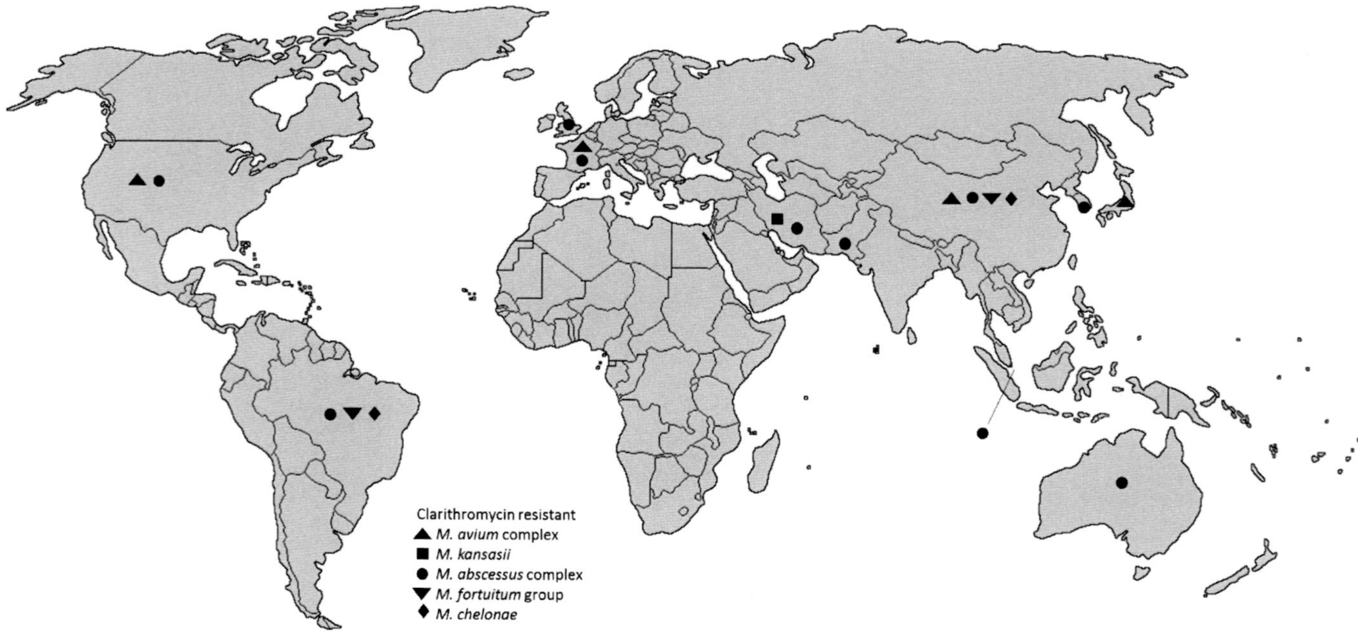

FIGURE 3.1 Reported clarithromycin resistance in selected mycobacteria species: 2013–17 (Chew et al., 2017; Zheng et al., 2017; Li et al., 2017; Zhao et al., 2014; Nie et al., 2014; Kim et al., 2015; Lee et al., 2014; Koh et al., 2016; Heidarieh et al., 2016; Ramis et al., 2015; Candido et al., 2014; Coolen et al., 2015; Bryant et al., 2016; Chua et al., 2015; Boyle et al., 2016; Luo et al., 2015; Kadota et al., 2016; Broda et al., 2013). Resistance to clarithromycin in *M. abscessus* is a recent finding from Pakistan and has not been published.

DETERMINATION OF CLINICAL SIGNIFICANCE

The site of isolation, number of positive cultures, clinical history, and immune status of patients are important factors to determine the clinical significance of a particular NTM isolate (Griffith et al., 2007). Although NTM frequently cause disease in the immunocompromised, NTM infections are reported in immune competent individuals as well. Possibility of NTM lung infection should especially be borne in mind in patients with postinfectious bronchiectasis presenting with new-onset pneumonia.

Healthcare-associated NTM infections are well documented in the literature. NTM have been associated with central line—associated bloodstream infections (El Helou et al., 2013; Al Yazidi et al., 2017), breast implants (Macadam et al., 2007; Romero et al., 2017), orthopedic implants (Eid et al., 2007), liposuction procedures (Hui et al., 2015; Al Soub et al., 2008; Kim and Mascola, 2010), injection abscesses (Yuan et al., 2009), and other healthcare-associated infections (Schnabel et al., 2016; Allen et al., 2017; Lee et al., 2016). NTM infections following cosmetic and tattooing procedures have also been reported (Falsey et al., 2013). However, in these cases, too, it is important to correlate NTM isolates with patient presentation and ensure clinical significance of the isolates. Pseudo-outbreaks involving NTM have also been reported in the literature (Guimaraes et al., 2016). ATS diagnostic criteria for NTM lung disease are helpful for making a diagnosis (Griffith et al., 2007). These criteria also appear in a modified form in the British Thoracic Society guidelines (Haworth et al., 2017).

MECHANISMS OF RESISTANCE

Mycobacteria have a hydrophobic and relatively impermeable cell wall which is the basis of their inherent resistance to several antibiotic compounds. All mycobacteria also possess beta-lactamases rendering penicillin ineffective and may also possess cephalosporinases. Mycobacteria are also rich in drug modifying enzymes and systems and in regulators and inducers of genes mediating susceptibility to antibiotics. Among these are the aminoglycoside modifying enzymes, chromosomal transcriptional regulator/promoter genes which activate antibiotic resistance mechanisms (e.g., *whiB7*) (Burian et al., 2012), erythromycin ribosomal methylase (*erm*) gene mediating inducible resistance to macrolides, a conserved DNA gyrase quinolone resistance determining region conferring low levels of susceptibility to quinolones (Guillemin et al., 1998), and several efflux pumps mediating resistance to aminoglycosides, tetracyclines, carbapenems, quinolones, and rifamycins (van Ingen et al., 2012).

Mutations in the mycobacterial genome result in acquired resistance against aminoglycosides, macrolides, and rifampicin. Mutations in the 16S ribosomal RNA gene *rrs* and in the 23S gene *rrl* are responsible for acquired aminoglycoside and macrolide resistance, respectively, in *M. abscessus*, *M. chelonae*, and *M. avium* complex (MAC). Acquired rifampicin resistance in *M. kansasii* and MAC is due to mutations in the *rpo*B gene.

Presence of extrachromosomal plasmid DNA has been implicated in resistance to aminoglycosides among MAC, albeit not confirmed (Franzblau et al., 1986). Due to the problems of eluting plasmid DNA in mycobacteria, this mechanism has been understudied; however, NTM are known to carry plasmids (Hull et al., 1984), and horizontal gene transfer through plasmids may be responsible for antibiotic resistance.

PHENOTYPIC DRUG SUSCEPTIBILITY TESTING FOR NONTUBERCULOUS MYCOBACTERIA

Drug susceptibility testing (DST) of NTM is relatively nascent when compared to that of other bacterial pathogens and even *Mycobacterium tuberculosis* complex (MTC). NTM are a diverse group and divided into two main groups; rapidly growing mycobacteria (RGM) and slow growing mycobacteria (SGM). Different methods have had to be tried, developed, and tested for each RGM and SGM.

Drug susceptibility test results are reported as one of three categories: sensitive, intermediate, or resistant. A sensitive result for an organism–antibiotic combination implies that the clinical infection with the test organism can be reliably treated with a standard recommended dosage of the antibiotic. A resistant result suggests that the organism will not be suppressed by the antibiotic at a specific infection site with any recommended dosage, while an intermediate category result indicates that suppression may be achieved if a higher dosage is used. These results help physicians modify antibiotic treatment regimens, where resistant antibiotics are discontinued to avoid the unnecessary toxicity and cost of treatment.

To ensure reliability and precision of susceptibility test results, quality control procedures are implemented that monitor the quality of reagents, test systems and technical competence of personnel carrying out the DST. This is largely achieved by regularly testing reference strains that may belong to the American Type Culture Collection (ATCC), or another strain repository.

Methods for Susceptibility Testing of Rapidly Growing Mycobacteria

Methods used for DST of other bacteria and MTC have been applied to RGM. These include disk diffusion, disk elution in agar, E-test, and testing of minimum inhibitory concentrations (MICs) by broth micro- and macrodilution. The only standardized method, validated in multicenter studies, is broth microdilution.

Broth microdilution method for RGM is performed with 0.5 McFarland standard of the organism inoculated into cation-adjusted Mueller Hinton Broth (CAMHB). For some RGM that are difficult to grow, CAMHB may be supplemented with oleic acid–albumin–dextrose–catalase (OADC). The inoculum is then introduced into a 96-well microtiter plate with antibiotic containing wells, achieving a final concentration of 5×10^5 CFU of bacteria per well. Antibiotic-free control wells are always included on the plate to compare and exclude contamination (CLSI, 2011). The MIC is the lowest concentration of an antibiotic able to prevent visible growth of the test organism. It is essential that each laboratory performing in-house MICs validate their methods to avoid errors of reporting that may be due to ineffective technique or reagents. For commercial methods that have been validated by the manufacturers and subsequently approved by the US Food and Drug Administration (FDA) or compliance with Conformité Européene standards (CE marked), laboratories should follow recommendations to verify that systems are performing as intended by the manufacturers. Quality control organisms to ensure reliability and correctness of methods for RGM include reference strains of *M. peregrinum* ATCC 700686, *Staphylococcus aureus* ATCC 29213, or *Enterococcus faecalis* ATCC 29212. Reference ranges for these organisms against antimycobacterial agents can be found in Clinical Laboratory Standards Institute (CLSI) documents M24-A2 and M100.

Methods for Susceptibility Testing of Slow Growing Mycobacteria

Broth microdilution and macrodilution are the only reliable methods for phenotypic DST of SGM. Broth macrodilution is best performed in the BACTEC 460 system; however, this system has been discontinued by the manufacturer, due to radioactivity and safety concerns, and can no longer be accessed. Broth microdilution remains the reference method for DST of SGM. The methods for inoculum preparation and testing are same as for RGM, except that CAMHB is always supplemented with 5%OADC. Quality control strains for SGM MIC testing include *M. avium* ATCC 700898 and *M. marium* ATCC 927 for MAC; *M. kansasii* ATCC 12478 for *M. kansasii*, MIC ranges for which can be found in CLSI document M24-A2.

As significance of SGM infections is increasingly being recognized, and, as new species are discovered, new methods need to be validated for each clinically significant species for which susceptibility testing may be required. Some species may require specialized methods; *M. hemophilum* is one species for which the recommendation (based on single center studies) is to use agar disk elution (CLSI, 2011), but no quality control strains or guidelines are available, and in-house validation is mandatory.

Commercial systems are available for DST of both RGM and SGM. Table 3.1 provides a brief description of these systems and their

TABLE 3.1 Commercial Systems for NTM DST

Test Name	Company	Description
Sensititre RAPMYCO	TREK Diagnostic Systems, Thermo Scientific, UK.	Dry broth microdilution, 96-well plates for rapid growing mycobacteria.
Sensitre SLOMYCO		Dry broth microdilution 96-well plates for slow growing mycobacteria.
		These may be used as standalone MIC plates or may be combined with Sensitire AutoReader for reading the endpoint MICs.

manufacturer. These systems are CE marked for in vitro diagnostics (IVD) use, however, not approved by the FDA and therefore not available for IVD use in the United States (where the commercial DST systems may be obtained for research use only).

Performing and reading MICs of RGM is easier compared to SGM. In our experience, reading the broth microdilution of SGM is particularly difficult owing to the fact that growth is faint and, hence, considerable time and expertise are required to read MICs. Automated readers for broth microdilution trays are available and are designed for specific plate formats, for example, Sensitire AutoReader (Trek Diagnostic Systems, ThermoFisher Scientific). This not only reads MIC values, but can also provide interpretations, transfer of data to laboratory information systems and report generation.

Susceptibility Interpretive Criteria (Breakpoints)

MIC cutoff values used to discriminate strains that are sensitive, intermediate, or resistant to a given antibiotic are referred to as breakpoints. Since the clinical significance of NTM was recognized, CLSI, previously National Committee on Clinical Laboratory Standards (NCCLS), developed and released breakpoints for susceptibility testing of some common NTM species in their document M24. Table 3.2 shows a list of NTM for which drug susceptibility breakpoints have been provided by CLSI.

Several NTM species are inherently resistant to a number of antimicrobial agents. As such, treatment regimens and antimicrobials to be tested vary depending on the species. Recommended antibiotics for individual NTM and their susceptibility breakpoints (MIC values expressed in μg/mL) as prescribed by CLSI are shown in Table 3.3. While the European Committee on

TABLE 3.2 List of NTMs for Which Drug Susceptibility Breakpoints Have Been Published by CLSI

SGM[a]	RGM[b]
M. avium complex	Common drug susceptibility criteria
M. kansasii	
M. marinum	
M. hemophilum	

NTM, Nontuberculous mycobacteria; *CLSI*, Clinical and Laboratory Standards Institute document M24-A2; *SGM*, slow growing mycobacteria; *RGM*, rapid growing mycobacteria.
[a]*Other than the SGM mentioned in this table, CLSI suggests broth microdilution method for certain nonfastidious, slow growing NTM species, for example,* M. terrae, M. xenopi, M. malmoense, *and* M. simiae. *The same methods, drugs, and interpretive criteria should be used as for* M. kansasii.
[b]*Common RGM species include Mycobacterium abscessus group (M. abscessus ssp. abscessus, M. abscessus ssp. massiliense, M. abscessus ssp. bolletti), M. fortuitum group (M. fortuitum ssp. fortuitum, M. fortuitum ssp. peregrinum, M. fortuitum ssp., M. fortuitum ssp. senegalense, M. fortuitum ssp. septicum, M. fortuitum ssp. alvei, M. fortuitum ssp. houstonense, M. fortuitum ssp. boenickei, M. fortuitum ssp. conceptionense, M. fortuitum ssp. porcinum, M. fortuitum ssp. neworleansense, M. fortuitum ssp. brisbanense, and M. fortuitum ssp. mageritense), and M. chelonae.*

Antimicrobial Susceptibility Testing (EUCAST) does not provide any guideline for susceptibility testing of NTM, published guidelines by the British Thoracic Society for NTM pulmonary disease (Haworth et al., 2017) mention CLSI for susceptibility breakpoints.

Breakpoints were traditionally derived in two ways: epidemiologically and clinically. Epidemiological cutoffs (ECOFFs) are established by testing a large number of wild-type (treatment-naïve and unexposed) bacteria to determine the MIC cutoff differentiating the bacterial population with acquired resistance. Clinically derived breakpoints are more relevant in identifying MICs that correlate with treatment success and are established on the basis of prospective clinical studies comparing MICs of infecting organisms with treatment outcomes. However, concentrations achievable at the site of infection based on the pharmacokinetics (PK) and pharmacodynamics (PD) of the antibiotic of interest may differ from in vitro MICs so PK/PD derived breakpoints are considered superior to clinical breakpoints alone or ECOFFs. However, this requires extensive PK–PD studies at various infection sites, dose–effect relationships, and finally, correlation of theoretically derived PK/PD breakpoints with clinical outcomes. For NTM, currently available breakpoints are ECOFFs, with very few instances of clinical studies; consequently, many available breakpoints have poor correlation with clinical outcomes. This poses a challenge for microbiologists and clinicians alike when advising/deciding on a regimen for a particular NTM infection based on in vitro susceptibility results. PK/PD approaches, however, are now being

TABLE 3.3 Drug Susceptibility Breakpoints (MIC Values Expressed in μg/mL) for NTM as Prescribed by CLSI (CLSI Document M24-A2)[a,b,c]

Drug	Susceptibility Category	Nontuberculous Mycobacteria			
		MAC[d]	MK[e]	MM	RGM[f]
AK	S				≤16
	I				32
	R		>32	>32	≥64
CIP	S				≤1
	I				2
	R		>2	>2	≥4
CLA[g]	S	≤8			≤2
	I	16			4
	R	≥32	>16	>16	≥8
DOX	S				≤1
	I				2–4
	R			>4	≥8
ETH	S				
	I				
	R		>4	>4	
FOX	S				≤16
	I				32–64
	R				≥128
IPM	S				≤4
	I				8–16
	R				≥32
LZD	S	≤8			≤8
	I	16			16
	R	≥32	>16		≥32
MEM	S				≤4
	I				8–16
	R				≥32

(*Continued*)

TABLE 3.3 (Continued)

Drug	Susceptibility Category	Nontuberculous Mycobacteria			
		MAC[d]	MK[e]	MM	RGM[f]
MIN	S				
	I				
	R			>4	
MOX	S	≤1			≤1
	I	2			2
	R	≥4	>2	>2	≥4
RFB	S				
	I				
	R		>2	>2	
RIF	S				
	I				
	R		>1	>1	
SXT	S				≤2/38
	I				
	R		>2/38	>2/38	≥4/76
TOB[h]	S				≤2
	I				4
	R				≥8

MIC, Minimum inhibitory concentration; *NTM*, nontuberculous mycobacteria; *CLSI*, Clinical & Laboratory Standards Institute; *MAC*, *Mycobacterium avium* complex; *MK*, *M. kansasii*; *MM*, *M. marinum*; *RGM*, rapidly growing mycobacteria; *AK*, amikacin; *CIP*, ciprofloxacin; *CLA*, clarithromycin; *DOX*, doxycycline; *ETH*, ethambutol; *FOX*, cefoxitin; *IPM*, imipenem; *LZD*, linezolid; *MEM*, meropenem; *MIN*, minocycline; *MOX*, moxifloxacin; *RFB*, rifabutin; *RIF*, rifampicin; *SXT*, trimethoprim-sulfamethoxazole; *TOB*, tobramycin.
[a]*Susceptibility method: broth microdilution using cation-adjusted Muller-Hinton broth (CAMHB). For slow growing NTM species, addition of oleic acid albumin dextrose and catalase (OADC) in CAMHB is advised.*
[b]*For* M. haemophilum, *agar disk elution method is described by CLSI. See CLSI document M24-A2 for details.*
[c]*CLSI suggests broth microdilution using CAMHB with OADC for certain other nonfastidious slow growing NTM species, for example,* M. terrae, M. xenopi, M. malmoense, *and* M. simiae. *Same methods, drugs and same interpretive criteria should be used as for* M. kansasii.
[d]*Linezolid and moxifloxacin breakpoints are tentative. For MAC, if performed by radiometric method at pH 6.8, the breakpoints for CLA are: susceptible ≤ 16, intermediate 32, resistant ≥ 64 μg/mL and if performed at pH 7.3–7.4, the breakpoints are: susceptible ≤ 4, intermediate 8–16, resistant ≥ 32 μg/mL.*
[e]*Susceptibility breakpoints for isoniazid and streptomycin are not described but may be useful clinically.*
[f]*Susceptibility breakpoints for IPM are tentative.*
[g]*Clarithromycin is class representative for azithromycin and roxithromycin.*
[h]*Tobramycin should not be reported for* M. abscessus *or* M. fortuitum *group.*

employed to determine therapeutic levels of antibiotics and antibiotic combinations. Until breakpoints based on species- and site-specific PK/PD data, in vitro MICs, and clinical outcome data, become available, this limitation of contemporary DST should be kept in mind.

Rapid Phenotypic Drug Susceptibility Testing for Nontuberculous Mycobacteria

Current phenotypic methods used to perform DST for NTMs are slow, time consuming, and require considerable technical resources. Newer methodologies are being developed. Matrix assisted laser desorption ionization (MALDI) time of flight mass spectrometry has become a standard technique for identification of microorganisms including mycobacteria. MALDI Biotyper antibiotic susceptibility test rapid assay is a recent technique described for susceptibility testing (Lange et al., 2014). This technique is being applied for susceptibility testing of mycobacteria (both *M. tuberculosis* and NTM) (Ceyssens et al., 2017). Although in its infancy, this technique has been shown to have good correlation for resistance detection when compared to standard phenotypic procedures.

GENOTYPIC DRUG SUSCEPTIBILITY TESTING FOR NONTUBERCULOUS MYCOBACTERIA

Targeted molecular detection of resistance genes associated with macrolide and aminoglycoside resistance have been validated for *M. abscessus* complex and MAC. A commercial line-probe method is available for the detection of resistance associated mutations in *erm* (inducible macrolide resistance), *rrl* (constitutive macrolide resistance), and *rrs* (aminoglycoside resistance) genes in NTM (Hain Lifescience GmbH). Sequencing approaches can also be used, but have not been standardized. A recent study identified some single nucleotide polymorphisms (SNPs) with possible correlation with inducible macrolide resistance in *M. abscessus* on whole genome sequencing (WGS) and also identified possible genetic correlates of phenotypic fluoroquinolone and amikacin resistance (Lipworth et al., 2018). WGS has the potential to detect additional SNPs that are missed by the line-probe assay. Further standardization and association studies of WGS for NTM resistance determination can lead to better utility of the method for prediction of NTM resistance in all species.

TREATMENT OF NONTUBERCULOUS MYCOBACTERIA INFECTIONS

1. SGM treatment regimens:
   a. MAC: CLSI provides drug susceptibility breakpoints (susceptible, intermediate and resistant) for clarithromycin, moxifloxacin and

linezolid. However, clinical correlation has only been made for clarithromycin. Clinical regimens used for treating MAC lung disease are based on clarithromycin (or azithromycin) with ethambutol and rifampicin (or rifabutin). Streptomycin or amikacin is also added in some advanced or previously treated disease for first 2–3 months of therapy. One prospective study compared a three-drug regimen (clarithromycin, ethambutol and rifampicin) with a two-drug regimen (clarithromycin and ethambutol) for MAC lung disease in HIV negative patients (Miwa et al., 2014). Treatment was given for 12 months and the primary endpoint was sputum culture conversion to negative after 12 months of treatment. Two-drug regimen was found to be noninferior to three-drug regimen. Authors suggested a larger and long-term clinical trial to compare the two treatment regimens. Adding moxifloxacin or clofazimine to the regimen was shown to be beneficial with improved outcomes in refractory cases (Jarand et al., 2016; Koh et al., 2013). Table 3.4 describes treatment regimens and durations currently recommended. Treatment outcomes were poor in patients infected with macrolide-resistant MAC isolates (Moon et al., 2016; Jo et al., 2014). In a recent meta-analysis, outcomes of several different regimens used for pulmonary MAC disease were poor (Pasipanodya et al., 2017a).

There is, thus, a need for better regimens based on clinical and PK/PD validation of susceptibility breakpoints for the other drugs used in the therapeutic regimens for MAC pulmonary disease. To initiate this process of clinical correlation, it is necessary to determine MIC distribution of various drugs used for treating MAC infections followed by investigations to explore whether therapeutic levels are achievable in the targeted tissue for a particular drug. MIC distribution of 229 clinical MAC isolates was determined against clarithromycin, rifampicin, rifabutin, ethambutol, amikacin, moxifloxacin, linezolid, and trimethoprim-sulfamethoxazole (Schon and Chryssanthou, 2017). This study established that MIC values were high for all of these drugs (except for clarithromycin) and that the therapeutic concentrations were not achievable in the body at recommended dosing. There may be value in determining synergy between various antibiotic combinations, and bactericidal effect of regimens, but methods for in vitro testing and endpoints for in vivo outcomes have not been developed and standardized.

b. *M. kansasii*: MIC values for classifying an *M. kansasii* isolate as resistant only have been provided by CLSI (Table 3.1). These include clarithromycin (also represents azithromycin) and rifampicin as primary agents, whereas, ethambutol, amikacin, ciprofloxacin, moxifloxacin, rifabutin, linezolid, and trimethoprim-sulfamethoxazole as secondary agents (CLSI, 2011). Isoniazid is used clinically to treat

TABLE 3.4 Antimicrobial Treatment Regimens for MAC Pulmonary Disease

Treatment Regimen	Duration	Comments	Susceptibility Breakpoints	Reference
Mild–moderate disease				
Azithromycin or clarithromycin and rifampicin and ethambutol.	Antibiotic treatment should continue for a minimum of 12 months after culture conversion.	3 times/week regimen. Macrolide alone or macrolide/ quinolone combination should not be used because of risk of development of macrolide resistance.	Available for clarithromycin, linezolid and moxifloxacin (CLSI, 2011).	Haworth et al. (2017), Griffith et al. (2006), Wallace et al. (2014), Park et al. (2017b)
Severe disease				
Azithromycin or clarithromycin and rifampicin and ethambutol and consider intravenous or nebulized amikacin for up to 3 months.	Antibiotic treatment should continue for a minimum of 12 months after culture conversion.	Daily regimen. Intermittent regimen should not be used. Macrolide alone or macrolide/ quinolone combination should not be used.		
		Treatment outcomes are worse for patients with NTM pulmonary disease with associated pleural effusion/ pleuritis due to NTM.		

(Continued)

TABLE 3.4 (Continued)

Treatment Regimen	Duration	Comments	Susceptibility Breakpoints	Reference
Clarithromycin-resistant MAC pulmonary disease				
Rifampicin and ethambutol and Isoniazid (+pyridoxine) or moxifloxacin and consider intravenous or nebulized amikacin for up to 3 months.	Antibiotic treatment should continue for a minimum of 12 months after culture conversion.	Daily regimen. Intermittent regimen should not be used. Macrolide alone or macrolide/quinolone combination should not be used. Amikacin therapy is effective in achieving culture conversion.		Haworth et al. (2017) Yagi et al. (2017) Olivier et al. (2014)

M. kansasii infections, however, susceptibility breakpoints for isoniazid have not been established. Regimens that are used to treat *M. kansasii* include a combination of clarithromycin (or azithromycin), rifampicin, and ethambutol. Alternate regimen replaces clarithromycin with isoniazid. Regimens and their recommended durations are shown in Table 3.5. There is emerging evidence that moxifloxacin has good in vitro activity, achieves inhibitory concentrations at the site of infection, and has the potential to shorten treatment duration for *M. kansasii* pulmonary disease, but larger studies are required (Srivastava et al., 2015; Padmapriyadarsini et al., 2016).

2. RGM treatment regimens:
   There are common drug susceptibility breakpoints provided by CLSI for RGM (CLSI, 2011). Correlation of DST results with the clinical outcomes are still lacking and it is important to embark on such a journey. It is not only the treatment outcomes of pulmonary infections, but also extrapulmonary infections caused by RGM that need to be correlated with the DST breakpoints.

Resistance to antimicrobial agents in NTMs may either be constitutive or inducible. This is a particular concern for macrolides. Inducible macrolide

TABLE 3.5 Antimicrobial Treatment Regimens for *M. kansasii* Pulmonary Disease

Regimen	Duration	Comments	Susceptibility Breakpoints	References
Rifampicin and ethambutol and isoniazid (with pyridoxine) or azithromycin or clarithromycin.	Antibiotic treatment should continue for a minimum of 12 months after culture conversion.	Daily regimen	Available for amikacin, ciprofloxacin, clarithromycin, doxycycline, cefoxitin, imipenem, meropenem, linezolid, moxifloxacin, cotrimoxazole, and tobramycin (CLSI, 2011).	Haworth et al. (2017)

resistance is mostly found in *M. abscessus* ssp. *abscessus*, but has also been reported in other RGM including *M. fortuitum* (Nash et al., 2005). When performing DST for RGM, it is therefore important to detect inducible macrolide resistance in the isolate (Nash 2003; Nash et al., 2006, 2009). While constitutive macrolide resistance can easily be detected in the routine 3-day susceptibility protocol, inducible macrolide-resistant isolates are susceptible at 3-days, but show resistance at 14 days. Therefore, to detect inducible macrolide resistance, it is imperative that the broth dilution MIC be incubated for a total of 14 days.

DST and its interpretation are inextricably linked to correct species (and, in some cases, correct subspecies) identification. *Mycobacterium abscessus* subspecies *abscessus* usually possess an *erm* (erythromycin ribosomal methylase) gene that makes them inducible resistant to macrolides. Whereas, most *M. abscessus* ssp. *massiliense* have a nonfunctional *erm* gene (C28 sequevar) and hence are susceptible to macrolides (Yoshida et al., 2018; Park et al., 2017a; Koh et al., 2017). Response to treatment regimen has also been reported to be better in patients with *M. abscessus* ssp. *Massiliense* lung disease than with *M. abscessus* ssp. *abscessus* lung disease (Lyu et al., 2014). However, even if these *M. abscessus* ssp. *massiliense* isolates develop macrolide resistance, outcomes have been reported to be poor (Choi et al., 2017).

Table 3.6 Regimens currently recommended for treatment of *M. abscessus* pulmonary disease, suggested durations, and comments on treatment outcomes for both macrolide susceptible and resistant infections.

TABLE 3.6 Antimicrobial Treatment Regimens for *M. abscessus* Complex Pulmonary Disease

Regimen	Duration	Comments	Susceptibility Breakpoints	References
Clarithromycin Susceptible or Inducible Macrolide Resistant Isolates				
Initial phase: Intravenous amikacin and intravenous tigecycline and where tolerated intravenous imipenem and where tolerated oral clarithromycin or oral azithromycin. *Continuation phase:* Nebulized amikacin and oral clarithromycin or azithromycin and 1–3 of the following antibiotics guided by drug susceptibility results and patient tolerance: oral clofazimine oral linezolid oral minocycline oral moxifloxacin oral cotrimoxazole.	*Initial phase:* ≥1 month *Continuation phase:* Antibiotic treatment should continue for a minimum of 12 months after culture conversion. Long-term suppressive antibiotic regimen may benefit those patients who fail to culture-convert.	Treatment outcomes are better in patients with clarithromycin susceptible isolates. These isolates are mostly *M. abscessus* ssp. *massiliense* which contain a nonfunctional *erm* (41) gene. Due to the poorer response rates in patients with inducible or constitutive macrolide-resistant isolates and the greater efficacy of antibiotics administered through the intravenous route, extending the duration of intravenous antibiotic therapy to 3–6 months in those that can	Available for amikacin, ciprofloxacin, clarithromycin, doxycycline, cefoxitin, imipenem, meropenem, linezolid, moxifloxacin, Cotrimoxazole and tobramycin (CLSI, 2011).	Haworth et al. (2017), Yoshida et al. (2018) Lyu et al. (2014) Choi et al. (2017), Choi et al. (2018), Pasipanodya et al. (2017b)

(*Continued*)

TABLE 3.6 (Continued)

Regimen	Duration	Comments	Susceptibility Breakpoints	References
		tolerate it may be the most appropriate treatment strategy in this subgroup of patients.		
		Substitute intravenous/nebulized amikacin with an alternative antibiotic if the *M. abscessus* is resistant to amikacin.		
		Start clofazimine during the initial phase of treatment if tolerated as steady-state serum concentrations may not be reached until ≥ 30 days of treatment.		
		Treatment success rates are higher for *M. abscessus* ssp. *massiliense* when compared to *M. abscessus* ssp. *abscessus*. This is due to a dysfunctional *erm* (41) gene in *M. abscessus* ssp. *massiliense* and thus a shorter duration of therapy may be sufficient.		

Constitutive Macrolide Resistant Isolates				
Initial phase: Intravenous amikacin and intravenous tigecycline and where tolerated intravenous imipenem. *Continuation phase:* Nebulized amikacin and 2–4 of the following antibiotics guided by drug susceptibility results and patient tolerance: oral clofazimine oral linezolid Oral minocycline oral moxifloxacin oral cotrimoxazole.	*Initial phase:* ≥1 month *Continuation phase:* Antibiotic treatment should continue for a minimum of 12 months after culture conversion. Long-term suppressive antibiotic regimen may benefit those patients who fail to culture-convert.	Outcomes are poor for macrolide-resistant isolates. Due to the poorer response rates in patients with inducible or constitutive macrolide-resistant isolates and the greater efficacy of antibiotics administered through the intravenous route, extending the duration of intravenous antibiotic therapy to 3–6 months in those that can tolerate it may be the most appropriate treatment strategy in this subgroup of patients. Substitute intravenous/nebulized amikacin with an alternative antibiotic if the *M. abscessus* is resistant to amikacin. Start clofazimine during the initial phase of treatment if tolerated as steady-state serum concentrations may not be reached until ≥30 days of treatment.		

Antimicrobial Treatment for *M. chelonae* Pulmonary Disease

There are no established regimens for this pathogen. It does not harbor *erm* (inducible macrolide resistance) gene. Suggested combinations include a macrolide and other antimicrobials based on susceptibility test results (Brown-Elliott et al., 2012). Tobramycin, rather than amikacin, is the preferred aminoglycoside for *M. chelonae* isolates which test susceptible on DST. Treatment should be continued for 12 months after culture conversion (Griffith et al., 2007).

Antimicrobial Treatment for *M. fortuitum* Pulmonary Disease

Treatment options should be based on in vitro DST and usually a combination of two active drugs is prescribed (Brown-Elliott et al., 2012). *M. fortuitum* group has been reported for the presence of *erm* gene and, thus, use of macrolides depends on a 14-day incubation of susceptibility test to detect this mechanism of resistance (Nash et al., 2005). Duration of treatment is guided by response to therapy and is usually continued for 12 months after culture conversion.

3. Treatment of pulmonary disease for NTM without standardized DST or breakpoints:

   CLSI does not provide separate drug susceptibility breakpoint tables for nonfastidious, slow growing NTM other than MAC, *M. kansasii*, and *M. marinum*; however, it provides a comment to use same methods, drugs and same breakpoints as for *M. kansasii* to interpret MIC values for certain nonfastidious SGM, for example, *M. terrae*, *M. xenopi*, *M. malmoense*, and *M. simiae*.

Antimicrobial Treatment for *M. malmoense* Pulmonary Disease

British thoracic society (BTS) (Haworth et al., 2017) recommends a combination of rifampicin, ethambutol, and clarithromycin (or azithromycin) daily for nonsevere disease. For severe pulmonary disease (acid-fast bacilli (AFB) smear positive specimen, lung cavitation on radiology, or severe symptoms/signs of systemic disease) it is advised to consider intravenous or nebulized amikacin for initial 3 months in addition to the above regimen. Treatment should be continued for a minimum of 12 months following culture conversion.

Antimicrobial Treatment of *M. xenopi* Pulmonary Disease

According to BTS (Haworth et al., 2017), a multidrug regimen of rifampicin, ethambutol, azithromycin, and moxifloxacin (or isoniazid) daily is recommended for nonsevere diseases. The above regimen should be supplemented with intravenous or nebulized amikacin for initial 3 months of treatment for severe disease (AFB smear positive specimen, lung cavitation on radiology or severe symptoms/signs of systemic disease). Treatment should be continued for a minimum of 12 months following culture conversion.

Antimicrobial Treatment of *M. simiae* Pulmonary Disease

Optimal therapy is still not defined for this NTM species. A combination of clarithromycin (or azithromycin), moxifloxacin (not ciprofloxacin), and trimethoprim/sulfamethoxazole is suggested (Brown-Elliott et al., 2012).

Treatment of *M. terrae* Complex Infections

Infections are rare and it can result in both pulmonary and extrapulmonary infections (tenosynovitis). A combination of clarithromycin (or azithromycin) and ethambutol plus other agents, such as fluoroquinolone and trimethoprim-sulfamethoxazole based on DST, is recommended (Brown-Elliott et al., 2012).

Treatment of *M. szulgai* Infections

Drug susceptibility criteria have not been established for *M. szulgai.* It has been reported to cause both pulmonary and extrapulmonary infections (skin, bone, cornea, and lymph nodes). Antituberculous drugs, including isoniazid, rifampicin, and pyrazinamide, are effective against this NTM in addition to clarithromycin and fluoroquinolones. A 3–4 drug combination till the culture remains negative for 12 months has been suggested. For extrapulmonary infections a 4–6 month therapy should be successful (Griffith et al., 2007).

Treatment of *M. hemophilum* Infections

M. hemophilum causes lymphadenitis in immunocompetent children. Classic lesion is a chronic draining skin sinus/ulcer. In immunocompromised individuals, chronic cutaneous sinus/ulcer, lymphadenitis, joint infections, and disseminated disease have been reported. Surgical excision of affected lymph nodes in immunocompetent patients is usually sufficient. A multidrug regimen consisting of clarithromycin, rifampicin, and ciprofloxacin is suggested by the ATS (Griffith et al., 2007).

Treatment of *M. marinum* Infections

M. marinum is associated with chronic granulomatous infections of skin and bone (swimming pool granuloma or fish tank granuloma). A two-drug combination of clarithromycin and ethambutol (or rifampicin) is recommended, with the addition of rifampicin if deep tissues are involved or osteomyelitis (Griffith et al., 2007).

Treatment of *M. ulcerans* Infections

M. ulcerans causes buruli ulcer, an endemic disease in regions around tropical rain forests. Response to antimicrobial therapy is poor. Surgery is the mainstay of treatment (Griffith et al., 2007).

CONCLUSION

DST of NTM is recommended only for clinically significant isolates. Only a limited number of methods and interpretive criteria have been validated, standardized, and clinically correlated. As sophisticated methods evolve and are correlated with PK/PD to derive clinically meaningful breakpoints, and genotypic markers are validated through association studies, it is hoped that NTM infections will become better manageable with improved outcomes.

REFERENCES

Al Soub, H., Al-Maslamani, E., Al-Maslamani, M., 2008. *Mycobacterium fortuitum* abdominal wall abscesses following liposuction. Indian J. Plast. Surg. 41 (1), 58–61.

Al Yazidi, L.S., Marais, B.J., Hazelton, B., Outhred, A., Kesson, A., 2017. Nontuberculous mycobacteria in children: a focus on bloodstream infections. Pediatr. Infect. Dis. J. 36 (4), 374–378.

Allen, K.B., Yuh, D.D., Schwartz, S.B., Lange, R.A., Hopkins, R., Bauer, K., et al., 2017. Nontuberculous *Mycobacterium* infections associated with heater-cooler devices. Ann. Thorac. Surg. 104 (4), 1237–1242.

Boyle, D.P., Zembower, T.R., Qi, C., 2016. Relapse versus reinfection of *Mycobacterium avium* complex pulmonary disease. Patient characteristics and macrolide susceptibility. Ann. Am. Thorac. Soc. 13 (11), 1956–1961.

Broda, A., Jebbari, H., Beaton, K., Mitchell, S., Drobniewski, F., 2013. Comparative drug resistance of *Mycobacterium abscessus* and *M. chelonae* isolates from patients with and without cystic fibrosis in the United Kingdom. J. Clin. Microbiol. 51 (1), 217–223.

Brown-Elliott, B.A., Nash, K.A., Wallace Jr., R.J., 2012. Antimicrobial susceptibility testing, drug resistance mechanisms, and therapy of infections with nontuberculous mycobacteria. Clin. Microbiol. Rev. 25 (3), 545–582.

Bryant, J.M., Grogono, D.M., Rodriguez-Rincon, D., Everall, I., Brown, K.P., Moreno, P., et al., 2016. Emergence and spread of a human-transmissible multidrug-resistant nontuberculous *Mycobacterium*. Science 354 (6313), 751–757.

Burian, J., Ramon-Garcia, S., Sweet, G., Gomez-Velasco, A., Av-Gay, Y., Thompson, C.J., 2012. The mycobacterial transcriptional regulator whiB7 gene links redox homeostasis and intrinsic antibiotic resistance. J. Biol. Chem. 287 (1), 299–310.

Candido, P.H., Nunes Lde, S., Marques, E.A., Folescu, T.W., Coelho, F.S., de Moura, V.C., et al., 2014. Multidrug-resistant nontuberculous mycobacteria isolated from cystic fibrosis patients. J. Clin. Microbiol. 52 (8), 2990–2997.

Ceyssens, P.J., Soetaert, K., Timke, M., Van den Bossche, A., Sparbier, K., De Cremer, K., et al., 2017. Matrix-assisted laser desorption ionization-time of flight mass spectrometry for combined species identification and drug sensitivity testing in mycobacteria. J. Clin. Microbiol. 55 (2), 624–634.

Chew, K.L., Cheng, J.W.S., Hudaa Osman, N., Lin, R.T.P., Teo, J.W.P., 2017. Predominance of clarithromycin-susceptible *Mycobacterium massiliense* subspecies: characterization of the *Mycobacterium abscessus* complex at a tertiary acute care hospital. J. Med. Microbiol. 66 (10), 1443–1447.

Choi, H., Jhun, B.W., Kim, S.Y., Kim, D.H., Lee, H., Jeon, K., et al., 2018. Treatment outcomes of macrolide-susceptible *Mycobacterium abscessus* lung disease. Diagn. Microbiol. Infect. Dis. 90 (4), 293–295.

Choi, H., Kim, S.Y., Lee, H., Jhun, B.W., Park, H.Y., Jeon, K., et al., 2017. Clinical characteristics and treatment outcomes of patients with macrolide-resistant *Mycobacterium massiliense* lung disease. Antimicrob. Agents Chemother. 61 (2).

Chua, K.Y., Bustamante, A., Jelfs, P., Chen, S.C., Sintchenko, V., 2015. Antibiotic susceptibility of diverse *Mycobacterium abscessus* complex strains in New South Wales, Australia. Pathology 47 (7), 678–682.

CLSI, 2011. Susceptibility. Testing of Mycobacteria, Nocardiae, and other aerobic Actinomycetes; Approved Standard-Second Edition. CLSI document M24-A2. Clinical and Laboratory Standards Institute, Wayne, PA.

Coolen, N., Morand, P., Martin, C., Hubert, D., Kanaan, R., Chapron, J., et al., 2015. Reduced risk of nontuberculous mycobacteria in cystic fibrosis adults receiving long-term azithromycin. J. Cyst. Fibros. 14 (5), 594–599.

Eid, A.J., Berbari, E.F., Sia, I.G., Wengenack, N.L., Osmon, D.R., Razonable, R.R., 2007. Prosthetic joint infection due to rapidly growing mycobacteria: report of 8 cases and review of the literature. Clin. Infect. Dis. 45 (6), 687–694.

El Helou, G., Viola, G.M., Hachem, R., Han, X.Y., Raad, I.I., 2013. Rapidly growing mycobacterial bloodstream infections. Lancet. Infect. Dis 13 (2), 166–174.

Falsey, R.R., Kinzer, M.H., Hurst, S., Kalus, A., Pottinger, P.S., Duchin, J.S., et al., 2013. Cutaneous inoculation of nontuberculous mycobacteria during professional tattooing: a case series and epidemiologic study. Clin. Infect. Dis. 57 (6), e143–e147.

Franzblau, S.G., Takeda, T., Nakamura, M., 1986. Mycobacterial plasmids: screening and possible relationship to antibiotic resistance in *Mycobacterium avium/Mycobacterium intracellulare*. Microbiol. Immunol. 30 (9), 903–907.

Griffith, D.E., Aksamit, T., Brown-Elliott, B.A., Catanzaro, A., Daley, C., Gordin, F., et al., 2007. An official ATS/IDSA statement: diagnosis, treatment, and prevention of nontuberculous mycobacterial diseases. Am. J. Respir. Crit. Care Med. 175 (4), 367–416.

Griffith, D.E., Brown-Elliott, B.A., Langsjoen, B., Zhang, Y., Pan, X., Girard, W., et al., 2006. Clinical and molecular analysis of macrolide resistance in *Mycobacterium avium* complex lung disease. Am. J. Respir. Crit. Care Med. 174 (8), 928–934.

Guillemin, I., Jarlier, V., Cambau, E., 1998. Correlation between quinolone susceptibility patterns and sequences in the A and B subunits of DNA gyrase in mycobacteria. Antimicrob. Agents Chemother. 42 (8), 2084–2088.

Guimaraes, T., Chimara, E., do Prado, G.V., Ferrazoli, L., Carvalho, N.G., Simeao, F.C., et al., 2016. Pseudooutbreak of rapidly growing mycobacteria due to *Mycobacterium abscessus* subsp *bolletii* in a digestive and respiratory endoscopy unit caused by the same clone as that of a countrywide outbreak. Am. J. Infect. Control. 44 (11), e221–e226.

Haworth, C.S., Banks, J., Capstick, T., Fisher, A.J., Gorsuch, T., Laurenson, I.F., et al., 2017. British Thoracic Society guidelines for the management of non-tuberculous mycobacterial pulmonary disease (NTM-PD). Thorax. 72 (Suppl. 2), ii1–ii64.

Heidarieh, P., Mirsaeidi, M., Hashemzadeh, M., Feizabadi, M.M., Bostanabad, S.Z., Nobar, M. G., et al., 2016. In vitro antimicrobial susceptibility of nontuberculous mycobacteria in Iran. Microb. Drug Resist. 22 (2), 172–178.

Hui, S.H., Noonan, L., Chavada, R., 2015. Post liposuction *Mycobacterium abscessus* surgical site infection in a returned medical tourist complicated by a paradoxical reaction during treatment. Infect. Dis. Rep. 7 (4), 6304.

Hull, S.I., Wallace Jr, R.J., Bobey, D.G., Price, K.E., Goodhines, R.A., et al., 1984. Presence of aminoglycoside acetyltransferase and plasmids in *Mycobacterium fortuitum*. Lack of correlation with intrinsic aminoglycoside resistance. Am. Rev. Respir. Dis. 129 (4), 614–618.

Jarand, J., Davis, J.P., Cowie, R.L., Field, S.K., Fisher, D.A., 2016. Long-term follow-up of *Mycobacterium avium* complex lung disease in patients treated with regimens including clofazimine and/or rifampin. Chest 149 (5), 1285–1293.

Jo, K.W., Kim, S., Lee, J.Y., Lee, S.D., Kim, W.S., Kim, D.S., et al., 2014. Treatment outcomes of refractory MAC pulmonary disease treated with drugs with unclear efficacy. J. Infect. Chemother. 20 (10), 602–606.

Kadota, T., Matsui, H., Hirose, T., Suzuki, J., Saito, M., Akaba, T., et al., 2016. Analysis of drug treatment outcome in clarithromycin-resistant *Mycobacterium avium* complex lung disease. BMC Infect. Dis. 16, 31.

Kim, M.J., Mascola, L., 2010. *Mycobacterium chelonae* wound infection after liposuction. Emerg. Infect. Dis. 16 (7), 1173–1175.

Kim, S.Y., Kim, C.K., Bae, I.K., Jeong, S.H., Yim, J.J., Jung, J.Y., et al., 2015. The drug susceptibility profile and inducible resistance to macrolides of *Mycobacterium abscessus* and *Mycobacterium massiliense* in Korea. Diagn. Microbiol. Infect. Dis. 81 (2), 107–111.

Koh, W.J., Hong, G., Kim, S.Y., Jeong, B.H., Park, H.Y., Jeon, K., et al., 2013. Treatment of refractory *Mycobacterium avium* complex lung disease with a moxifloxacin-containing regimen. Antimicrob. Agents Chemother. 57 (5), 2281–2285.

Koh, W.J., Jeong, B.H., Jeon, K., Kim, S.Y., Park, K.U., Park, H.Y., et al., 2016. Oral macrolide therapy following short-term combination antibiotic treatment of *Mycobacterium massiliense* lung disease. Chest 150 (6), 1211–1221.

Koh, W.J., Jeong, B.H., Kim, S.Y., Jeon, K., Park, K.U., Jhun, B.W., et al., 2017. Mycobacterial characteristics and treatment outcomes in *Mycobacterium abscessus* lung disease. Clin. Infect. Dis. 64 (3), 309–316.

Lange, C., Schubert, S., Jung, J., Kostrzewa, M., Sparbier, K., 2014. Quantitative matrix-assisted laser desorption ionization-time of flight mass spectrometry for rapid resistance detection. J. Clin. Microbiol. 52 (12), 4155–4162.

Lee, H., Hwang, D., Jeon, M., Lee, E., Kim, T., Yu, S.N., et al., 2016. Clinical features and treatment outcomes of septic arthritis due to *Mycobacterium massiliense* associated with intra-articular injection: a case report. BMC Res. Notes 9 (1), 443.

Lee, S.H., Yoo, H.K., Kim, S.H., Koh, W.J., Kim, C.K., Park, Y.K., et al., 2014. The drug resistance profile of *Mycobacterium abscessus* group strains from Korea. Ann. Lab. Med. 34 (1), 31–37.

Li, G., Pang, H., Guo, Q., Huang, M., Tan, Y., Li, C., et al., 2017. Antimicrobial susceptibility and MIC distribution of 41 drugs against clinical isolates from China and reference strains of nontuberculous mycobacteria. Int. J. Antimicrob. Agents 49 (3), 364–374.

Lipworth, S.I.W., Hough, N., Leach, L., Morgan, M., Jeffrey, K., Andersson, M., et al., 2018. Whole genome sequencing for predicting *Mycobacterium abscessus* drug susceptibility. bioRxiv. Jan 1, 251918.

Luo, R.F., Curry, C., Taylor, N., Budvytiene, I., Banaei, N., 2015. Rapid detection of acquired and inducible clarithromycin resistance in *Mycobacterium abscessus* group by a simple real-time PCR assay. J. Clin. Microbiol. 53 (7), 2337–2339.

Lyu, J., Kim, B.J., Kim, B.J., Song, J.W., Choi, C.M., Oh, Y.M., et al., 2014. A shorter treatment duration may be sufficient for patients with *Mycobacterium massiliense* lung disease than with *Mycobacterium abscessus* lung disease. Respir. Med. 108 (11), 1706–1712.

Macadam, S.A., Mehling, B.M., Fanning, A., Dufton, J.A., Kowalewska-Grochowska, K.T., Lennox, P., et al., 2007. Nontuberculous mycobacterial breast implant infections. Plast. Reconstr. Surg. 119 (1), 337–344.

Miwa, S., Shirai, M., Toyoshima, M., Shirai, T., Yasuda, K., Yokomura, K., et al., 2014. Efficacy of clarithromycin and ethambutol for *Mycobacterium avium* complex pulmonary disease. A preliminary study. Ann. Am. Thorac. Soc. 11 (1), 23–29.

Moon, S.M., Park, H.Y., Kim, S.Y., Jhun, B.W., Lee, H., Jeon, K., et al., 2016. Clinical characteristics, treatment outcomes, and resistance mutations associated with macrolide-resistant *Mycobacterium avium* complex lung disease. Antimicrob. Agents Chemother. 60 (11), 6758–6765.

Nash, K.A., 2003. Intrinsic macrolide resistance in Mycobacterium smegmatis is conferred by a novel erm gene, erm(38). Antimicrob Agents Chemother. 47 (10), 3053–3060.

Nash, K.A., Andini, N., Zhang, Y., Brown-Elliott, B.A., Wallace Jr., R.J., 2006. Intrinsic macrolide resistance in rapidly growing mycobacteria. Antimicrob Agents Chemother. 50 (10), 3476–3478.

Nash, K.A., Brown-Elliott, B.A., Wallace Jr., R.J., 2009. A novel gene, erm(41), confers inducible macrolide resistance to clinical isolates of Mycobacterium abscessus but is absent from Mycobacterium chelonae. Antimicrob Agents Chemother. 53 (4), 1367–1376.

Nash, K.A., Zhang, Y., Brown-Elliott, B.A., Wallace Jr., R.J., 2005. Molecular basis of intrinsic macrolide resistance in clinical isolates of *Mycobacterium fortuitum*. J. Antimicrob. Chemother. 55 (2), 170–177.

Nie, W., Duan, H., Huang, H., Lu, Y., Bi, D., Chu, N., 2014. Species identification of *Mycobacterium abscessus* spp. *abscessus* and *Mycobacterium abscessus* spp. *bolletii* using rpoB andhsp65, and susceptibility testing to eight antibiotics. Int. J. Infect. Dis. 25, 170–174.

Olivier, K.N., Shaw, P.A., Glaser, T.S., Bhattacharyya, D., Fleshner, M., Brewer, C.C., et al., 2014. Inhaled amikacin for treatment of refractory pulmonary nontuberculous mycobacterial disease. Ann. Am. Thorac. Soc. 11 (1), 30–35.

Padmapriyadarsini, C., Nair, D., Gomathi, N.S., Velayudham, B., 2016. Pulmonary *Mycobacterium kansasii* disease in immunocompetent host: treatment outcomes with short-course chemotherapy. Indian J. Med. Microbiol. 34 (4), 516–519.

Park, J., Cho, J., Lee, C.H., Han, S.K., Yim, J.J., 2017a. Progression and treatment outcomes of lung disease caused by *Mycobacterium abscessus* and *Mycobacterium massiliense*. Clin. Infect. Dis. 64 (3), 301–308.

Park, S., Jo, K.W., Lee, S.D., Kim, W.S., Shim, T.S., 2017b. Clinical characteristics and treatment outcomes of pleural effusions in patients with nontuberculous mycobacterial disease. Respir. Med. 133, 36–41.

Pasipanodya, J.G., Ogbonna, D., Deshpande, D., Srivastava, S., Gumbo, T., 2017a. Meta-analyses and the evidence base for microbial outcomes in the treatment of pulmonary

Mycobacterium avium-intracellulare complex disease. J. Antimicrob. Chemother 72 (Suppl. 2), i3–i19.

Pasipanodya, J.G., Ogbonna, D., Ferro, B.E., Magombedze, G., Srivastava, S., Deshpande, D., et al., 2017b. Systematic review and meta-analyses of the effect of chemotherapy on pulmonary *Mycobacterium abscessus* outcomes and disease recurrence. Antimicrob. Agents Chemother 61 (11).

Ramis, I.B., Cnockaert, M., von Groll, A., Nogueira, C.L., Leao, S.C., Andre, E., et al., 2015. Antimicrobial susceptibility of rapidly growing mycobacteria using the rapid colorimetric method. Eur. J. Clin. Microbiol. Infect. Dis. 34 (7), 1403–1413.

Romero, F.A., Powell, E.A., Babady, N.E., Kaltsas, A., Figueroa, C.J., Pulitzer, M., et al., 2017. Nontuberculous mycobacterial infections after silicone breast implant reconstruction emphasize a diversity of infecting mycobacteria. Open Forum Infect. Dis. 4 (4), ofx189.

Schnabel, D., Esposito, D.H., Gaines, J., Ridpath, A., Barry, M.A., Feldman, K.A., et al., 2016. Multistate US Outbreak of rapidly growing mycobacterial infections associated with medical tourism to the Dominican Republic, 2013–2014(1). Emerg. Infect. Dis. 22 (8), 1340–1347.

Schon, T., Chryssanthou, E., 2017. Minimum inhibitory concentration distributions for Mycobacterium avium complex-towards evidence-based susceptibility breakpoints. Int J Infect Dis 55, 122–124.

Srivastava, S., Pasipanodya, J., Sherman, C.M., Meek, C., Leff, R., Gumbo, T., 2015. Rapid drug tolerance and dramatic sterilizing effect of moxifloxacin monotherapy in a novel hollow-fiber model of intracellular *Mycobacterium kansasii* disease. Antimicrob. Agents Chemother. 59 (4), 2273–2279.

van Ingen, J., Boeree, M.J., van Soolingen, D., Mouton, J.W., 2012. Resistance mechanisms and drug susceptibility testing of nontuberculous mycobacteria. Drug Resist. Updat. 15 (3), 149–161.

Wallace Jr, R.J., Brown-Elliott, B.A., McNulty, S., Philley, J.V., Killingley, J., et al., 2014. Macrolide/azalide therapy for nodular/bronchiectatic *Mycobacterium avium* complex lung disease. Chest 146 (2), 276–282.

Yagi, K., Ishii, M., Namkoong, H., Asami, T., Iketani, O., Asakura, T., et al., 2017. The efficacy, safety, and feasibility of inhaled amikacin for the treatment of difficult-to-treat nontuberculous mycobacterial lung diseases. BMC Infect. Dis. 17 (1), 558.

Yoshida, S., Tsuyuguchi, K., Kobayashi, T., Tomita, M., Inoue, Y., Hayashi, S., et al., 2018. Association between sequevar and antibiotic treatment outcome in patients with *Mycobacterium abscessus* complex infections in Japan. J. Med. Microbiol. 67 (1), 74–82.

Yuan, J., Liu, Y., Yang, Z., Cai, Y., Deng, Z., Qin, P., et al., 2009. *Mycobacterium abscessus* post-injection abscesses from extrinsic contamination of multiple-dose bottles of normal saline in a rural clinic. Int. J. Infect. Dis. 13 (5), 537–542.

Zhao, X., Wang, Y., Pang, Y., 2014. Antimicrobial susceptibility and molecular characterization of *Mycobacterium intracellulare* in China. Infect. Genet. Evol. 27, 332–338.

Zheng, H.W., Pang, Y., He, G.X., Song, Y.Y., Zhao, Y.L., 2017. Comparing the genotype and drug susceptibilities between *Mycobacterium avium* and *Mycobacterium intracellulare* in China. Biomed. Environ. Sci. 30 (7), 517–525.

Chapter 4

Future Nontuberculous Mycobacteria DST and Therapeutic Interventions

Sven Hoffner[1] and Diane Ordway[2]

[1]*Department of Public Health Sciences, Karolinska Institute, Stockholm, Sweden,* [2]*Mycobacteria Research Laboratory, Department of Microbiology, Immunology and Pathology, Colorado State University, Fort Collins, CO, United States*

CURRENT CHALLENGES OF DRUG SUSCEPTIBILITY TESTING

The biggest challenge of nontuberculous mycobacterial (NTM) therapeutic intervention is the commonly held view that NTM share similar pathogenesis, drug-susceptible testing, and treatment as *Mycobacterium tuberculosis* (*M. tuberculosis*). One significant difference between NTM and *M. tuberculosis* is that identification of *M. tuberculosis* in a clinical specimen reflects only one species resulting in the causative agent of tuberculosis (TB) (Matteelli et al., 2016; Haworth et al., 2017; Haworth and Floto, 2017; Lewinsohn et al., 2017). This is a disease that requires immediate treatment with an active combination therapy to ensure bacterial transmission is stopped and a cure is achieved. TB drug susceptibility testing (DST) is for detection of drug resistance profiles that allow clinicians to treat patients effectively, thus leading to an improved patient clinical outcome (Lewinsohn et al., 2017).

In contrast, laboratory detection of NTM in a clinical sample is insufficient justification for immediate medical treatment because NTM infection does not always lead to NTM disease (Matteelli et al., 2016; Haworth and Floto, 2017). Of the 150 recognized NTM species, only a small proportion are associated with human clinical NTM disease, thus requiring DST. Therefore, additional information is required before DST is used. First, identification of the specific species of NTM is needed. Second, the patient's clinical findings should be sufficiently significant to warrant combination drug therapy. Further complicating matters, NTM DST is often not capable of early identification of drug resistance profiles, leading to ineffective

Nontuberculous Mycobacteria (NTM). DOI: https://doi.org/10.1016/B978-0-12-814692-7.00004-8

clinical treatment of patients and worsening clinical outcomes (Lewinsohn et al., 2017).

A second important difference between DST of NTM and *M. tuberculosis* is the over 50-year history *M. tuberculosis* has of standardized international recommendations on performing reliable DST for all classical TB drugs (Matteelli et al., 2016; Haworth and Floto, 2017). However, in the case of DST for NTM, the situation is complicated by the difficulty of initial identification of optimal compounds for testing in the *in vitro* DST assays and how to interpret the results of these assays to ensure successful clinical treatment outcomes. Evidence suggests DST assays recommended for *M. tuberculosis* should not be used for NTM (Matteelli et al., 2016; Haworth and Floto, 2017). Due to a large number of NTM species, it is imperative to identify individual NTM species responsible for the clinical infection prior to recommendation of a combination treatment regimen (Brown-Elliott and Wallace, 2017; Bryant et al., 2016).

Vast differences exist in the ability to detect drug resistance in NTM compared to *M. tuberculosis* since there is a better understanding of the molecular causes of drug resistance in TB (Brown-Elliott and Wallace, 2017; Orme and Ordway, 2014). Drug therapy of NTM infections can be extremely challenging (for an outstanding review on this topic, see Orme and Ordway, 2014; van Ingen et al., 2012). A significant factor is antibiotic inactivation, with many mechanisms of resistance such as β-lactamases, aminoglycoside phosphotransferases, and aminoglycoside acetyltransferases. Also, the p55 efflux pump confers resistance to tetracyclines and aminoglycosides, and multiple other efflux pumps likely cause antimicrobial resistance and persistence (van Ingen et al., 2013a). Finally, the erythromycin ribosomal methylase (erm) gene found in almost all *M. abscessus* subsp. *abscessus* and *M. fortuitum* strains results in methylation of the 23S rRNA, rendering the bacteria resistant to macrolides (van Ingen et al., 2012). Complicating DST is that induction of this gene is not detected by DST because as this requires prolonged incubation for up to 2 weeks. Commercially available DST assays are available for TB that are capable of detection of specific resistance to the major antitubercular drugs in just a few hours (Mboowa et al., 2014; Cirillo et al., 2017). Conversely, NTM drug resistance mechanisms are poorly understood, and limited comprehension exists of the role of rapid molecular testing of NTM drug resistance (Luetkemeyer et al., 2014; Lan et al., 2011). In spite of this, efforts to develop new molecular testing approaches for DST of NTM are underway (Orme and Ordway, 2014; Luetkemeyer et al., 2014; Lan et al., 2011).

THE ROLE OF TESTING DRUG COMBINATION THERAPY

Bacterial infections are normally treated with one antibiotic substance, and in vitro susceptibility testing determines the extent this drug inhibits the infecting bacteria when grown under optimal laboratory conditions. However, NTM infection resulting in clinical disease requires combination drug treatment to counteract the development of drug resistance. Mycobacterial infections are notable examples of such infections and should always necessitate combination drug treatment with two or more drugs (Lewinsohn et al., 2017; van Ingen et al., 2013a). The recommended therapy for infections caused by fully drug-susceptible *M. tuberculosis* includes four "first-line drugs" (rifampicin (Rif), isoniazid (Inh), ethambutol (Emb), and pyrazinamide). If we are sure that the combined drugs act independently, not showing any antagonistic, synergistic, or additive effects, it should be acceptable to test the drugs separately—exposing the infecting bacteria in the laboratory to the drugs used in the combined drug therapy separately. This single-drug testing approach for combination treatment for TB predicts the clinical usefulness of each drug (Lewinsohn et al., 2017).

On the contrary, it is well recognized that NTM single-drug *in vitro* susceptibility testing results often do not predict clinical outcome of the patient combination drug therapy (Soni et al., 2016). Thus, a more reliable *in vitro* DST approach may be to use drug combinations rather than single drugs for DST. Experimental studies comparing single-DST results with combination drugs *in vitro* susceptibility testing against a drug-resistant *M. avium* strain showed improved patient clinical outcome based on the DST using combinations drug testing compared to single-DST (Hoffner et al., 1987). Additional studies, showed other possible drug interactions exist when using standard anti-NTM compounds commonly used in combined drug therapy against NTM infections (Hoffner et al., 1989a). The results obtained from these studies demonstrated a synergistic interaction, independent of the number of drugs studied (Zimmer et al., 1982; Hoffner et al., 1989a).

Based on these reports, where the radiometric Bactec 460 system from Becton Dickinson (MD, United States) was used to measure the daily bacterial growth index (metabolic activity), a model for studying combined drug effects for up to four combined drugs and a definition of drug synergy was suggested. In each experiment, the daily growth was followed for combinations of the studied drugs (two drugs or more) and compared to single-drug suspensions (Hoffner et al., 1989a). In subsequent studies that used the identical methods, different two-drug combinations were evaluated using a growth index value of <0.5 as an indicator of combination drug synergy. A study evaluated 20 clinical isolates of the *M. avium* complex that were exposed to the following anti-TB drugs Emb (5 mg/L), Inh (0.2 mg/L), Rif (2 mg/L) and Streptomycin (Sm) (4 mg/L), and their combinations. While the tested isolates were all classified as resistant to the drugs tested individually, a significant antibacterial synergism was evident with drug combinations (Hoffner et al., 1989a). Table 4.1 shows the number and percentage of the 20 *M. avium* complex isolates showing a growth inhibition indicating drug synergy. Notably, two of the drugs stand out—Emb most often showing a synergistic interaction, and Inh rarely showing any synergy. Based on these promising initial results, subsequent studies were carried out with additional drugs and species of NTM. The main goal of these studies was to evaluate the interaction between Emb and fluorinated quinolones (Fqs). The tested Fqs were ciprofloxacin, ofloxacin, and levofloxacin, investigated for their synergy against 100 clinical *M. avium* complex isolates (Hoffner et al., 1989a). These studies illustrated Emb combinations with Rif and Sm and showed a substantial synergistic effect that was also demonstrated when in combination with the Fq compounds. More importantly, a pronounced synergistic effect was evident for all three Fq compounds when tested in combination with Emb rather than singly.

Conversely, all *M. avium* strains tested were resistant to Emb with less than 7% of the strains being susceptible to ciprofloxacin, and no less than 76% being susceptible to the combination of these two compounds. In addition to Rif, Sm, and Fq, other drugs showed similar synergism when combined with Emb. Among these drugs we saw β-lactams (Abate and Hoffner, 1997), erythromycin, tobramycin, and fusidic acid (Hoffner et al., 1989b). Importantly, these studies show that *in vitro* testing drugs singly rather than in combination would noticeably underestimate the potential synergy of a

TABLE 4.1 Drug Combination and Synergistic Interaction (Number of Strains/Percentage) Showing Synergy

Ethambutol + isoniazid	1 (5%)
Ethambutol + rifampicin	20 (100%)
Rifampicin + streptomycin	20 (100%)
Isoniazid + ethambutol	1 (5%)
Isoniazid + rifampicin	0 (0%)
Isoniazid + streptomycin	0 (0%)
Rifampicin + ethambutol	20 (100%)
Rifampicin + isoniazid	0 (0%)
Rifampicin + streptomycin	5 (25%)
Streptomycin + ethambutol	20 (100%)
Streptomycin + isoniazid	0 (0%)
Streptomycin + rifampicin	5 (25%)

combined therapy with Emb and Fq. These significant findings support the need for further studies and future changes in our current DST algorithms used for NTM.

The initial series of investigations on *in vitro* combination drug synergy for *M. avium* was expanded to evaluate other NTM species to verify whether they also show *in vitro* combination drug synergy. These studies documented that this synergistic effect was not limited to *M. avium* complex, but was also apparent among the NTM species *M. kansasii* and *M. malmoense* which both showed similar patterns of synergy with Emb (Hjelm et al., 1992; Hoffner et al., 1993; van Ingen et al., 2013b). Synergistic effects of drug combinations with Emb and other antibacterial agents were shown for a number of combinations, indicating that the potentiated impact of Emb were not a specific effect for just one or a few combinations of drugs, but rather a more general characteristic. These studies suggested that this drug may contribute to successful therapy of many NTM infections (Kallenius et al., 1989).

One important question is: are these findings clinically meaningful and could the *in vitro* DST results of combined drugs rather than single drugs better predict the clinical efficacy of a specific drug combination? Reports have illustrated that patients with *M. avium* infection initially show *in vitro* susceptibility to the combination of Emb and Rif, but some patients failed when treated with a combination of these drugs. Interestingly, in these nonresponders, resistance to the combined drugs had occurred and could be identified when Emb and Rif were used in combination for *in vitro* testing (Hoffner et al., 1994). The hypothesis for the mechanism of Emb in potentiating a number of other drugs against several different species of NTM was related to its interaction with the cell wall of the bacteria. Studies evaluating the microcalorimetric potential show a significant interaction between Emb and the cell wall of *M. avium* (Hoffner et al., 1990). Overall drug synergistic effects and their clinical relevance need to be examined in tandem using single and combination drugs *in vitro* DST while evaluating patient clinical outcomes. Future drug discovery and identification of effective compounds against NTM will undoubtedly require expansion of the current DST for NTM, including tests of combinations of new compounds with existing anti-NTM drugs.

FUTURE THERAPEUTIC APPROACHES

The recommended available antibiotic treatment regimens for pulmonary *M. avium* complex and *M. abscessus* subspecies are demonstrated in Table 4.2. Clinical studies have shown that only 70% of *M. avium* complex patients and 50% of *M. abscessus* subspecies patients who successfully complete the recommended treatment result in cured NTM disease (van Ingen et al., 2013a). Moreover, those patients that have refractory disease have recurrence rates as high as 25%–45% (Lande et al., 2018). The need for not only improved combination therapies but also novel approaches, to tackle reinfection of pulmonary NTM disease has led to multiple preclinical and clinical studies expanding the potential for building a successful treatment regimen.

SYNERGISTIC TREATMENT REGIMENS

Synergistic drug combinations have been viewed by many to be a promising approach to treat multiple chronic bacterial diseases (Yin et al., 2014). Nonetheless, chronic combination treatment regimens can interact in unpredictable ways and show not only drug synergy but also drug antagonism. Drug development aimed at optimizing a synergistic increase in drug efficacy and reducing drug resistance is a highly pursued approach of combinational drug development (Zheng et al., 2018). Some approaches to suppress the phenotype of multiple drug resistance have included the pairing of antibiotic–antibiotic combinations, adding a nonantibiotic adjuvant molecule to target resistance mechanisms directly or to indirectly target resistance by blocking bacterial signaling pathways (enzymes, and/or two-component systems) (Zheng et al., 2018). Synergistic drug combinations have the potential to be extremely efficacious by attaining more specific drug targets and reducing the development of drug resistance (Worthington and Melander, 2013).

In recent years, many insights into the process of achieving drug synergy aimed at NTM have been gained, resulting in increasing awareness that this approach can successfully lead to a patient cure (van Ingen et al., 2013a). Recent preclinical animal models (Orme and Ordway, 2014; Bernut et al., 2017; Obregon-Henao et al., 2015) comparing the antimycobacterial activity of clarithromycin, clofazimine, bedaquiline, and clofazimine–bedaquiline combinations against *M. abscessus* treatment of GKO and SCID mice demonstrated clofazimine and bedaquiline were the most effective in decreasing the *M. abscessus* organ burden. This study highlights the importance of synergistic compounds capable of enhanced efficacy against *M. abscessus*.

TABLE 4.2 Recommended Treatment Regimens

Species	Recommended Treatment Regimen
Mycobacterium avium complex	Rifampicin or rifabutin and ethambutol and macrolide
Mycobacterium abscessus	Initially: cefoxitin or imipenem and amikacin and macrolide and tigecycline
	Secondarily: macrolide and two oral drugs based on ATS and inhaled amikacin

Additional studies using the zebrafish model demonstrated bedaquiline was highly efficacious against *M. abscessus* infection and treatment was sufficient to protect the infected larvae from *M. abscessus*-induced killing (Dupont et al., 2017). However, clofazimine alone was bacteriostatic for both *M. abscessus* and *M. avium*. Clofazimine-amikacin was synergistic against *M. abscessus* and *M. avium* (Dupont et al., 2017). Evaluation of time kill kinetic assays (Ferro et al., 2016) resulted in finding that addition of clofazimine to a regimen of amikacin and clarithromycin led to suppression of the emergence of resistance.

The mechanism of clofazimine and bedaquiline combination treatment synergy has been evaluated in an animal model against drug-resistant strains of *M. tuberculosis* (Lechartier and Cole, 2015; Shang et al., 2011a). These studies demonstrated clofazimine appears to act as a prodrug, which is reduced by NADH dehydrogenase (NDH-2), leading to the release of reactive oxygen species upon reoxidation by O_2 (Lechartier and Cole, 2015). Importantly, clofazimine likely competes with menaquinone (MK-4), a main cofactor in the mycobacterial electron transfer chain, for its reduction by NDH-2 (Lechartier and Cole, 2015). Additional studies investigated the effect of MK-4 supplementation on the action of clofazimine against *M. tuberculosis* and discovered a direct competition between clofazimine and MK-4 for the cidal effect of clofazimine in nonreplicating and replicating mycobacteria (Lechartier and Cole, 2015). Additional studies are required to elucidate the synergistic mechanisms of clofazimine and bedaquiline drug combination treatments against NTM.

Preclinical murine models evaluating the efficacy of bedaquiline against *M. avium complex* species infected mice showed reduced organ bacterial burden; however, this efficacy was not bactericidal (Lounis et al., 2009). Preclinical studies using bedaquiline for treatment failure of lung disease caused by *M. avium* (six patients) or *M. abscessus* (four patients) evaluated bedaquiline treatment between one and eight years (Philley et al., 2015). Eighty percent of patients enrolled in the study had macrolide-resistant isolate (eight out of ten) (Philley et al., 2015). All the patients were treated with the same bedaquiline dosage (400 mg daily) and received the best available companion drugs (mean, 5.0 drugs) (Philley et al., 2015). All patients completed six months of therapy and remain on bedaquiline. After six months of therapy, 60% of patients (six out of ten) had a microbiologic response, with 50% (five out of ten) having one or more negative cultures (Philley et al., 2015). However, additional preclinical studies evaluating the efficacy of bedaquiline treatment in patients with *Mycobacterium intracellulare* pulmonary disease showed seven out of 13 patients with an initial culture conversion resulted in relapse (Alexander et al., 2017). Subsequent sequencing revealed nonsynonymous mmpT5 mutations in isolates from all seven relapse cases, including two pretreatment isolates, supporting the need for further optimization of treatment regimens to prevent the emergence of mmpT5 variants and relapse (Alexander et al., 2017). In spite of this, bedaquiline and clofazimine combination treatments in the absence of rifamycin may increase drug exposure enhancing efficacy against NTM. Frequently standard treatment for NTM pulmonary disease is not possible because of drug intolerance, antibiotic resistance, or progression of the disease. Second-line treatment such as inhaled amikacin, clofazimine, bedaquiline, and delamanid are considered, regardless of only a few studies to guide their use (Lande et al., 2018).

POTENTIAL NEW TREATMENTS

Recent preclinical studies evaluating the use of oxazolidinone treatment for *M. avium complex* species and *M. abscessus* species have discovered new drugs to improve current treatment regimens. Oxazolidinones are synthetic antibacterial agents with a distinctive mechanism of action having a wide spectrum of activity, oral bioavailability, and well-established structureactivity relationship (Jadhavar et al., 2015). Oxazolidinones act by inhibiting translation at the initiation phase of protein synthesis (Jadhavar et al., 2015). Combination treatment with tedizolid and clofazimine has the potential to improve treatment efficacy and reduce the development of drug resistance. Tedizolid has superior bioavailability, longer half-life, and accumulates in macrophages in the lung (Brown-Elliott and Wallace, 2017; Raju et al., 2016) compared to linezolid. A recent screening of T compounds for activity against NTM delivered three top hits from the oxazolidinones, suggesting the potential for repositioning this class of protein synthesis inhibitors to replace linezolid which suffers from low potency (Low et al., 2017). The usefulness of oxazolidinones against drug-resistant TB is already established (Singh and Silakari, 2018). Several oxazolidinones are currently in development for their possible use in TB therapy (Singh and Silakari, 2018). The oxazolidinones linezolid, sutezolid and toxicity, especially myelosuppression, has been a significant limiting factor for their development (Singh and Silakari, 2018).

Growing evidence suggests targeting biofilm formation is essential to combat NTM disease (Orme and Ordway, 2014; Chan et al., 2010). Glycopeptidolipids are a class of glycolipids produced by several nontuberculous-causing members of the mycobacterium genus that are constituents of the biofilm (Schorey and Sweet, 2008). NTM biofilm formation results in drug resistance to most antibiotics, and more recalcitrant to host immunity (Schorey and Sweet, 2008; Xiang et al., 2014; Hall-Stoodley et al., 2006). Preclinical studies evaluating the use of new drugs for treatment for *M. avium complex* species and *M. abscessus* species have also discovered inhibitors of the trehalose monomycolate transporter MmpL3, suggesting that this transmembrane protein may be an attractive target for NTM (Low et al., 2017). The MmpL family of proteins translocates complex glycol-lipids and siderophores across the cell envelope of mycobacteria and plays an important role in the biogenesis of the outer membrane of NTM (Belardinelli et al., 2016). Recent *in vitro* studies illustrating indolecarboxamides lead compounds, 6 and 12, exhibited strong activity *in vitro* against a wide panel of *M. abscessus* isolates and in infected macrophages (Low et al., 2017). The indolecarboxamides strongly inhibited the transport of trehalose monomycolate, resulting in the loss of trehalose dimycolate production and abrogating mycolylation of arabinogalactan constituents of the biofilm. These studies supported an unexploited chemical structure class active against *M. abscessus* infections with promising translational development possibilities for the treatment of NTM in patients (Kozikowski et al., 2017).

Two β-lactams, cefoxitin, and imipenem, are part of the recommended treatment for pulmonary infections with *M. abscessus* (Lefebvre et al., 2016; Dubee et al., 2015). *M. abscessus* has been shown to produce a broad-spectrum β-lactamase, BlaMab, indicating that the combination of β-lactams with a BlaMab inhibitor may increase treatment efficacy (Lefebvre et al., 2016). Preclinical studies showed avibactam efficiently inactivated BlaMab by the formation of a covalent adduct. Also, inhibition of BlaMab by avibactam was observed in both infected macrophages and zebrafish (Lefebvre et al., 2016). Assessment of the killing and intracellular activities of β-lactams indicates that imipenem is superior to cefoxitin at clinically achievable drug concentrations. Inhibition of BlaMab could increase the efficacy of imipenem and extend the spectrum of drugs potentially useful to treat pulmonary infections (Dubee et al., 2015).

INHALATION TREATMENT REGIMENS

Inhalation treatment targets the airways, quickly lowering the dose of the drug required for optimal efficacy. Also, a low incidence of systemic side effects normally occurs with a rapid onset of drug action. A randomized trial adding liposomal amikacin by inhalation to patients with *M. avium* complex species pulmonary disease that was failing previous treatment regimens resulted in a significantly higher rate of culture conversion compared to the placebo group (Olivier et al., 2017). A second, preclinical trial was conductive to evaluate inhaled amikacin therapy in patients with difficult-to-treat NTM lung diseases (Yagi et al., 2017). A total of 26 cases which consisted of 23 *M. avium* complex and 3 *M. abscessus* complex infections cases, were included in this study (Yagi et al., 2017). All patients received combination therapy, including inhaled amikacin therapy (Yagi et al., 2017). Twenty-three of the 26 patients (88.5%), including *M. avium* complex patients (91.3%) and *M. abscessus* patients (66.7%), were administered inhaled amikacin therapy for >3 months (Yagi et al., 2017). Ten of the 23 patients (43.5%) who received amikacin inhalation, *M. avium* complex (38.1%) and *M. abscessus* patients (100%), showed sputum conversion. Seven of the 23 patients, including, 5/21 *M. avium* complex and 2/2 *M. abscessus* patients, showed improvements in high-resolution computed tomography imaging of the chest (Yagi et al., 2017). These clinical studies support the need for the development of new combination inhaled therapies.

Initial phase 1 preclinical studies evaluating the efficacy and safety of ciprofloxacin dry powder for inhalation (DPI) in patients with noncystic fibrosis bronchiectasis, two or more exacerbations in the previous year with a predefined bacteria in sputum have been completed (De Soyza et al., 2018). These studies resulted in unfavorable outcomes for ciprofloxacin DPI 28 days on/off since, although ciprofloxacin was found to be safe, the differences in the treated and placebo groups were not significant (De Soyza et al., 2018). A subsequent phase 2 trial aimed to evaluate the efficacy of Ciprofloxacin DPI in prolonging the time to first exacerbation and reducing the frequency of exacerbations with two treatment regimens (28 days on/off and 14 days on/off) in patients with NCFB using a stringent definition of exacerbation (Aksamit et al., 2017). Ciprofloxacin DPI showed efficacy; demonstrating reduced bacterial burden and a trend towards fewer exacerbations suggesting microbiological clinical efficacy of Ciprofloxacin. Although the phase 1 and 2 trails targeted reduced bacterial burden in seven different pathogens, the RESPIRE studies could provide insight into the potential use of other anti-NTM compounds formulated in a dry powder inhalation treatment.

INTERMITTENT TREATMENT REGIMENS

The chronic treatment of patients with pulmonary nontuberculous disease often results in toxicity and poor clinical outcome. Multiple clinical studies (Jeong et al., 2015; Jhun et al., 2018; Wallace et al., 2014) have been conducted using an intermittent treatment regimen (Jeong et al., 2015). All three clinical studies (Jeong et al., 2015; Jhun et al., 2018; Wallace et al., 2014) showed use of an intermittent three-times-weekly therapy with a macrolide, rifampin, and Emb for patients with noncavitary nodular bronchiectatic *M. avium* complex pulmonary disease lead to prolonged culture conversion in 70%–85% of patients. Intermittent therapy leads to less treatment regimen changes and improved tolerance. Regardless of the success of intermittent therapy in patients with noncavitary nodular bronchiectatic *M. avium* complex pulmonary disease, the rate of reinfection continues to be high (Jeong et al., 2015; Jhun et al., 2018; Wallace et al., 2014).

THERAPEUTIC VACCINE APPROACHES

The development of highly efficacious new treatment regimens against NTM is achievable in the near future. However, NTM patients have a high chance of being reinfected with NTM. Reinfection of NTM patients has been shown to be due to environmental sources and/or human transmission (Shang et al., 2011b; Bryant et al., 2016). Thus, to control opportunistic NTM and to cease subsequent reinfection, a therapeutic vaccine is urgently needed against NTM. Vaccine strategies represent one of the fighting answers against multi-drug resistance NTM in different of clinical settings.

Many approaches can be used to develop vaccines against NTM. For example, similar virulence factors are expressed during infection and disease progression with other pathogens. These cross-reactive antigens produced by other pathogens may allow development of NTM-preventive vaccines, a therapeutic vaccine given in combination with effective treatment and a post-exposure vaccine to prevent reinfection shown in Fig. 4.1 (Le Moigne et al., 2016). New studies using reverse vaccinology have identified an *M. abscessus* phospholipase C (MA-PLC) as a potential vaccine target (Le Moigne et al., 2015). The protective response generated by vaccination with plasmid DNA encoding the MA-PLC formulated with a tetra functional block copolymer 704, was tested in CF (ΔF508) mice against aerosolized smooth and rough (hypervirulent) variants of *M. abscessus* (Le Moigne et al., 2015). The MA-PLC DNA vaccination on 0, 21, and 42 days elicited a strong antibody response. A significant protective effect was obtained against aerosolized *M. abscessus* (S variant) in ΔF508 mice, but not in wild-type FVB littermates (Le Moigne et al., 2015). Interestingly, high IgG titers against MA-PLC protein were measured in cystic fibrosis patients infected with *M. abscessus* (Le Moigne et al., 2015).

Vaccine development against NTM must consider the impact of prior *Bacillus* Calmette–Guérin (BCG) vaccination in countries endemic with *M. tuberculosis*. Studies have shown that NTM-based preexisting immunity could lead to lack of BCG vaccine efficacy (Poyntz et al., 2014). Additional

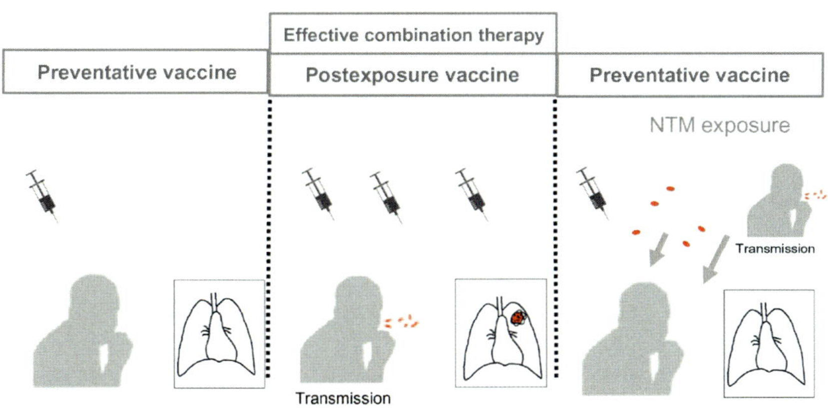

FIGURE 4.1 Potential vaccine development to prevent NTM infection. Therapeutic vaccine strategies and postexposure vaccination to stop NTM reinfection.

studies have shown that if BCG is given before NTM exposure in a murine model, there is some protection against *M. kansasii* and *M. avium* (Orme and Collins, 1985). Additional studies have shown BCG was not protective against *M. intracellulare* and *M. simiae*, possibly because BCG lacks cross-reactive antigens against these NTM species (Orme and Collins, 1985). An interesting study in Sweden supports BCG vaccination playing a role in protection again localized NTM disease (Romanus et al., 1995). This study tracked NTM case rates after their BCG vaccination program ended in 1975 and was only given to select groups with increased risk (Romanus et al., 1995). The report shows that the average yearly incidence of NTM disease for children under five years of age increased from 0.06 during the period 1969–74 to a maximum level of 5.7 during 1981–85.

Moreover, this study also showed that in individuals born in Sweden in the period 1975–85 the overall incidence rate before five years of age was at 26.8 per 100,000 non-BCG-vaccinated children and at 4.6 in those who were BCG-vaccinated; ratio 5:9 (95% confidence limits 1.6, 48.5). In this report, 83% of individuals with NTM disease had *M. avium-intracellulare* (Romanus et al., 1995). The differences in countries' BCG vaccination programs must be taken into consideration in any attempt to develop a vaccine against NTM.

The most important question that exists in developing a vaccine against NTM is what immune responses are required to induce long-lived, protective immunity against NTM. Interestingly, studies of BCG vaccinated mice infected with *M. ulcerans* developed transient $CD4^+$ TH1 protective responses (Fraga et al., 2012). Unexpected cross-reactivity of NTM/*Mtb*-specific $CXCR3^+CCR6^+$ memory T cells in non-TB exposed healthy control donors has been observed (Lindestam Arlehamn et al., 2015), including three of the four antigens (Rv3619c, Rv3620c, and Rv2608) that make up the ID93 fusion protein. Rv3619c and Rv3620c were shown to be TB-reactive (notably, these antigens are not included in BCG) whereas Rv2608 was cross-reactive in individuals with both latent TB infection and nonexposed household contacts (Lindestam Arlehamn et al., 2015).

In addition to a new anti-NTM vaccine, a potent adjuvant is likely to play a role in a robust vaccine response against NTM (Coler et al., 2011). Immunity against *M. ulcerans* infection has been conferred by early IFN-gamma induction as shown by the delayed progression of infection, and reduced bacterial burden in wild-type compared to GKO mice (Bieri et al., 2016). Other data demonstrate promising protective responses in mice, including increased survival, decreased pathology, and reduced bacterial load compared to BCG, using a recombinant BCG expressing *M. ulcerans* Ag85A in the Buruli ulcer mouse model (Hart et al., 2015). Additional research is required to develop vaccines against NTM infections (Checkley et al., 2011; Lee et al., 2016).

SUMMARY

NTM DST and therapeutic interventions have made significant progress over the years. However, additional progress is needed to combat the current emergence of NTM. In spite of considerable progress in the identification of new antibiotics and novel treatment approaches we still have no new regimens against NTM. The need still exists for future combination therapies to improve efficacy, reduce toxicity, limit drug resistance, and ultimately stop subsequent reinfection. More clinically useful methods or identified drug resistance mechanisms using DST are a critical component in our progress toward controlling future NTM clinical disease. The development of new vaccine approaches and shorter combination therapies that improve efficacy will lead to a transformation of current DST approaches.

REFERENCES

Abate, G., Hoffner, S.E., 1997. Synergistic antimycobacterial activity between ethambutol and the beta-lactam drug cefepime. Diagn. Microbiol. Infect. Dis. 28 (3), 119–122.

Aksamit, T., et al., 2017. The RESPIRE trials: two phase III, randomized, multicentre, placebo-controlled trials of ciprofloxacin dry powder for inhalation (ciprofloxacin DPI) in non-cystic fibrosis bronchiectasis. Contemp. Clin. Trials 58, 78–85.

Alexander, D.C., et al., 2017. Emergence of mmpT5 variants during bedaquiline treatment of *Mycobacterium intracellulare* lung disease. J. Clin. Microbiol. 55 (2), 574–584.

Belardinelli, J.M., et al., 2016. Structure–function profile of MmpL3, the essential mycolic acid transporter from *Mycobacterium tuberculosis*. ACS Infect. Dis. 2 (10), 702–713.

Bernut, A., et al., 2017. The diverse cellular and animal models to decipher the physiopathological traits of *Mycobacterium abscessus* infection. Front Cell Infect. Microbiol. 7, 100.

Bieri, R., et al., 2016. Interferon-gamma is a crucial activator of early host immune defense against *Mycobacterium ulcerans* infection in mice. PLoS Negl. Trop. Dis. 10 (2), e0004450.

Brown-Elliott, B.A., Wallace Jr., R.J., 2017. In vitro susceptibility testing of tedizolid against nontuberculous mycobacteria. J. Clin. Microbiol. 55 (6), 1747–1754.

Bryant, J.M., et al., 2016. Emergence and spread of a human-transmissible multidrug-resistant nontuberculous *Mycobacterium*. Science 354 (6313), 751–757.

Chan, E.D., et al., 2010. Host immune response to rapidly growing mycobacteria, an emerging cause of chronic lung disease. Am. J. Respir. Cell Mol. Biol. 43 (4), 387–393.

Checkley, A.M., et al., 2011. Identification of antigens specific to non-tuberculous mycobacteria: the MCE family of proteins as a target of T cell immune responses. PLoS One 6 (10), e26434.

Cirillo, D.M., Miotto, P., Tortoli, E., 2017. Evolution of phenotypic and molecular drug susceptibility testing. Adv. Exp. Med. Biol. 1019, 221–246.

Coler, R.N., et al., 2011. Development and characterization of synthetic glucopyranosyl lipid adjuvant system as a vaccine adjuvant. PLoS One 6 (1), e16333.

De Soyza, A., et al., 2018. RESPIRE 1: a phase III placebo-controlled randomised trial of ciprofloxacin dry powder for inhalation in non-cystic fibrosis bronchiectasis. Eur. Respir. J. 51 (1).

Dubee, V., et al., 2015. beta-Lactamase inhibition by avibactam in *Mycobacterium abscessus*. J. Antimicrob. Chemother. 70 (4), 1051–1058.

Dupont, C., et al., 2017. Bedaquiline inhibits the ATP synthase in *Mycobacterium abscessus* and is effective in infected zebrafish. Antimicrob. Agents Chemother. 61 (11).

Ferro, B.E., et al., 2016. Clofazimine prevents the regrowth of *Mycobacterium abscessus* and *Mycobacterium avium* type strains exposed to amikacin and clarithromycin. Antimicrob. Agents Chemother. 60 (2), 1097–1105.

Fraga, A.G., et al., 2012. Cellular immunity confers transient protection in experimental Buruli ulcer following BCG or mycolactone-negative *Mycobacterium ulcerans* vaccination. PLoS One 7 (3), e33406.

Hall-Stoodley, L., et al., 2006. *Mycobacterium marinum* biofilm formation reveals cording morphology. FEMS Microbiol. Lett. 257 (1), 43–49.

Hart, B.E., Hale, L.P., Lee, S., 2015. Recombinant BCG expressing *Mycobacterium ulcerans* Ag85A imparts enhanced protection against experimental Buruli ulcer. PLoS Negl. Trop. Dis. 9 (9), e0004046.

Haworth, C.S., Floto, R.A., 2017. Introducing the new BTS Guideline: management of nontuberculous mycobacterial pulmonary disease (NTM-PD). Thorax 72 (11), 969–970.

Haworth, C.S., et al., 2017. British Thoracic Society Guideline for the management of nontuberculous mycobacterial pulmonary disease (NTM-PD). BMJ Open Respir. Res. 4 (1), e000242.

Hjelm, U., et al., 1992. Susceptibility of *Mycobacterium kansasii* to ethambutol and its combination with rifamycins, ciprofloxacin and isoniazid. Eur. J. Clin. Microbiol. Infect. Dis. 11 (1), 51–54.

Hoffner, S.E., Svenson, S.B., Kallenius, G., 1987. Synergistic effects of antimycobacterial drug combinations on *Mycobacterium avium* complex determined radiometrically in liquid medium. Eur. J. Clin. Microbiol. 6 (5), 530–535.

Hoffner, S.E., et al., 1989a. In-vitro synergistic activity between ethambutol and fluorinated quinolones against *Mycobacterium avium* complex. J. Antimicrob. Chemother. 24 (3), 317–324.

Hoffner, S.E., et al., 1989b. Studies on the mechanisms of the synergistic effects of ethambutol and other antibacterial drugs on *Mycobacterium avium* complex. Acta Leprol. 7 (Suppl. 1), 195–199.

Hoffner, S.E., Svenson, S.B., Beezer, A.E., 1990. Microcalorimetric studies of the initial interaction between antimycobacterial drugs and *Mycobacterium avium*. J. Antimicrob. Chemother. 25 (3), 353–359.

Hoffner, S.E., Hjelm, U., Kallenius, G., 1993. Susceptibility of *Mycobacterium malmoense* to antibacterial drugs and drug combinations. Antimicrob. Agents Chemother. 37 (6), 1285–1288.

Hoffner, S.E., et al., 1994. *Mycobacterium avium* complex develop resistance to synergistically active drug combinations during infection. Eur. Respir. J. 7 (2), 247–250.

van Ingen, J., et al., 2012. Resistance mechanisms and drug susceptibility testing of nontuberculous mycobacteria. Drug Resist. Updat. 15 (3), 149–161.

van Ingen, J., et al., 2013a. Drug treatment of pulmonary nontuberculous mycobacterial disease in HIV-negative patients: the evidence. Expert Rev. Anti-Infect. Ther. 11 (10), 1065–1077.

van Ingen, J., et al., 2013b. Synergistic activity of rifampicin and ethambutol against slow-growing nontuberculous mycobacteria is currently of questionable clinical significance. Int. J. Antimicrob. Agents. 42 (1), 80–82.

Jadhavar, P.S., et al., 2015. Oxazolidinones as anti-tubercular agents: discovery, development and future perspectives. Curr. Med. Chem. 22 (38), 4379–4397.

Jeong, B.H., et al., 2015. Intermittent antibiotic therapy for nodular bronchiectatic *Mycobacterium avium* complex lung disease. Am. J. Respir. Crit. Care Med. 191 (1), 96–103.

Jhun, B.W., et al., 2018. Intermittent antibiotic therapy for recurrent nodular bronchiectatic *Mycobacterium avium* complex lung disease. Antimicrob. Agents Chemother. 62 (2).

Kallenius, G., Svenson, S.B., Hoffner, S.E., 1989. Ethambutol: a key for *Mycobacterium avium* complex chemotherapy? Am. Rev. Respir. Dis. 140 (1), 264.

Kozikowski, A.P., et al., 2017. Targeting mycolic acid transport by indole-2-carboxamides for the treatment of *Mycobacterium abscessus* infections. J. Med. Chem. 60 (13), 5876–5888.

Lan, R., et al., 2011. *Mycobacterium tuberculosis* and non-tuberculous mycobacteria isolates from HIV-infected patients in Guangxi, China. Int. J. Tuberc. Lung Dis. 15 (12), 1669–1675.

Lande, L., George, J., Plush, T., 2018. *Mycobacterium avium* complex pulmonary disease: new epidemiology and management concepts. Curr. Opin. Infect. Dis. 31 (2), 199–207.

Le Moigne, V., et al., 2015. Bacterial phospholipases C as vaccine candidate antigens against cystic fibrosis respiratory pathogens: the *Mycobacterium abscessus* model. Vaccine 33 (18), 2118–2124.

Le Moigne, V., Gaillard, J.L., Herrmann, J.L., 2016. Vaccine strategies against bacterial pathogens in cystic fibrosis patients. Med. Mal. Infect. 46 (1), 4–9.

Lechartier, B., Cole, S.T., 2015. Mode of action of clofazimine and combination therapy with benzothiazinones against *Mycobacterium tuberculosis*. Antimicrob. Agents Chemother. 59 (8), 4457–4463.

Lee, S.J., et al., 2016. *Mycobacterium abscessus* D-alanyl-D-alanine dipeptidase induces the maturation of dendritic cells and promotes Th1-biased immunity. BMB Rep.

Lefebvre, A.L., et al., 2016. Bactericidal and intracellular activity of beta-lactams against *Mycobacterium abscessus*. J. Antimicrob. Chemother. 71 (6), 1556–1563.

Lewinsohn, D.M., et al., 2017. Official American Thoracic Society/Infectious Diseases Society of America/Centers for Disease Control and Prevention Clinical Practice Guidelines: diagnosis of tuberculosis in adults and children. Clin. Infect. Dis. 64 (2), 111–115.

Lindestam Arlehamn, C.S., et al., 2015. Immunological consequences of intragenus conservation of *Mycobacterium tuberculosis* T-cell epitopes. Proc. Natl. Acad. Sci. U.S.A. 112 (2), E147–E155.

Lounis, N., et al., 2009. ATP synthase inhibition of *Mycobacterium avium* is not bactericidal. Antimicrob. Agents Chemother. 53 (11), 4927–4929.

Low, J.L., et al., 2017. Screening of TB actives for activity against nontuberculous mycobacteria delivers high hit rates. Front. Microbiol. 8, 1539.

Luetkemeyer, A.F., et al., 2014. Evaluation of two line probe assays for rapid detection of *Mycobacterium tuberculosis*, tuberculosis (TB) drug resistance, and non-TB mycobacteria in HIV-infected individuals with suspected TB. J. Clin. Microbiol. 52 (4), 1052–1059.

Matteelli, A., et al., 2016. WHO strategies for the programmatic management of drug-resistant tuberculosis. Expert Rev. Respir. Med. 10 (9), 991–1002.

Mboowa, G., Namaganda, C., Ssengooba, W., 2014. Rifampicin resistance mutations in the 81 bp RRDR of rpoB gene in *Mycobacterium tuberculosis* clinical isolates using Xpert(R) MTB/RIF in Kampala, Uganda: a retrospective study. BMC Infect. Dis. 14, 481.

Obregon-Henao, A., et al., 2015. Susceptibility of *Mycobacterium abscessus* to antimycobacterial drugs in preclinical models. Antimicrob. Agents Chemother. 59 (11), 6904–6912.

Olivier, K.N., et al., 2017. Randomized trial of liposomal amikacin for inhalation in nontuberculous mycobacterial lung disease. Am. J. Respir. Crit. Care Med. 195 (6), 814–823.

Orme, I.M., Collins, F.M., 1985. Prophylactic effect in mice of BCG vaccination against nontuberculous mycobacterial infections. Tubercle 66 (2), 117–120.

Orme, I.M., Ordway, D.J., 2014. Host response to nontuberculous mycobacterial infections of current clinical importance. Infect. Immun. 82 (9), 3516–3522.

Philley, J.V., et al., 2015. Preliminary results of bedaquiline as salvage therapy for patients with nontuberculous mycobacterial lung disease. Chest 148 (2), 499–506.

Poyntz, H.C., et al., 2014. Non-tuberculous mycobacteria have diverse effects on BCG efficacy against *Mycobacterium tuberculosis*. Tuberculosis (Edinb.) 94 (3), 226–237.

Raju, R.M., et al., 2016. Leveraging advances in tuberculosis diagnosis and treatment to address nontuberculous mycobacterial disease. Emerg. Infect. Dis. 22 (3), 365–369.

Romanus, V., et al., 1995. A typical mycobacteria in extrapulmonary disease among children. Incidence in Sweden from 1969 to 1990, related to changing BCG-vaccination coverage. Tuber. Lung Dis. 76 (4), 300–310.

Schorey, J.S., Sweet, L., 2008. The mycobacterial glycopeptidolipids: structure, function, and their role in pathogenesis. Glycobiology 18 (11), 832–841.

Shang, S., et al., 2011a. Activities of TMC207, rifampin, and pyrazinamide against *Mycobacterium tuberculosis* infection in guinea pigs. Antimicrob. Agents Chemother. 55 (1), 124–131.

Shang, S., et al., 2011b. Increased virulence of an epidemic strain of *Mycobacterium massiliense* in mice. PLoS One 6 (9), e24726.

Singh, P.K., Silakari, O., 2018. Current status of O-heterocycles: a synthetic and medicinal overview. ChemMedChem.

Soni, I., et al., 2016. Challenges facing the drug discovery pipeline for non-tuberculous mycobacteria. J. Med. Microbiol. 65 (1), 1–8.

Wallace Jr., R.J., et al., 2014. Macrolide/azalide therapy for nodular/bronchiectatic *Mycobacterium avium* complex lung disease. Chest 146 (2), 276–282.

Worthington, R.J., Melander, C., 2013. Combination approaches to combat multi-drug resistant bacteria. Trends in Biotechnol. 31 (3), 177–184.

Xiang, X., et al., 2014. *Mycobacterium biofilms*: factors involved in development, dispersal, and therapeutic strategies against biofilm-relevant pathogens. Crit. Rev. Eukaryot. Gene Expr. 24 (3), 269–279.

Yagi, K., et al., 2017. The efficacy, safety, and feasibility of inhaled amikacin for the treatment of difficult-to-treat non-tuberculous mycobacterial lung diseases. BMC Infect. Dis. 17 (1), 558.

Yin, N., et al., 2014. Synergistic and antagonistic drug combinations depend on network topology. PLoS One 9 (4), e93960.

Zheng, W., Sun, W., Simeonov, A., 2018. Drug repurposing screens and synergistic drug-combinations for infectious diseases. Br. J. Pharmacol. 175 (2), 181–191.

Zimmer, B.L., DeYoung, D.R., Roberts, G.D., 1982. In vitro synergistic activity of ethambutol, isoniazid, kanamycin, rifampin, and streptomycin against *Mycobacterium avium-intracellulare* complex. Antimicrob. Agents Chemother. 22 (1), 148–150.

Chapter 5

Nontuberculous Mycobacterial Diseases in Humans

Lars-Olof. Larsson[1], Rutger Bennet[2], Margareta Eriksson[2], Bodil Jönsson[3] and Malin Ridell[4]

[1]Division of Respiratory Medicine, Department of Medicine, Karolinska University Hospital, Stockholm, Sweden, [2]Astrid Lindgren Children's Hospital, Karolinska University Hospital, Stockholm, Sweden, [3]Department of Infectious Medicine, Institute of Biomedicine, University of Gothenburg, Gothenburg, Sweden, [4]Department of Microbiology and Immunology, Institute of Biomedicine, University of Gothenburg, Gothenburg, Sweden

ETIOLOGICAL AGENTS

The genus *Mycobacterium* encompasses not only the organisms responsible for tuberculosis and leprosy, but also a large number of species occurring in water, soil, and animals, etc. These mycobacteria are designated nontuberculous mycobacteria (NTM) or environmental mycobacteria. They are sometimes called anonymous or atypical mycobacteria, but this terminology is misleading. Most mycobacterial species are saprophytes, but there are some opportunistic ones, that is, organisms capable of causing diseases in humans and animals (Wayne and Sramek, 1992; Tortoli, 2003; Primm et al., 2004; Falkinham, 2016). The mycobacteria are aerobic, nonspore-forming bacteria that are stained acid-fast (Grange, 1996; Wayne, 1984). They grow more slowly than most other bacteria, but there is also a great variation in growth rate between different species. Based on this variation in growth rate, mycobacteria are divided into slow growers and rapid growers (Wayne and Sramek, 1992; Wayne, 1984). A characteristic feature of the mycobacteria is their unique cell wall being rich in complex lipids; the assembly of which results in a robust cell envelope (Daffé, 2008; Minnikin et al., 2015). The lipid composition varies between species and can be used systematically, that is, to differentiate and identify them (Dobson et al., 1985). The slow growth and lipid rich cell envelope make mycobacteria more resistant to antibiotics, disinfectants, and hostile environmental conditions than most other microorganisms (Falkinham, 2016).

The genus *Mycobacterium* encompasses a large number of species, but the majority of the NTM infections in humans are caused by a relatively

Nontuberculous Mycobacteria (NTM). DOI: https://doi.org/10.1016/B978-0-12-814692-7.00005-X

small number of species. The species most frequently associated with human diseases are those of the *Mycobacterium avium* complex (MAC). These organisms are slow-growing and consist of, for example, *M. avium*, and *Mycobacterium intracellulare*, capable of causing lung infections, lymphadenitis, and disseminating infections (Falkinham, 2002). *Mycobacterium chimaera* is also referred to as the MAC and has been identified as causing outbreaks after open-chest surgery (Sax et al., 2015). Other slow-growing NTM associated with human infections particularly pulmonary ones are *Mycobacterium kansasii*, *Mycobacterium xenopi*, and *Mycobacterium malmoense* (Wayne and Sramek, 1992; Grange, 1996; Griffith et al., 2007). The latter is particularly slow growing; it takes 8 weeks or more to obtain visible colonies on solid media. Other examples of slow growers are *Mycobacterium scrofulaceum*, *Mycobacterium hemophilum*, *Mycobacterium marinum*, and *Mycobacterium ulcerans* (Wayne and Sramek, 1992; Griffith et al., 2007). The first two cause, for example, skin lesions and lymphadenitis (Griffith et al., 2007; Wolinsky, 1979). The latter two cause skin infections and grow best at 25–33°C. *M. marinum* is acquired from sea water, swimming pools, and aquariums and is the common cause of fish tank granuloma in which the infection is introduced through injuries when handling raw or living fish (Griffith et al., 2007). *M. ulcerans* causes the so-called Buruli (Africa) or Barnesdale (Australia) ulcers, and is one of the most prevalent species after *M. tuberculosis* and *M. leprae* to cause infection in immunocompetent individuals (Griffith et al., 2007; van der Werf et al., 2005). These ulcers are large, indolent, progressive, necrotic lesions of the skin, penetrating into underlying tissue. They occur in a tropical climate and mainly affect children and young adults. In contrast to all other mycobacterial species known to infect humans, this species has a defined virulence factor which is secreted. This factor is a potent macrolide toxin, mycolactone, encoded by a plasmid. It causes tissue damage and inhibits immune responses (Adusumilli et al., 2005; Portaels et al., 2009; Walsh et al., 2011).

The environment is rich in rapidly growing NTM species, most of which never, or only exceptionally, are associated with human disease. Some species can, however, be pathogenic for humans giving pulmonary, skin, and disseminating infections; good examples are *Mycobacterium fortuitum* (Wayne and Sramek, 1992) and *Mycobacterium abscessus* (Jonsson et al., 2007; Qvist et al., 2015; Bryant et al., 2016). The latter encompasses the subspecies *abscessus*, *bolletii*, and *massiliense* (Sassi and Drancourt, 2014). Infections with these organisms have increased dramatically and patients with cystic fibrosis (CF) are particularly at risk (Jonsson et al., 2007; Qvist et al., 2015; Bryant et al., 2016).

EPIDEMIOLOGY AND TRANSMISSION

Epidemiological data concerning NTM infections are not readily available (Kendall and Winthrop, 2013; Prevots and Marras, 2015) since NTM diseases are not generally reported to the public health authorities. However, recent reports have shown increasing incidence in Europe and the United States (McShane and Glassroth, 2015; Cowman et al., 2016). In developing countries, the incidence of NTM infections is particularly unknown, since epidemiological studies have only rarely been performed in these settings. In addition, many patients with NTM pulmonary disease are classified and treated as having multidrug-resistant tuberculosis when laboratory resources are insufficient.

NTM infection is acquired from the environment and not transmitted from person to person. There are, however, some indications that *M. abscessus* might be transmitted between CF patients (Bryant et al., 2016), but this phenomenon is regarded as an exception (Jonsson et al., 2007). Exposure to NTM is common given their ubiquitous presence in the environment. Waterborne aerosol transmission is the most common route since NTM are normal inhabitants of surface water and plumbing systems. NTM are common in tap water since they are far more resistant to chlorine than other bacteria, and pulmonary NTM infections have been traced to, for example, shower heads (Falkinham, 2016). Contact transmission occurs also and skin lesions have been shown to be caused by inoculation in the skin by spiky vegetation (Walsh et al., 2011). Another possibility is ingestion of NTM; for example, by newly harvested vegetables or when children play with sand and soil. NTM infections are observed not only in humans, but also in wild animals and in domesticized stock (Bercouvier and Vincent, 2001). The reservoirs for NTM are numerous and particularly frequent in moist, warm, and humid surroundings.

In children, NTM incidence has been calculated to be 1−5/100,000 out of which 80% are younger than 5 years old while other studies showed that 7% of preschool children and 20% of the school children had a positive sensitin test suggesting exposure to NTM (Lind et al., 1991; Larsson et al., 1991; Lopez-Varela et al., 2015). Consequently, it can be speculated that mild cases (small adenitis) are never diagnosed and that there also are subclinical cases.

PATHOGENESIS AND IMMUNITY

The cellular and physiologic conditions causing NTM diseases are poorly understood (Honda et al., 2015). The complex interaction of an array of lipid molecules and associated macromolecules are in all probability the most important factors responsible for mycobacterial virulence (Minnikin et al., 2015). Presence of specific virulence factors has been difficult to prove and apart from the mycolactone of *M. ulcerans* no toxins—neither endotoxins nor exotoxins—have been discovered among the mycobacteria. Recently it has been shown that hydrophobicity has a role in pathogenesis and that rough (hydrophobic) strains of, for example, *M. avium*, *M. kansasii*, and

M. abscessus are more pathogenic than smooth (less-hydrophobic) strains of the same species (Jonsson et al., 2013; Jankute et al., 2017). The shift from smooth to rough involves relatively few genes and it has been discussed whether the recent increase in severe cases of NTM infections are caused by such a shift.

Similarly to tuberculosis, professional phagocytes like macrophages are crucial for controlling NTM infection and the mycobacteria can survive and grow inside the infected phagocytes (Awuh and Flo, 2017). Cytokines such as IL-12, TNF-α, and IFN-γ are especially important in antimycobacterial immune response and regulation (Wu and Holland, 2015). Antibodies seem to play a minor role in NTM infections in humans, although some studies indicate that they contribute to the protective immunity (Glatman-Freedman, 2005). The formation of granulomas is a defense mechanism and encapsulates the mycobacteria, but in immunocompromised patients the granuloma formation may be incomplete (Wu and Holland, 2015). In addition, the granuloma may erupt and viable mycobacteria can escape and disseminate, resulting in relapse or aggravation of the infection.

PREDISPOSING FACTORS FOR NONTUBERCULOUS MYCOBACTERIA DISEASES

Patients with NTM infections often have a known predisposing condition, such as impaired resistance to infection or preexisting disease (Honda et al., 2015). Defects in the pathways of the cytokines IL-12, TNF-α, IFN-γ, HIV infections, and CF are well-known risk factors as well as immunosuppressive treatment (Knoll et al., 2012; Jones and Havlir, 2002; Chan and Iseman, 2013; Brode et al., 2015). Immunosuppressive treatment increases the risk of progression to disease by NTM in an infected individual (Brode et al., 2015). It has further been discussed whether or not inhalation of glucocorticoids may facilitate infection by NTM (Andrejak et al., 2013). Screening for NTM infection in these patients would, therefore, be of clinical importance since these patients are at risk. Comparative skin tests with tuberculins and sensitins may be a valuable option (Larsson et al., 1993).

NTM infections are also strongly associated with preexisting pulmonary diseases, (Griffith et al., 2007; Wu and Holland, 2015) good examples being chronic obstructive pulmonary disease (COPD), and bronchiectasis (Szymanski et al., 2015; Mirsaeidi et al., 2013; Sexton and Harrison, 2008). COPD is one of the most frequent NTM-associated conditions, and the number of cases is increasing. A history of excessive smoking and/or occupational exposure to airborne particles is common among these patients (Wolinsky, 1979). It has furthermore been shown that 9%–50% of patients with bronchiectasis may have active NTM lung infections (Mirsaeidi et al., 2013; Tanaka et al., 1997; Faverio et al., 2016).

During the past few years it has been shown that previous tuberculosis increases the risk of getting tuberculosis again (Borgdorff and van Soolingen, 2013), and it is also clear that the risk of getting NTM infections is higher in patients who have had tuberculosis (Sexton and Harrison, 2008).

Another further major risk factor is CF, an autosomal recessive systemic disease that involves the respiratory tract, among others (Skolnik et al., 2016). The prevalence of pulmonary NTM disease in patients with CF ranges from 4% to 14% (Skolnik et al., 2016) and *M. abscessus* occurs particularly frequently (Jonsson et al., 2007; Bryant et al., 2016; Griffith, 2003).

Many patients with NTM infections have no recognized predisposition or preexisting disease. Nonsmoking, postmenopausal, tall, and slender women are overrepresented among these NTM cases (Chan and Iseman, 2009).

Predisposing conditions are usually not found in children (Eriksson et al., 2001). Children, thus, differ from adults concerning such conditions. In a Swedish study, it was found that children with NTM lymphadenitis had been healthy before they were diagnosed (Eriksson et al., 2001). However, chronic lung disease and immunosuppression, etc., can predispose children to NTM disease also.

BACTERIOLOGICAL DIAGNOSIS

Correct identification of the etiologic agent, that is, the NTM species, is critical for diagnosis, management, and treatment of patients. Diagnosis of mycobacterioses may be delayed due to nonspecific symptoms and overlapping symptoms of underlying diseases (Griffith et al., 2007). Such a delay may result in disease progression and prolong symptom burdens. Furthermore, in an advanced disease the treatment is more complicated and with more side effects.

The most conclusive criterion for NTM infection is a positive culture together with clinical symptoms. Positive culture results, with the same NTM species from at least two separate samples, are needed for the diagnosis. The presence of a NTM in one single sample might reflect a transient contaminant since NTM are frequent in the environment. Ordinary tap water often contains entirely saprophytic NTM, such as *Mycobacterium terrae* and *Mycobacterium gordonae*, and their presence in clinical specimens are generally considered as colonization from the environment (Wayne and Sramek, 1992; Griffith et al., 2007).

Routine methods for microbiological diagnosis of NTM starts with microscopy using acid-fast stain preferable with the fluorochrome method. If the bacterial load is >10,000 bacteria/mL, the samples are smear positive (Forbes et al., 2018). In this case, tuberculosis must be excluded which can be done with a specific polymerase chain reaction (PCR) method validated for smear-positive samples (direct *M. tuberculosis* complex PCR). If this PCR is negative, the patient is infected by another mycobacterial species

(Jönsson et al., 2015). There is still no direct molecular identification method available for detection of NTM in clinical samples, except for MAC, in some laboratories. Such tests are under development for several other NTM species, but so far they are not clinically validated (British Thoracic Society NTM, 2017; Kim et al., 2008; Perry et al., 2014).

In general, NTM species should be identified. Commercially available molecular assays rapidly identify most of the clinically important MAC. If an isolate is not possible to identify with these assays, Sanger sequencing of the *16S RNA* gene, or/and the *hsp*65-gene, usually can identify the isolate. In some laboratories, species identification with the more laborious high-performance liquid chromatography of mycolic acids are in use (Forbes et al., 2018). Concerning the *M. abscessus* complex identification to subspecies level is important since the subspecies *M. abscessus* ssp. *abscessus* and *M. abscessus* ssp. *bolletii* might have a gene variant that gives inducible macrolide resistance undetectable in the phenotypic susceptibility tests. *M. abscessus* ssp. *massiliense* have this gene variant in a defect form and, therefore, does not have this special form of macrolide resistance (Mougari et al., 2017).

Some patients, particularly those who are immunocompromised, may be infected with more than one strain. They can have, for example, two different species or strains of NTM (Levy-Frebault et al., 1987) or an NTM and *M. tuberculosis* (Yilmaz et al., 2017).

Drug susceptibility testing (DST) is particularly challenging since in vitro results are only in congruence with the clinical outcome for some species and some antibiotics. Routine susceptibility testing for MAC isolates is recommended for clarithromycin only, and for *M. kansasii* only for rifampicin. Routine susceptibility testing of rapidly growing mycobacteria (*M. fortuitum*, *M. abscessus* complex, and *M. chelonae*) is reliable for several different antibiotics, for example, amikacin, clarithromycin, and linezolid. Concerning *M. marinum*, susceptible testing is only recommended in cases of treatment failure. Results of susceptibility testing of *M. malmoense* are variable and several studies have reported a lack of consistency and correlation between clinical response and in vitro results. The correlation between clinical response and in vitro results of susceptibility tests of more uncommon NTM species is unknown.

CLINICAL DIAGNOSIS AND PRESENTATION OF NONTUBERCULOUS MYCOBACTERIA DISEASE IN CHILDREN

Peripheral lymphadenitis in young children may be due to infection by NTM. In addition, NTM have also been isolated and designated as the causative agent in skin lesions, otomastoiditis, and pulmonary disease. These NTM infections have generally occurred in previously healthy children, though preexisting lung disease and immunosuppression can predispose to

NTM infection as in adults. Fine needle aspiration of samples from lesions is usually performed to obtain diagnostic material. The diagnosis is confirmed by mycobacterial culture which excludes the primary differential diagnosis tuberculosis. Tuberculin skin test (TST) and IGRA may be supportive in the diagnostic procedure. The IGRAs are quite specific for *M. tuberculosis*, but do not differentiate between latent infection and active disease and could only be supportive for the exclusion or confirmation of tuberculosis (Hermansen et al., 2014). The TST is unspecific and positive in about half of children with NTM lymphadenitis, which is why this test can be useful, particularly for those who are not BCG vaccinated.

Enlarged lymph nodes are a common clinical finding in children. Differential diagnoses to lymphadenitis by NTM infections are pyogenic lymphadenitis and lymphadenitis caused by *M. tuberculosis*.

The clinical presentation of lymphadenitis due to NTM is usually a cervical adenitis, most often in the submandibular region with purple color and firm or fluctuant consistency. The adenitis is part of a cluster of nodes, with the palpable ones being 2–3 cm in size. The lymphadenitis usually progresses in volume over a period of several weeks. The adenitis is cold, nontender, or marginally sore before rupture, which relatively frequently occurs. The child has no constitutional symptoms and routine laboratory values are normal. NTM lymphadenitis is most commonly caused by slow-growing NTM species such as *M. avium* and *M. haemophilum* (Eriksson et al., 2001). *M. avium* is the most commonly identified species, though *M. haemophilum* has been reported as a causative pathogen in Israel and the Netherlands.

Skin and soft tissue infections by NTM occur occasionally in previously healthy children (Gonzalez-Santiago and Drage, 2015). The most common species in Europe and the United States causing these infections in children is *M. marinum*, usually acquired through contact with aquarium water. Crater-like indurations are seen on areas of the body where the temperature is lower, such as fingers, elbows, or knees. Spontaneous recovery is the rule. In Australia and West Africa, *M. ulcerans* is the dominant NTM causing skin and soft tissue infections. In a study from Australia, 65% out of 25 cases were caused by *M. ulcerans* (O'Brien et al., 2018). *M. ulcerans* causes ulcers which are more likely to develop into permanent disfiguring skin lesions than those caused by other NTM species.

Otomastoiditis due to infection by NTM is a rare condition that has achieved increasing interest lately. In a Swedish study, 15 children and one adult were found over a 13-year period, which corresponds to about one per million children per year (Lundman et al., 2015). Even though these children, except one, were considered previously healthy, all had a history of ventilation tubes for treatment of recurrent otitis. Signs were secretion, polyposis, and granulomatous tissue. There were usually minor constitutional symptoms. Serious intracranial complications have been described, such as sinus thrombosis and brain abscesses. The most common causative species in this study was *M. abscessus*.

CLINICAL DIAGNOSIS AND PRESENTATION OF NONTUBERCULOUS MYCOBACTERIA DISEASE IN ADULTS

The most common NTM manifestation among adults are pulmonary infections, though all organs can be infected. The diagnosis of NTM pulmonary disease is based on four criteria: symptoms, radiological findings, microbiological findings, and the exclusion of other diagnoses (Griffith et al., 2007; British Thoracic Society NTM, 2017). The symptoms are described as local and/or general. The local symptoms are chronic or recurrent cough, sputum production, and dyspnea. The general symptoms are fatigue, weight loss, fever, and chest pain. The symptoms are generally nonspecific and mimic other pulmonary diseases, which is why diagnosis often is delayed.

Radiological findings may support the diagnosis of NTM disease with fibrocavitary or nodular-bronchiectatic infiltrates. The radiological findings can, however, not be distinctly separated from the findings due to tuberculosis, Non-NTM bronchiectasis, or certain malignancies. A conclusive diagnosis cannot finally be settled without positive NTM cultures. The microbiological diagnosis of NTM is settled by culture from two or more separate smears or a positive culture from one or more bronchial wash or lavage. Diagnosis can be further settled by a biopsy with mycobacterial histopathological features, also including a positive culture of the biopsy or of sputum or bronchial washing. As a final requirement for diagnoses, other diagnoses such as tuberculosis and malignancy should be excluded. It should be noted that besides the microbiological identification of NTM not all of the clinical findings listed need to be present for a positive diagnosis (Griffith et al., 2007; British Thoracic Society NTM, 2017).

It is of outmost importance to consider coexisting diseases, which may facilitate the development of NTM infection, such as immune deficiency or malignancy (Griffith et al., 2007). In the majority of patients with pulmonary NTM disease, the NTM infection dominates though often associated with preexisting diseases. The clinical presentation and the diagnostic findings vary between patients and is dependent on individual characteristics and comorbidities.

The clinical manifestations in the fully developed disease have a negative impact on many aspects of quality of life, that is, as in several chronic diseases (Mehta and Marras, 2011).

There are two main clinical presentations of NTM pulmonary disease; the fibrocavitary form and the nodular-bronchiectatic form. The fibrocavitary form of NTM pulmonary disease is generally associated with MAC, *M. kansasii*, and *M. abscessus* complex, and less frequently with *M. malmoense* and *M. xenopi* (Stout, 2006; Wagner and Young, 2004; Weiss and Glassroth, 2012). This form is the most progressive (Weiss and Glassroth, 2012) and untreated cases may cause extensive lung destruction and respiratory failure within 1–2 years (Griffith et al., 2007). Fibrocavitary disease is often observed in Caucasian male smokers over 50 years of age. Preexisting underlying structural lung diseases, such as tuberculosis, COPD, bronchiectasis, and pneumoconiosis, are common in fibrocavitary disease. Symptoms include coughing, which is often productive with purulent sputum, dyspnea, and constitutional symptoms (Weiss and Glassroth, 2012). Apical fibrocavitary lesions are typical. Radiographic findings are characterized by areas of increased opacity and cavitations, usually localized in the upper lobes, and calcification may be observed. Fibrosis with volume loss and traction bronchiectasis as well as apical pleural thickening are frequently found during radiological examinations (Bonaiti et al., 2015). The findings do not differentiate from those seen in active tuberculosis.

The number of patients with nodular-bronchiectatic form has increased and is presently the most common form of disease (Weiss and Glassroth, 2012). It usually develops without any known preexisting lung disease (Jarzembowski and Young, 2008). This form of disease is most frequently associated with MAC (Stout, 2006; Wagner and Young, 2004; Weiss and Glassroth, 2012). The nodular-bronchiectatic form tends to have a much slower progression than the fibrocavitary form, occurring over months and years or remaining in a stabile phase (Weiss and Glassroth, 2012). A long-term follow up may, therefore, be necessary to demonstrate clinical or radiographic changes (Griffith et al., 2007). The symptoms are nonspecific, such as chronic cough, dyspnea, and often a rich production of sputa (Weiss and Glassroth, 2012). General symptoms as fatigue, fever, and weight loss; hemoptysis and chest pains occur less commonly (McShane and Glassroth, 2015). Nodular bronchiectasis may be visible on a chest radiograph, but are more apparent on high-resolution computed tomography (HRCT). The abnormalities are primarily found in the mid- and lower-lung field. Up to 90% of patients have associated multifocal bronchiectasis, and many patients have clusters of small (5 mm) predominantly peripheral nodules in associated areas of the lung. Small cavitary lesions may also occur in advanced stages of the disease (Griffith et al., 2007; McShane and Glassroth, 2015; Weiss and Glassroth, 2012).

A further major risk factor is CF, due to the highly viscous airway secretions. These patients have an impaired mucociliary clearance and are therefore often infected. In CF patients with progressive NTM pulmonary disease, lung transplantation is generally not feasible (Weill et al., 2015; Gilljam et al., 2010).

NTM can cause diseases that are not proper infections. Exposure to aerosols containing NTM may cause hypersensitivity pneumonitis which appears to involve components of both inflammation and infection; the inflammation is predominant and the infection is of less importance (Falkinham, 2002; Griffith et al., 2007; Marras et al., 2005). These conditions result in unique features that differ distinctly from other NTM lung diseases (Griffith et al., 2007). The disease is sometimes called "hot-tub lung" since some patients become ill after using or cleaning hot tubs. NTM are tough organisms and are able to survive and thrive in harsh environments, including disinfection (Falkinham, 2002; Griffith et al., 2007). Inadequate ventilation, wet storage of filters, and aerosol-producing procedures are the main risk factors in the hot-tub setting (Fjallbrant et al., 2013) in which the causative species generally are MAC (Marras et al., 2005).

Symptom onset is subacute and dyspnea, cough, and fever are the most common symptoms. Occasionally, hypoxemic respiratory failure requires hospitalization or intensive care (Griffith et al., 2007). Chest radiographs and/or chest CT scans are abnormal. Findings include diffuse infiltrates with prominent nodularity throughout all lung fields and ground glass opacities are often present on HRCT chest scans, Pulmonary function testing demonstrates a varying degree of ventilation and diffusion capacity reduction (Griffith et al., 2007). Some patients improve after corticosteroids or simple removal of the exposure, while others improve only after antibiotic therapy (Marras et al., 2005). Prognosis of patients who are not infected, but exposed to large amounts of NTM, is generally good (Griffith et al., 2007).

THERAPY FOR NONTUBERCULOUS MYCOBACTERIA INFECTIONS IN CHILDREN

In a previously healthy child with NTM lymphadenitis there is a spontaneous resolution and surgical resection and/or drug treatment are generally not necessary. However recovery may take 6 months or more. It is important that the family can meet an experienced physician who can give adequate information and reassurance about the benign condition that requires acceptance during several months. There are several published retrospective studies and some comparative studies concerning therapy and they have also been summarized in meta analyses (Lindeboom, 2011; Lopez-Varela et al., 2015; Tebruegge et al., 2016; Zimmermann et al., 2017; Zimmermann et al., 2015). Surgical resection and prolonged antibiotic treatment are used as therapeutic options. Surgical resections are often advocated due to a presumed more rapid cure rate. There are, however, known complications such as nerve dysfunction of the facial nerve branches (Claesson et al., 2011; Lindeboom, 2012). The palsy is often believed to be temporary, however remaining symptoms can be observed, for example, when the child is tired or has an infection. Curettage has been suggested as an alternative to complete resection with less side effects. According to a recent study, complete cure following surgery was established within 3 weeks, but with complications in 16%, and curettage gave cure in 11 weeks without complications (Lindeboom, 2012). Some cases were treated with a rifamycin and/or a macrolide, but it is questionable whether this treatment shortens the duration until cure (Lindeboom, 2011). The duration of treatment has usually been 6 months.

Skin infection with *M. marinum* also heals spontaneously. Tetracycline can be used for faster resolution of symptoms. Early antibiotic treatment should be used for lesions caused by *M. ulcerans* to avoid deformities. Mastoiditis caused by *M. abscessus* requires both intensive medical treatment and invasive procedures.

There are special considerations for children regarding the pharmacokinetics of antimycobacterial drugs. Treatment is complicated by the shortage of child-friendly products.

THERAPY FOR NONTUBERCULOUS MYCOBACTERIA INFECTIONS IN ADULTS

Therapy should be person-centered. The unique conditions of each individual patient should be considered and evaluated. In addition to the medical therapy, adjunctive measures such as physiotherapy and nutritional advice should be given. In some cases, surgery might be part of the treatment and has a beneficial effect for long-term outcome. The medical therapy should be based on evidence-based regimes where adherence, monitoring, and evaluation of the treatment are key factors. The therapeutic arsenal is limited and adherence to treatment guidelines is often low since the treatment regimens are complex and lengthy and side effects are common. The decision to treat depends also on the aim of the therapy, since eradication is not always possible and stability may be the only expected achievement.

The primary drug treatment endpoint for eradication is 12 months of negative sputum cultures while on therapy (Griffith et al., 2007).

When the microbiologic finding reveals a positive smear for acid-fast bacilli and there is no history of a previous pulmonary disease by NTM (Griffith et al., 2007) an individual should receive tuberculosis treatment pending the confirmation of the diagnosis. The identification of the NTM species does however not always prompt the initiation of treatment. The decision to start treatment should be based on the clinical presentation, the expected course, and progression of disease and the individual risk factors, such as preexisting lung disease, genetic, or iatrogenetic conditions. The bacterial load and clinical significance of the infecting species are also important factors as basis for treatment decisions (Griffith et al., 2007; Thomson and Yew, 2009). Low-virulence NTM, such as *M. fortuitum*, and environmental ones, such as *M. gordonae* and *M. terrae*, are often associated with contamination and usually do not warrant treatment. *M. kansasii* tends to be more associated with tuberculosis-like illness, but is also generally easier to treat. Therefore, the threshold for treatment with antimicrobials is somewhat lower than, for example, with *M. abscessus* complex which is notoriously difficult to eradicate (Griffith et al., 2007; Thomson and Yew, 2009).

Few studies exist that guide the treatment of pulmonary disease by NTM. The ATS/IDSA guidelines are based largely on expert opinion and offer the most ample information on treatment of pulmonary disease by NTM caused by MAC, *M. kansasii*, and *M. abscessus* complex. The current evidence base is composed of single-center observational studies and randomized–controlled trials from the British Thoracic Society (Research Committee of the British Thoracic Society, 2001; Jenkins et al., 2008). The treatment regimens differ by species, with the most important distinction being between slow growing mycobacteria and rapid growing ones. For most slow growers, the recommended regimen consists of rifampin (or rifabutin), ethambutol, and a macrolide antibiotic and is given for 12 months following culture conversion; amikacin or streptomycin can be added in the initial 3 months in cases of severe disease (Griffith et al., 2007). As *M. kansasii* is clinically more closely related to *M. tuberculosis* than to other NTM species (Evans et al., 1996), pulmonary disease caused by *M. kansasii* is usually treated with a combination of isoniazid, rifampin, and ethambutol (Griffith et al., 2007). However, isoniazid should not be used for the treatment of disease by other NTM species than *M. kansasii*. For the rapidly growing mycobacteria, regimens are primarily based on in vitro DST results. For *M. abscessus* complex, the most notorious causative agent of disease among rapid growers, these regimens often include a macrolide antibiotic, amikacin and either cefoxitin, imipenem, or tigecycline (van Ingen et al., 2012).

Accurate identification of the disease-causing strain in *M. abscessus* complex lung disease is critical before treatment initiation since the subspecies differ in their susceptibility to antibiotics and treatment outcomes. *M. abscessus* spp. *massiliense* is susceptible to macrolides, and treatment outcomes are better than for *M. abscessus* spp. *abscessus*, which has a gene that confers inducible macrolide resistance (Nash et al., 2009; Koh et al., 2011). Generally, the *M. abscessus* complex subspecies are susceptible to amikacin, cefoxitin, and clarithromycin in vitro (Griffith et al., 2007). However, the correlation between in vitro and in vivo susceptibility is only established for macrolides and the clinician should be aware of the limitations of DST for the *M. abscessus* complex (Maurer et al., 2014). The initial treatment over 2–4 months consists of one or two parenteral antibiotics and an oral macrolide if the species is macrolide susceptible (Griffith et al., 2007; Kasperbauer and De Groote, 2015). The most active parenteral agents are amikacin, cefoxitin, imipenem, and tigecycline (Griffith et al., 2007). After the initial treatment phase and cessation of parenteral therapy, an additional oral antibiotic is recommended, for example, a fluoroquinolone, linezolid, or clofazimine (Kasperbauer and De Groote, 2015).

The most favorable option of curing local *M. abscessus* complex lung disease is optimized multidrug chemotherapy and surgical resection of localized disease (Griffith et al., 2007). The prognosis is generally discouraging in cases where a medical and joint surgical intervention is not possible (Griffith et al., 1993, 2007). Suppressive therapy, including periodic parenteral antibiotic or oral macrolide therapy, may be offered to control the disease symptoms in these patients (Griffith et al., 2007).

In some cases of NTM disease, surgery may be appropriate, but there are no established criteria for patient selection. Patients whose pulmonary disease is predominantly localized to one lung and/or one lobe who can tolerate surgical resection may be candidates for surgery under some circumstances. These include poor response to drug therapy, the development of macrolide-resistant disease or the presence of significant disease-related complications such as hemoptysis. Surgery should be performed in centers with expertise in both medical and surgical management of mycobacterial diseases because lung resection surgery for mycobacterial disease is potentially associated with significant morbidity and mortality. Overall, surgical resection of limited (focal) disease in a patient with adequate cardiopulmonary reserve to withstand partial or complete lung resection can be successful in combination with multidrug treatment regimens for treating MAC lung disease (Griffith et al., 2007; Wallace et al., 2014).

Use of adjunctive therapies such as physiotherapy and nutritional advice in addition to antibiotics should be part of the treatment. The patients should play a key role in daily disease management and they should be made aware that smoking cessation is important for improved airway function. Modalities offering increased mucus clearance, especially in patients with bronchiectasis, include several methods such as inhalation of hypertonic saline. An individual physical training program in close cooperation with a physiotherapist is essential for the wellbeing of patients and helps preserve their physical capacity (Bouchard et al., 2012; Pollock et al., 2001). Physical function can be monitored with an exercise test, for example, the 6-min walk test (Singh et al., 2014). Involuntary weight loss is a negative prognostic factor and the nutritional status should be evaluated in cooperation with a dietitian (Berthon and Wood, 2015). Severe vitamin D deficiency may predispose for development or progression of NTM disease (McShane and Glassroth, 2015).

PROGNOSIS

Treatment for NTM pulmonary disease can be disappointing. The prognosis is related to the treatment response. Sustained clinical improvement and culture conversion is not achievable for all patients. Cure is usually defined as sustained culture conversion >12 months after treatment (Griffith et al., 2007). Cure rates differ by species, disease severity and line of treatment (first line, relapse/reinfection) and range from 30%–50% in *M. abscessus* complex to 50%–70% in MAC (van Ingen et al., 2012). Fibrocavitary disease by NTM is associated with greater mortality compared to other forms of pulmonary NTM (Hayashi et al., 2012). Patients are considered treatment failures if they have not had response (microbiologic, clinical or radiographic) after 6 months of appropriate therapy or achieved conversion of sputum to culture negative after 12 months of appropriate therapy

(Griffith et al., 2007). Macrolides have become the mainstay of treatment regimens for MAC. Infection with a macrolide-resistant strain or acquisition of macrolide resistance during treatment hinders adequate therapy and is a negative prognostic factor (Griffith et al., 2007). The likelihood of developing macrolide resistance appears to be relatively low when the treatment regimen complies with the ATS/IDSA-recommended 3-drug therapy (Griffith et al., 2006; Adjemian et al., 2014).

Even with treatment that is considered successful, there is a possibility of recurrence, either in the form of relapse or reinfection. Among patients who initially have sputum conversion but subsequently develop positive cultures for MAC after discontinuing therapy, many are reinfected by new MAC strains rather than having a relapse with their initial MAC strain (Griffith et al., 2007). The rate of recurrence can be considerable (Wallace et al., 2014).

CONCLUSIONS

There is now a broad spectrum of factors which promote NTM as a causative agent of disease in humans. The healthcare provider needs insight and understanding of the current dynamic of NTM. A consensus rationale for clinical management and plans for clinical development are needed since diseases caused by NTM represent a growing problem and further research is of major importance. Current disease management is far from satisfactory and needs to be improved (Mirsaeidi et al., 2015; Griffith, 2015).

Little is known concerning the interaction between NTM and humans in spite of the fact that all humans are repeatedly exposed to NTM during their whole life. Recent analyses indicate that *M. tuberculosis* has developed and evolved from being an environmental saprophyte to a parasite (Jankute et al., 2017). Can a similar pernicious process occur with any NTM species? On the other hand, can repeated NTM exposure be beneficial for humans by activating the immune system?

REFERENCES

Adjemian, J., Prevots, D.R., Gallagher, J., Heap, K., Gupta, R., Griffith, D., 2014. Lack of adherence to evidence-based treatment guidelines for nontuberculous mycobacterial lung disease. Ann. Am. Thorac. Soc. 11, 9–16.

Adusumilli, S., Mve-Obiang, A., Sparer, T., Meyers, W., Hayman, J., Small, P.L., 2005. *Mycobacterium ulcerans* toxic macrolide, mycolactone modulates the host immune response and cellular location of *M. ulcerans* in vitro and in vivo. Cell Microbiol. 7, 1295–1304.

Andrejak, C., Nielsen, R., Thomsen, V.O., Duhaut, P., Sorensen, H.T., Thomsen, R.W., 2013. Chronic respiratory disease, inhaled corticosteroids and risk of non-tuberculous mycobacteriosis. Thorax 68, 256–262.

Awuh, J.A., Flo, T.H., 2017. Molecular basis of mycobacterial survival in macrophages. Cell Mol. Life Sci. 74, 1625–1648.

Bercouvier, H., Vincent, V., 2001. Mycobacterial infections in domestic and wild animals due to *Mycobacterium marinum*, *M. fortuitum*, *M. chelonae*, *M. porcinum*, *M. farcinogenes*, *M. smegmatis*, *M. scrofulaceum*, *M. xenopi*, *M. kansasii*, *M. simiae* and *M. genavense*. Rev. Sci. Tech. 20, 265–290.

Berthon, B.S., Wood, L.G., 2015. Nutrition and respiratory health—feature review. Nutrients 7, 1618–1643.

Bonaiti, G., Pesci, A., Marruchella, A., Lapadula, G., Gori, A., Aliberti, S., 2015. Nontuberculous mycobacteria in noncystic fibrosis bronchiectasis. Biomed. Res. Int. 2015, 197950.

Borgdorff, M.W., van Soolingen, D., 2013. The re-emergence of tuberculosis: what have we learnt from molecular epidemiology? Clin. Microbiol. Infect. 19, 889–901.

Bouchard, C., Blair, S., Haskell, W., 2012. Physical Activity and Health, second edition. Human Kinetics, Champaign, IL.

Brode, S.K., Jamieson, F.B., Ng, R., et al., 2015. Increased risk of mycobacterial infections associated with anti-rheumatic medications. Thorax 70, 677–682.

British Thoracic Society NTM Guideline Development Group British Thoracic Society Guidelines for the Diagnosis and Management of Non-tuberculous Mycobacterial Pulmonary Disease (NTM-PD), 2017. Under consultation. <https://www.brit-thoracic.org.uk/document-library/clinical-information/non-tuberculous-mycobacteria/ntm-guideline/bts-guidelines-for-the-diagnosis-and-management-of-ntm-pd/> (accessed 03.03.17).

Bryant, J.M., Grogono, D.M., Rodriguez-Rincon, D., et al., 2016. Emergence and spread of a human-transmissible multidrug-resistant nontuberculous mycobacterium. Science 354, 751–757.

Chan, E., Iseman, M., 2009. Slender old woman appear to be more susceptible to non tuberculous mycobacterial disease. Gender 7, 5–18.

Chan, E.D., Iseman, M.D., 2013. Underlying host risk factors for nontuberculous mycobacterial lung disease. Semin. Respir. Crit. Care Med. 34, 110–123.

Claesson, G., Bennet, R., Eriksson, M., Petrini, B., 2011. Nerve dysfunction following surgical treatment of cervical non-tuberculous mycobacterial lymphadenitis in children. Acta. Paediatr. 100, 299–302.

Cowman, S., Burns, K., Benson, S., Wilson, R., Loebinger, M.R., 2016. The antimicrobial susceptibility of non-tuberculous mycobacteria. J. Infect. 72, 324–331.

Daffé, M., 2008. The global architecture of the mycobacterial cell envelope. In: Daffé, M., Reyat, J. (Eds.), The Mycobacterial Cell Envelope. ASM Press, Washington, DC, pp. 3–11.

Dobson, G., Minnikin, D., Minnikin, S., et al., 1985. Systematic analyses of complex mycobacterial lipids. In: Goodfellow, M., Minnikin, D. (Eds.), Clinical Methods in Bacterial Systematics. Academic Press, London, pp. 237–265.

Eriksson, M., Bennet, R., Danielsson, N., 2001. Non-tuberculous mycobacterial lymphadenitis in healthy children: another "lifestyle disease?" Acta. Paediatr. 90, 1340–1342.

Evans, S.A., Colville, A., Evans, A.J., Crisp, A.J., Johnston, I.D., 1996. Pulmonary *Mycobacterium kansasii* infection: comparison of the clinical features, treatment and outcome with pulmonary tuberculosis. Thorax 51, 1248–1252.

Falkinham, J.O., 2002. 3rd nontuberculous mycobacteria in the environment. Clin. Chest Med. 23, 529–551.

Falkinham, J.O., 2016. 3rd current epidemiologic trends of the nontuberculous mycobacteria (NTM). Curr. Environ. Health Rep. 3, 161–167.

Faverio, P., Stainer, A., Bonaiti, G., et al., 2016. Characterizing non-tuberculous mycobacteria infection in bronchiectasis. Int. J. Mol. Sci. 17, 1913.

Fjallbrant, H., Akerstrom, M., Svensson, E., Andersson, E., 2013. Hot tub lung: an occupational hazard. Eur. Respir. Rev. 22, 88–90.

Forbes, B.A., Hall, G.S., Miller, M.B., et al., 2018. Practice guidelines for clinical microbiology laboratories: mycobacteria. Clin. Microbiol. Rev. 31.

Gilljam, M., Schersten, H., Silverborn, M., Jonsson, B., Ericsson Hollsing, A., 2010. Lung transplantation in patients with cystic fibrosis and *Mycobacterium abscessus* infection. J. Cystic Fibrosis 9, 272–276.

Glatman-Freedman, A., 2005. Role of antibody-mediated immunity in host defence against *Mycobacterium tuberculosis*. In: Cole, S., Eisenach, K., McMurray, D., Jacobs, J. (Eds.), Tuberculosis and the Tubercle Bacillus. ASM Press, Washington, DC.

Gonzalez-Santiago, T.M., Drage, L.A., 2015. Nontuberculous mycobacteria: skin and soft tissue infections. Dermatol. Clin. 33, 563–577.

Grange, J., 1996. Mycobacteria and Human Disease. Oxford University Press, New York.

Griffith, D.E., 2003. Emergence of nontuberculous mycobacteria as pathogens in cystic fibrosis. Am. J. Respir. Crit. Care Med. 167, 810–812.

Griffith, D.E., 2015. Nontuberculous mycobacterial disease research. An end to the beginning. Am. J. Respir. Crit. Care Med. 192, 535–537.

Griffith, D.E., Girard, W.M., Wallace Jr., R.J., 1993. Clinical features of pulmonary disease caused by rapidly growing mycobacteria. An analysis of 154 patients. Am. Rev. Respir. Dis. 147, 1271–1278.

Griffith, D.E., Brown-Elliott, B.A., Langsjoen, B., et al., 2006. Clinical and molecular analysis of macrolide resistance in *Mycobacterium avium* complex lung disease. Am. J. Respir. Crit. Care Med. 174, 928–934.

Griffith, D.E., Aksamit, T., Brown-Elliott, B.A., et al., 2007. An official ATS/IDSA statement: diagnosis, treatment, and prevention of nontuberculous mycobacterial diseases. Am. J. Respir. Crit. Care Med. 175, 367–416.

Hayashi, M., Takayanagi, N., Kanauchi, T., Miyahara, Y., Yanagisawa, T., Sugita, Y., 2012. Prognostic factors of 634 HIV-negative patients with *Mycobacterium avium* complex lung disease. Am. J. Respir. Crit. Care Med. 185, 575–583.

Hermansen, T.S., Thomsen, V.O., Lillebaek, T., Ravn, P., 2014. Non-tuberculous mycobacteria and the performance of interferon gamma release assays in Denmark. PLoS One 9, e93986.

Honda, J.R., Knight, V., Chan, E.D., 2015. Pathogenesis and risk factors for nontuberculous mycobacterial lung disease. Clin. Chest Med. 36, 1–11.

van Ingen, J., Boeree, M.J., van Soolingen, D., Mouton, J.W., 2012. Resistance mechanisms and drug susceptibility testing of nontuberculous mycobacteria. Drug Resist. Updat. 15, 149–161.

Jankute, M., Nataraj, V., Lee, O.Y., et al., 2017. The role of hydrophobicity in tuberculosis evolution and pathogenicity. Sci. Rep. 7, 1315.

Jarzembowski, J.A., Young, M.B., 2008. Nontuberculous mycobacterial infections. Arch. Pathol. Lab. Med. 132, 1333–1341.

Jenkins, P.A., Campbell, I.A., Banks, J., Gelder, C.M., Prescott, R.J., Smith, A.P., 2008. Clarithromycin vs ciprofloxacin as adjuncts to rifampicin and ethambutol in treating opportunist mycobacterial lung diseases and an assessment of *Mycobacterium vaccae* immunotherapy. Thorax 63, 627–634.

Jones, D., Havlir, D.V., 2002. Nontuberculous mycobacteria in the HIV infected patient. Clin. Chest. Med. 23, 665–674.

Jonsson, B., Ridell, M., Wold, A.E., 2013. Phagocytosis and cytokine response to rough and smooth colony variants of *Mycobacterium abscessus* by human peripheral blood mononuclear cells. APMIS 121, 45–55.

Jönsson, B., Lönnermark, E., Ridell, M., 2015. Evaluation of the Cobas TaqMan MTB test for detection of *Mycobacterium tuberculosis* complex. Infect. Dis. 47, 231–236.

Jonsson, B.E., Gilljam, M., Lindblad, A., Ridell, M., Wold, A.E., Welinder-Olsson, C., 2007. Molecular epidemiology of *Mycobacterium abscessus*, with focus on cystic fibrosis. J. Clin. Microbiol. 45, 1497–1504.

Kasperbauer, S.H., De Groote, M.A., 2015. The treatment of rapidly growing mycobacterial infections. Clin. Chest Med. 36, 67–78.

Kendall, B.A., Winthrop, K.L., 2013. Update on the epidemiology of pulmonary nontuberculous mycobacterial infections. Semin. Respir. Crit. Care Med. 34, 87–94.

Kim, S., Park, E.M., Kwon, O.J., et al., 2008. Direct application of the PCR restriction analysis method for identifying NTM species in AFB smear-positive respiratory specimens. Int. J. Tuberc. Lung Dis. 12, 1344–1346.

Knoll, B.M., Kappagoda, S., Gill, R.R., et al., 2012. Non-tuberculous mycobacterial infection among lung transplant recipients: a 15-year cohort study. Transpl. Infect. Dis. 14, 452–460.

Koh, W.J., Jeon, K., Lee, N.Y., et al., 2011. Clinical significance of differentiation of *Mycobacterium massiliense* from *Mycobacterium abscessus*. Am. J. Respir. Crit. Care Med. 183, 405–410.

Larsson, L.O., Skoogh, B.E., Bentzon, M.W., et al., 1991. Sensitivity to sensitins and tuberculin in Swedish children. II. A study of preschool children. Tubercle 72, 37–42.

Larsson, L.O., Bentzon, M.W., Lind, A., et al., 1993. Sensitivity to sensitins and tuberculin in Swedish children. Part 5: A study of school children in an inland rural area. Tuber. Lung Dis. 74, 371–376.

Levy-Frebault, V., Pangon, B., Bure, A., Katlama, C., Marche, C., David, H.L., 1987. *Mycobacterium simiae* and *Mycobacterium avium–M. intracellulare* mixed infection in acquired immune deficiency syndrome. J. Clin. Microbiol. 25, 154–157.

Lind, A., Larsson, L.O., Bentzon, M.W., et al., 1991. Sensitivity to sensitins and tuberculin in Swedish children. I. A study of schoolchildren in an urban area. Tubercle 72, 29–36.

Lindeboom, J.A., 2011. Conservative wait-and-see therapy versus antibiotic treatment for nontuberculous mycobacterial cervicofacial lymphadenitis in children. Clin. Infect. Dis. 52, 180–184.

Lindeboom, J.A., 2012. Surgical treatment for nontuberculous mycobacterial (NTM) cervicofacial lymphadenitis in children. J. Oral. Maxillofac. Surg. 70, 345–348.

Lopez-Varela, E., Garcia-Basteiro, A.L., Santiago, B., Wagner, D., van Ingen, J., Kampmann, B., 2015. Non-tuberculous mycobacteria in children: muddying the waters of tuberculosis diagnosis. Lancet Respir. Med. 3, 244–256.

Lundman, L., Edvardsson, H., Angeby, K., 2015. Otomastoiditis caused by non-tuberculous mycobacteria: report of 16 cases, 3 with infection intracranially. J. Laryngol. Otol. 129, 644–655.

Marras, T.K., Wallace Jr., R.J., Koth, L.L., Stulbarg, M.S., Cowl, C.T., et al., 2005. Hypersensitivity pneumonitis reaction to *Mycobacterium avium* in household water. Chest 127, 664–671.

Maurer, F.P., Bruderer, V.L., Ritter, C., Castelberg, C., Bloemberg, G.V., Bottger, E.C., 2014. Lack of antimicrobial bactericidal activity in *Mycobacterium abscessus*. Antimicrob. Agents Chemother. 58, 3828–3836.

McShane, P.J., Glassroth, J., 2015. Pulmonary disease due to nontuberculous mycobacteria: current state and new insights. Chest 148, 1517–1527.

Mehta, M., Marras, T.K., 2011. Impaired health-related quality of life in pulmonary nontuberculous mycobacterial disease. Respir. Med. 105, 1718–1725.

Minnikin, D.E., Lee O.-C., Wu, H.H.T., et al., 2015. Pathophysiological implications of cell envelope structure of *Mycobacterium tuberculosis* and related taxa. In: Wellman, R. (Ed.), TuberculosisExpanding Knowledge, InTechOpen, pp. 145–175.

Mirsaeidi, M., Hadid, W., Ericsoussi, B., Rodgers, D., Sadikot, R.T., 2013. Non-tuberculous mycobacterial disease is common in patients with non-cystic fibrosis bronchiectasis. Int. J. Infect. Dis. 17, e1000–e1004.

Mirsaeidi, M., Farnia, P., Sadikot, R., Hsueh, P.R., Aliberti, S., 2015. Nontuberculous mycobacteria: epidemiologic, mycobacteriologic, and clinical aspects. Biomed. Res. Int. 2015, 523697.

Mougari, F., Loiseau, J., Veziris, N., et al., 2017. Evaluation of the new GenoType NTM-DR kit for the molecular detection of antimicrobial resistance in non-tuberculous mycobacteria. J. Antimicrob. Chemother. 72, 1669–1677.

Nash, K.A., Brown-Elliott, B.A., Wallace Jr., R.J., 2009. A novel gene, erm(41), confers inducible macrolide resistance to clinical isolates of *Mycobacterium abscessus* but is absent from *Mycobacterium chelonae*. Antimicrob. Agents Chemother. 53, 1367–1376.

O'Brien, D.P., Friedman, N.D., McDonald, A., et al., 2018. Wound healing: natural history and risk factors for delay in Australian patients treated with antibiotics for *Mycobacterium ulcerans* disease. PLoS Negl. Trop. Dis. 12, e0006357.

Perry, M.D., White, P.L., Ruddy, M., 2014. Potential for use of the Seegene Anyplex MTB/NTM real-time detection assay in a regional reference laboratory. J. Clin. Microbiol. 52, 1708–1710.

Pollock, J.M., Buddle, B.M., Andersen, P., 2001. Towards more accurate diagnosis of bovine tuberculosis using defined antigens. Tuberculosis 81, 65–69.

Portaels, F., Silva, M.T., Meyers, W.M., 2009. Buruli ulcer. Clin. Dermatol. 27, 291–305.

Prevots, D.R., Marras, T.K., 2015. Epidemiology of human pulmonary infection with nontuberculous mycobacteria: a review. Clin. Chest. Med. 36, 13–34.

Primm, T.P., Lucero, C.A., Falkinham, J.O., 2004. 3rd health impacts of environmental mycobacteria. Clin. Microbiol. Rev. 17, 98–106.

Qvist, T., Gilljam, M., Jönsson, B., et al., 2015. Epidemiology of nontuberculous mycobacteria among patients with cystic fibrosis in Scandinavia. J. Cystic Fibrosis 14, 46–52.

Research Committee of the British Thoracic Society, 2001. First randomised trial of treatments for pulmonary disease caused by *M avium intracellulare*, *M. malmoense*, and *M. xenopi* in HIV negative patients: rifampicin, ethambutol and isoniazid versus rifampicin and ethambutol. Thorax 56, 167–72.

Sassi, M., Drancourt, M., 2014. Genome analysis reveals three genomospecies in *Mycobacterium abscessus*. BMC Genomics 15, 359.

Sax, H., Bloemberg, G., Hasse, B., et al., 2015. Prolonged outbreak of *Mycobacterium chimaera* infection after open-chest heart surgery. Clin. Infect. Dis. 61, 67–75.

Sexton, P., Harrison, A.C., 2008. Susceptibility to nontuberculous mycobacterial lung disease. Eur. Respir. J. 31, 1322–1333.

Singh, S.J., Puhan, M.A., Andrianopoulos, V., et al., 2014. An official systematic review of the European Respiratory Society/American Thoracic Society: measurement properties of field walking tests in chronic respiratory disease. Eur. Respir. J. 44, 1447–1478.

Skolnik, K., Kirkpatrick, G., Quon, B.S., 2016. Nontuberculous mycobacteria in cystic fibrosis. Curr. Treat. Options Infect. Dis. 8, 259–274.

Stout, J.E., 2006. Evaluation and management of patients with pulmonary nontuberculous mycobacterial infections. Expert Rev. Anti. Infect. Ther. 4, 981–993.

Szymanski, E.P., Leung, J.M., Fowler, C.J., et al., 2015. Pulmonary nontuberculous mycobacterial infection. A multisystem, multigenic disease. Am. J. Respir. Crit. Care Med. 192, 618–628.

Tanaka, E., Amitani, R., Niimi, A., Suzuki, K., Murayama, T., Kuze, F., 1997. Yield of computed tomography and bronchoscopy for the diagnosis of *Mycobacterium avium* complex pulmonary disease. Am. J. Respir. Crit. Care Med. 155, 2041–2046.

Tebruegge, M., Pantazidou, A., MacGregor, D., et al., 2016. Nontuberculous mycobacterial disease in children—epidemiology, diagnosis & management at a tertiary center. PLoS One 11, e0147513.

Thomson, R.M., Yew, W.W., 2009. When and how to treat pulmonary nontuberculous mycobacterial diseases. Respirology 14, 12–26.

Tortoli, E., 2003. Impact of genotypic studies on mycobacterial taxonomy: the new mycobacteria of the 1990s. Clin. Microbiol. Rev. 16, 319–354.

Wagner, D., Young, L.S., 2004. Nontuberculous mycobacterial infections: a clinical review. Infection 32, 257–270.

Wallace Jr., R.J., Brown-Elliott, B.A., McNulty, S., et al., 2014. Macrolide/azalide therapy for nodular/bronchiectatic *Mycobacterium avium* complex lung disease. Chest 146, 276–282.

Walsh, D.S., Portaels, F., Meyers, W.M., 2011. Buruli ulcer: advances in understanding *Mycobacterium ulcerans* infection. Dermatol. Clin. 29, 1–8.

Wayne, L.G., 1984. Mycobacterial speciation. In: Kubica, G.P., Wayne, L.G. (Eds.), The Mycobacteria a Sourcebook. Marcel Dekker Inc, New York.

Wayne, L.G., Sramek, H.A., 1992. Agents of newly recognized or infrequently encountered mycobacterial diseases. Clin. Microbiol. Rev. 5, 1–25.

Weill, D., Benden, C., Corris, P.A., et al., 2015. A consensus document for the selection of lung transplant candidates: 2014—an update from the Pulmonary Transplantation Council of the International Society for Heart and Lung Transplantation. J. Heart Lung Transplant. 34, 1–15.

Weiss, C.H., Glassroth, J., 2012. Pulmonary disease caused by nontuberculous mycobacteria. Expert Rev. Respir. Med. 6, 597–612.

van der Werf, T.S., Stienstra, Y., Johnson, R.C., et al., 2005. *Mycobacterium ulcerans* disease. Bull. World Health Organ. 83, 785–791.

Wolinsky, E., 1979. Nontuberculous mycobacteria and associated diseases. Am. Rev. Respir. Dis. 119, 107–159.

Wu, U.I., Holland, S.M., 2015. Host susceptibility to non-tuberculous mycobacterial infections. Lancet Infect. Dis. 15, 968–980.

Yilmaz, N., Ucar, E.Y., Saglam, L., 2017. *Mycobacterium tuberculosis* and nontuberculous mycobacteria coinfection of the lungs. Turk Thorac J. 18, 23–26.

Zimmermann, P., Tebruegge, M., Curtis, N., Ritz, N., 2015. The management of non-tuberculous cervicofacial lymphadenitis in children: a systematic review and meta-analysis. J. Infect. 71, 9–18.

Zimmermann, P., Curtis, N., Tebruegge, M., 2017. Nontuberculous mycobacterial disease in childhood—update on diagnostic approaches and treatment. J. Infect. 74 (Suppl. 1), S136–S142.

Chapter 6

Nontuberculous Mycobacterial Lung Disease

Ruxana T. Sadikot
Section of Pulmonary, Critical Care Medicine, Emory University, Atlanta, GA, United States

INTRODUCTION

Nontuberculous mycobacteria (NTM) refer to mycobacteria other than *Mycobacterium tuberculosis* and *Mycobacterium leprae*. NTM are gram-positive, acid-fast, aerobic bacilli that are ubiquitous in the environment. They are isolated from shower heads, water distribution systems, bird droppings, soil, dust, milk, and laboratory equipment. Unlike tuberculosis (TB), the exact prevalence is difficult to estimate as NTM is not reportable (Brode et al., 2014). Additionally, NTM is acquired from the environment and is not transmitted from person-to-person, except in select situations such as cystic fibrosis (CF). The incidence and prevalence of NTM lung disease (NTM-LD) is rising worldwide and accounts for most of clinical cases of NTM (Adjemian et al., 2012; Ringshausen et al., 2016). This increased incidence has been attributed to environmental and climate changes, showering, genetic mutations, nosocomial infections, and improved molecular testing. However, it is also possible that the true number of cases may not have changed dramatically, but rather there are simply more cases being diagnosed due to greater utilization of chest CT scan (prompting sputum cultures in those with abnormalities), more reliance on more sensitive molecular diagnostic techniques, and greater awareness of NTM-LD by clinicians.

NTM are usually classified based on their growth rate. Rapid growers form visible colonies within 7 days on solid media, whereas slow growers take longer to appear in culture. With advancement in molecular techniques, new NTM species have been described and now 160 different NTM species have been identified, with more being added on a regular basis (Claeys and Robinson, 2018). Among these recognized NTM species, the most frequent human pathogens are the slow growing *Mycobacterium avium* complex (MAC), *Mycobacterium kansasii*, and rapidly growing *Mycobacterium abscessus* complex. MAC includes *M. avium* and *M. intracellulare* and are

Nontuberculous Mycobacteria (NTM). DOI: https://doi.org/10.1016/B978-0-12-814692-7.00006-1

the main pathogens causing pulmonary NTM infections. *M. abscessus* complex, encompasses the subspecies *abscessus*, *bolletii*, and *massiliense* (Esteban and García-Coca, 2018). Other NTM infections such as *Mycobacterium fortuitum*, *xenopi*, *simiae*, *chelonae*, and *szulgai* are less common causes of NTM-LD. Several anecdotal reports of other less common species are described as case reports or small case series.

The diagnosis is based on an association of clinical, microbiological and radiological criteria. In immunocompetent hosts, structural lung diseases, such as bronchiectasis, chronic obstructive pulmonary disease (COPD), pneumoconiosis, sarcoidosis and CF, appear to increase the risk of NTM infection or colonization (Holland, 2001). Although rare NTM-LD is also known to occur in immunocompromised individuals including patients who are on biologics.

CLINICAL MANIFESTATIONS OF NONTUBERCULOUS MYCOBACTERIA LUNG DISEASE

Pulmonary symptoms of NTM-LD are usually nonspecific and comprise dry or productive cough, dyspnea, hemoptysis, fever, weight loss, and malaise. The onset may be acute or insidious, subacute or chronic. Lung NTM disease typically presents in three major clinical phenotypes: (1) fibro-cavitary, (2) nodular bronchiectatic form, and (3) hypersensitivity pneumonitis. The increased association of NTM with preexisting lung diseases is being more frequently recognized because of the use of high-resolution computed tomography (HRCT) and improved molecular approaches to diagnose NTM (Honda et al., 2015; Holland, 2001; Griffith and Aksamit, 2012; Porvaznik et al., 2017).

The fibro-cavitary form is characterized by cavitary lesions, occurring predominantly in the upper lobes, and usually develops in older males with underlying lung disease such as chronic obstructive airways disease and pneumoconiosis and is often associated with alcohol misuse disorder. The fibro-cavitary form of NTM-LD is usually progressive with worse prognosis, especially if untreated. The high morbidity and mortality in this group results from chronic respiratory failure (Griffith and Aksamit, 2012; Porvaznik et al., 2017).

On the other hand, the nodular bronchiectatic form occurs predominantly in nonsmoking, postmenopausal women and presents as bilateral bronchiectasis with multiple nodules and tree-in-bud opacities on HRCT. It often presents without underlying lung disease. The diagnosis may be delayed because of the insidious presentation.

Hypersensitivity pneumonitis or extrinsic allergic alveolitis, also known as "hot-tub lung" is an inflammatory reaction mounted from a hypersensitivity response rather than a true infection. This is associated with inhalation of a large amount of NTM. It is important to recognize and distinguish this presentation as the management involves removal of the inciting agent and steroids (Griffith and Aksamit, 2012; Porvaznik et al., 2017).

Other clinical and radiological presentations, such as reticular nodular shadowing, consolidation, and pleural effusions, can also occur, although in general these are uncommon. In a recent study. Park et al. characterized patients with NTM pleuritis and pleural effusions. They found that the laboratory characteristics of patients with NTM pleural disease were like that of TB with elevation of adenosine deaminase and lymphocytic exudative effusion. However, the treatment outcome in general was worse in patients with pleural disease compared to those with NTM-LD (Park et al., 2017). In the immunocompromised patient, mediastinal involvement, unilateral interstitial infiltrates, segmental pneumonia, pleural effusion, and disseminated disease have been reported.

CHRONIC OBSTRUCTIVE AIRWAYS DISEASE, BRONCHIECTASIS, AND NONTUBERCULOUS MYCOBACTERIA

Host predispositions to NTM-LD are less well established, with the exception of COPD and preexisting bronchiectasis (Marras et al., 2016; Andrejak et al., 2013; Prevots et al., 2010). Char et al. examined the histology of lung tissue from COPD patients following lung volume reduction with a focus on evidence of mycobacterial infection. In this study, they found that 10% of specimens had histological evidence of mycobacterial infection, as identified by presence of granulomas that were necrotizing. These patients also had higher exacerbation rates, with lower FEV1 and diffusion capacity (Char et al., 2014). In a recent study from Germany, Ringshausen et al., found that chronic airway diseases, COPD/emphysema, and bronchiectasis were present in most case patients diagnosed with NTM-LD (Ringshausen et al., 2016).

It is unclear whether the association between COPD and NTM-LD is due to structural and functional abnormalities of the diseased lung, medications used to treat airways disease, both factors combined, or other factors (Andrejak et al., 2013). In a population-based nested case−control study using health databases in Ontario, Brode et al. studied the risk of developing NTM and TB in patients who were on inhaled corticosteroids. Their study suggests that use of inhaled corticosteroid is associated with an increased risk of NTM-LD, but not TB. Since the use of inhaled corticosteroid has increased in patients with COPD, the rising incidence of NTM in this patient population may reflect this change. However, further studies are needed to establish this association. Together, these studies suggest that NTM infections in patients with COPD are probably underestimated and should be kept in mind, particularly when patients have frequent exacerbations and display accelerated respiratory function decline, before implementing inhaled corticosteroids or macrolides (Brode et al., 2017).

The association between non-CF bronchiectasis and NTM has been more extensively studied. It has been shown that 9%–63% of patients with bronchiectasis may have active NTM lung infections (Faverio et al., 2016; Bonaiti et al., 2015; Mirsaeidi et al., 2013; Aksamit et al., 2017). Further, investigators in the United States and Europe have formed registries of adult patients with non-CF bronchiectasis. In the United States, the Bronchiectasis Research Registry enrolled 1826 patients from 13 clinical sites and found that 63% of patients had either a history of NTM-LD or sputum culture positive for an NTM at the time of enrollment (Aksamit et al., 2017). The European Multi-Center Bronchiectasis Audit and Research Collaboration registry have enrolled up to 10,000 adult patients with non-CF bronchiectasis, including data on NTM disease and colonization. The results from this registry are awaited and should provide valuable data about the association between non-CF bronchiectasis and NTM.

NONTUBERCULOUS MYCOBACTERIA IN PATIENTS WITH CYSTIC FIBROSIS

The effect of CF on patient airways is highly complex and influenced by environmental stress, including bacterial infections. Multifarious interplay between the host, pathogens, and the environment shapes the course of the disease. In CF, the absence of functional cystic fibrosis transmembrane conductance regulator (CFTR) impairs hydration of the mucus, ciliary function, and bacterial clearance, providing a favorable environment for infection and colonization of microbes. Due to the high viscosity of airway secretions, CF patients have impaired mucociliary clearance, predisposing them to bacterial infections (Skolnik et al., 2016). Although *Pseudomonas aeruginosa* and *Staphylococcus aureus* continue to predominate colonization and infections in CF patients, recent data suggests that emerging pathogens such as NTM are increasing in these patients. The prevalence of NTM infection in patients with CF ranges between 1.3% and 32.7%. The 2015 Cystic Fibrosis Foundation Annual Report noted that almost 12% of patients had a positive sputum culture for nontuberculous mycobacterial species (Marras et al., 2016). Unlike in other patients with NTM, in patients with CF, there is some evidence of person-to-person transmission of NTM (Bryant et al., 2016).

Within the CF population, multiple potential risk factors for NTM pulmonary infection have been investigated although many have not been consistently reproduced. Clinical characteristics, including baseline lung function, have generally not been linked to NTM risk in CF. However, coexistence with other pathogens can worsen the outcome of patients with NTM-LD. In a recent study, Kamata et al. showed that MAC-LD patients with chronic *P. aeruginosa* infection exhibited significantly higher respiratory scores, indicating impaired health-related quality of life. Chronic *P. aeruginosa* infection with NTM was significantly associated with the severity of bronchiectasis (Kamata et al., 2017).

The diagnosis of nontuberculous mycobacteria-pulmonary disease (NTM-PD) in CF is challenging, as it requires both microbiological and clinical evidence in the setting of coexisting airway infections and progressive lung disease. Although in some individuals NTM can accelerate the progression of CF lung disease, in others NTM may remain indolent for years, or appear transiently in sputum cultures. CF patients infected with *M. abscessus* are associated with increased morbidity and mortality, as well as with a rapid decline in lung function (Qvist, 2016). In fact, NTM-LD infection with *M. abscessus* complex may impact eligibility for lung transplantation in patients with CF. The dilemma faced by clinicians is to accurately identify those patients who are likely to benefit from therapy, while avoiding unnecessary treatment in those with indolent infections (Nick et al., 2016).

NONTUBERCULOUS MYCOBACTERIA IN IMMUNOCOMPROMISED PATIENTS

Immunosuppressive conditions associated with NTM disease are usually HIV, hematological, and lymphoproliferative malignancy, stem-cell and solid-organ transplantations or inflammatory disorders treated with biologicals (Henkle and Winthrop, 2015; Honda et al., 2015, 2018). Among TNF-α antagonists, treatment with infliximab and adalimumab are more commonly associated with the development of NTM infections rather than etanercept (Winthrop et al., 2009). In a small retrospective study, Meier et al. described patients with previous lung cancer who developed NTM (Meier et al., 2017). They found that symptomatic improvement with MAC antimicrobial therapy was lower in this population, however, it might have been confounded by comorbidities. Whether patients with lung cancer are more prone to develop NTM and are less responsive to treatment, need further studies.

NTM infections in transplant patients are uncommon, but are associated with marked morbidity and mortality. Transmission can occur through donor-derived infection, through de novo infection, or as a reactivation of latent disease in the recipient. There is increasing evidence that transplant patients may be at greater risk for acquiring nosocomial NTM (Pena and Klesney-Tait, 2017; Meije et al., 2014). Water systems in hospitals, and even hemodialysis centers, have been reported to have high rates of mycobacterial colonization. Among solid-organ transplant recipients, heart and lung transplant patients appear to have the highest rate of infection with NTM. In general, the mortality associated with NTM infection in solid-organ transplant patients is increased compared with nonimmunocompromised patients. In thoracic, solid-organ transplant patients, NTM are of special concern because of the potential for anastomotic infections in the perioperative period. In lung transplantation, surgical wound infection with *M. abscessus*, which is inherently resistant to most antimicrobials, causes significant morbidity and mortality. Based on this, patients with CF with *M. abscessus* are often eliminated from transplantation (Pena and Klesney-Tait, 2017). In Hematopoietic stem cell transplantation stem cell transplantation (HSCT), the incidence of infections due to NTM ranges from 0.4% to 4.9%, with MAC being the most common species (Doucette and Fishman, 2004).

Hematopoietic stem-cell transplant recipients are also at risk for a variety of opportunistic infections. While typical bacterial infections remain the most frequent cause of infection, NTM are becoming increasingly recognized as pathogens after stem-cell transplant (Cordonnier et al., 2004; Knoll, 2014). In general, the strategy for diagnosis of NTM-LD in patients who undergo stem-cell transplant is similar to immunocompetent individuals, although a greater emphasis on microbiology is required. Most data suggest that radiological and microbiological improvement can be achieved in these patients although survival was less favorable (Hirama et al., 2018).

Pneumonia is the most common presentation for immunocompromised patients with NTM infection; disseminated infections are more likely to occur in immunocompromised patients. Clinical course varies depending on the species. *M. abscessus* tends to cause more aggressive disease, often spreading to adjacent structures such as pleura and mediastinum. The combination of agents selected for treatment will depend on the subspecies of NTM and the location of the infection. Clinicians must have a high level of suspicion and engage a multidisciplinary team of physicians, in the diagnosis and treatment of these challenging infections (Henkle and Winthrop, 2015).

Diagnosis and treatment of NTM-LD: A diagnosis of NTM-PD must meet both clinical and microbiologic criteria as outlined by ATS/IDSA guidelines, with the exclusion of other etiologies such as TB. Recently, these guidelines were revised by the British Thoracic Society as outlined in Box 6.1. A clinical algorithm as published by BTS is shown in Fig. 6.1 (Haworth et al., 2017; van Ingen et al., 2018; Griffith et al., 2007).

Clinical criteria include persistent pulmonary symptoms and radiographic changes (nodular and/or cavitary opacities on radiograph or bronchiectasis with nodules on CT scan), and microbiologic criteria which require persistent growth of NTM in respiratory specimens, such as sputum (Adelman and Addrizzo-Harris, 2018). When both expectorated or induced sputum samples are nondiagnostic or negative but a high index of suspicion for NTM-LD remains, bronchoscopy should be pursued for bronchoalveolar lavage. Until recently, the phylogenetic relationships of mycobacteria species were largely based on 16S rRNA sequencing; however, the increased availability and cost-effectiveness of whole genome sequencing (WGS) has led to more mycobacteria species being sequenced, as well as more robust comparative genomics to base phylogenetic relationships on (Haworth et al., 2017; Abubakar et al., 2017).

BOX 6.1 Clinical and Microbiological Criteria for Diagnosing Nontuberculous Mycobacterial Lung Disease

Clinical (both required)

1. Pulmonary symptoms, nodular or cavitary opacities on chest radiograph, or a high-resolution computed tomography scan that shows multifocal bronchiectasis with multiple small nodules.

   and

2. Appropriate exclusion of other diagnoses.

Microbiological

1. Positive culture results from at least two separate expectorated sputum samples. If the results are nondiagnostic, consider repeat sputum AFB smears and cultures.

   or

2. Positive culture result from at least one bronchial wash or lavage.

   or

3. Transbronchial or other lung biopsy with mycobacterial histopathological features (granulomatous inflammation or AFB) and positive culture for NTM or biopsy showing mycobacterial histopathologic features (granulomatous inflammation or AFB) and one or more sputum or bronchial washings that are culture positive for NTM.

AFB, Acid-fast bacilli; *NTM*, nontuberculous mycobacteria.

Modified with permission from BMJ Open Respir. Res. 4(1), e000242–52. Reprinted with permission of the American Thoracic Society. Copyright © 2017 American Thoracic Society. Cite: An Official ATS/IDSA Statement: Diagnosis, treatment, and prevention of nontuberculous mycobacterial diseases. Am. J. Respir. Crit. Care Med. 2007, 175(4), 367–416. The American Journal of Respiratory and Critical Care Medicine is an official journal of the American Thoracic Society.

NTM-LD treatment is challenging due to long, complicated, and difficult to tolerate antibiotic regimens. Treatment regimens differ by species, with the most important distinction being between slow growing and rapid growing mycobacteria. As a general principle, multiple drugs should be used in the treatment of NTM to reduce the risk of developing resistance with monotherapy. Agents used for treatment include macrolides, rifampin, ethambutol, streptomycin, amikacin, moxifloxacin, clofazimine, and bedaquiline (Haworth et al., 2017).

For most slow growing NTM, the recommended regimen consists of oral drugs including rifampin (or rifabutin), ethambutol, and a macrolide antibiotic, given for 12 months which should be monitored for culture conversion. In severe cases, amikacin or streptomycin can be added for the initial 3 months to reduce the bacterial burden (Kasperbauer and DeGroote, 2015; Griffith et al., 2007). *M. kansasii* is clinically more closely related to

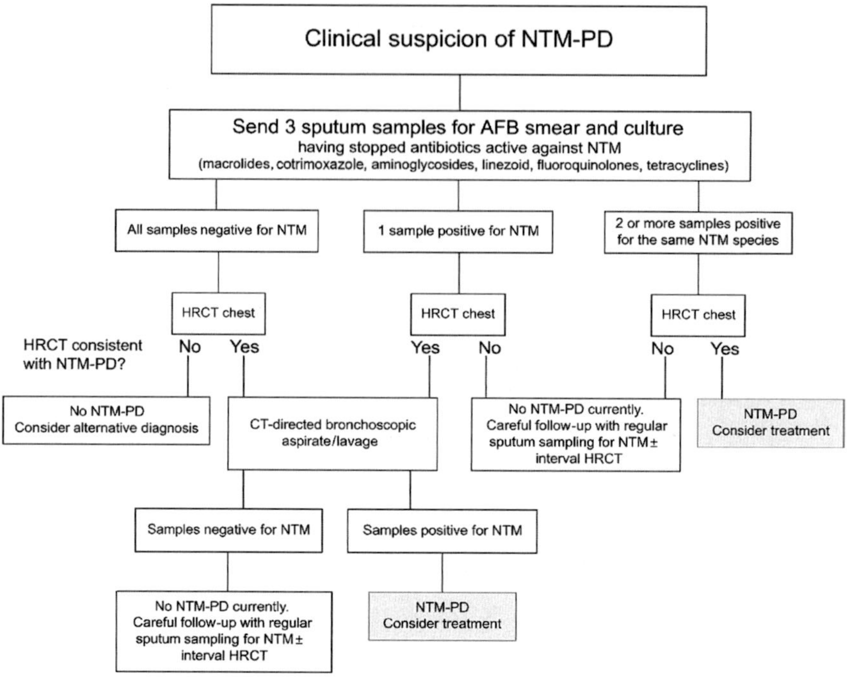

FIGURE 6.1 A suggested algorithm for the investigation of individuals with clinical suspicion of NTM-PD. *AFB*, Acid-fast bacilli; *HRCT*, high-resolution CT; *NTM-PD*, nontuberculous mycobacterial pulmonary disease.

M. tuberculosis than to other NTM species. Lung disease caused by *M. kansasii* is usually treated with a combination of isoniazid, rifampin, and ethambutol. For the rapid growing NTM, regimens are primarily based on in vitro drug susceptibility testing data. *M. abscessus* lung disease is one of the most difficult NTM infections for treatment, because of its natural resistance to most available antimycobacterial agents. For Mycobacterium abscessus (MAB), treatment regimens often include a macrolide antibiotic and amikacin in combination with other intravenous drugs often selected from cefoxitin, imipenem, or tigecycline. Duration of treatment should follow recommendations from the ATS/IDSA (Griffith et al., 2007) and BTS guidelines (Haworth et al., 2017). Follow-up of patients with sputum conversion and cultures have also been defined, as shown in Box 6.2.

Some studies suggest a high risk of microbiologic recurrence up to 50%, after successful completion of antimycobacterial therapy. Alternative therapies that are being revisited include antimycobacterial drugs, like clofazimine, which have demonstrated efficacy and tolerability for treatment-refractory NTM infections. Bedaquiline, which is approved for the treatment of multidrug-resistant TB, may have a role in treatment of refractory NTM-PD as well. Other therapies include newer formulations of amikacin for inhalation

BOX 6.2 Definitions for Microbiological Outcomes

- Culture conversion: three consecutive negative mycobacterial sputum cultures collected over a minimum of 3 months, with the time of conversion being the date of the first of the three negative mycobacterial cultures. In patients unable to expectorate sputum, a single negative mycobacterial culture of a CT-directed bronchial wash is indicative of culture conversion.
- Recurrence: two positive mycobacterial cultures following culture conversion. If available, genotyping may help distinguish relapse from reinfection.
- Refractory disease: failure of the culture to convert after 12 months of nontuberculous mycobacterial treatment.

with a view that this form may be better tolerated than parenteral administration (Adelman and Addrizzo-Harris, 2018; Abubakar et al., 2017).

In addition to antimicrobial therapy, avoidance of potential sources of exposure (e.g., aerosolized water, soil, and biofilms) is necessary to prevent reinfection. Although large studies are lacking, nonpharmacological treatments such as airway clearance have been utilized in the management of non-CF bronchiectasis. Chest physical therapy (PT) can be provided as manual chest PT, high-frequency chest wall oscillation using a vest or an oscillating device, such as a flutter device, that can help mobilize mucus (Flude et al., 2012). In a recent retrospective review, Basavaraj et al. found that select patients with NTM may have clinical improvement with chest PT without being subjected to prolonged antibiotic therapy (Basavaraj et al., 2017). This study also highlighted the significance of association of GERD and asymptomatic reflux in patients with NTM (61%) as previously reported by Koh et al. (2007) suggesting that treatment of reflux may help alleviate some of the symptoms.

Surgical resection can also play an important adjunctive role in the management of NTM-LD, especially if the disease is localized. Indications for surgery include disease refractory to treatment with persistence of bacteria, hemoptysis, or severe localized lesions in the lungs, such as cavity or bronchiectasis. It is important that the patient is evaluated for surgical suitability and is treated with combination therapy for 2–3 months prior to surgery to reduce the bacterial burden (Mitchell et al., 2008; Mitchell, 2015).

It is also essential to monitor patients closely for drug toxicity, given the use of a combination of drugs that are administered for a prolonged period. Liver, renal, and bone marrow toxicity with rifampin-based agents need close monitoring of serum chemistry. Macrolide use often warrant monitoring for cardiotoxicity especially in elderly patients who are on multiple medications. Change of NTM species may also occur during, or after, treatment for NTM-LD. In a recent study by Lee et al. (2017) changes from MAC to *M. avium* ssp. *avium* (MAA) were accompanied by symptomatic and radiographic worsening, as well as inducible resistance to clarithromycin.

SUMMARY AND CONCLUSIONS

NTM-LD is increasing globally and is often seen in immune-competent individuals. However, there has also been a rise in lung diseases such as COPD and bronchiectasis which are associated with NTM infections. NTM vary widely in their pathogenicity and clinical relevance. In addition to the host, the environment seems to play a critical role in NTM prevalence and individual susceptibility. To be diagnosed with NTM-LD, patients should meet the clinical and microbiologic criteria; however, the decision to initiate treatment should be individualized depending on the analysis of risks and benefits. The accurate identification of the NTM species is critical for diagnosis and management of patients with NTM-LD. Diagnosis of NTM-PD is often delayed due to nonspecific symptoms and inadequate use of ATS/IDSA guidelines and criteria. This delay potentially leads to disease progression, since treatment is also deferred. WGS of NTM has increased our understanding of the genetic evolution of the mycobacteria family with improved diagnostics. The available therapeutic arsenal for treatment of NTM is limited and evidence-based approaches are lacking because of the lack of controlled trials. Treatment of NTM-LD poses challenges because of long, complicated antibiotic regimens which are often difficult to tolerate. Multicenter studies are much needed to provide more robust evidence-based recommendations for the treatment and outcome analysis of NTM-LD. Clinicians should be aware of the aspects of NTM-LD, including the need for accurate diagnosis, the availability of advanced molecular genotyping to identify the species, and the benefits and limitations of recommended treatments.

REFERENCES

Abubakar, I., Gupta, R.K., Rangaka, M.X., Lipman, M., 2017. Update in tuberculosis and nontuberculous mycobacteria. Am. J. Respir. Crit. Care. Med. 197 (10), 1248–1253.

Adelman, M.H., Addrizzo-Harris, D.J., 2018. Management of nontuberculous mycobacterial pulmonary disease. Curr. Opin. Pulm. Med. 24 (3), 212–219.

Adjemian, J., Olivier, K.N., Seitz, A.E., Holland, S.M., Prevots, D.R., 2012. Prevalence of nontuberculous mycobacterial lung disease in U.S. Medicare beneficiaries. Am. J. Respir. Crit. Care. Med. 185, 881–886.

Aksamit, T.R., O'Donnell, A.E., Barker, A., 2017. Bronchiectasis research registry consortium. Adult bronchiectasis patients: a first look at the United States bronchiectasis research registry. Chest 151 (5), 982–992.

Andrejak, C., Nielsen, R., Thomson, R., Dahaut, P., Sorensen, H.T., Thomsen, R.W., 2013. Chronic respiratory disease, inhaled corticosteroids and risk of non-tuberculosis mycobacteriosis. Thorax 68 (3), 256–262.

Basavaraj, A., Segal, L., Samuels, J., Feintuch, J., Feintuch, J., Alter, K., et al., 2017. Effects of chest physical therapy in patients with non-tuberculous mycobacteria. Int. J. Respir. Pulm. Med. 4 (1), 65–72.

Bonaiti, G., Pesci, A., Marruchell, A., Lapadula, G., Gori, A., 2015. Nontuberculous mycobacteria in noncystic fibrosis bronchiectasis. Biomed. Res. Int. 197950.

Brode, S.K., Daley, C.L., Marras, T.K., 2014. The epidemiologic relationship between tuberculosis and non-tuberculous mycobacterial disease: a systematic review. Int. J. Tuberc. Lung. Dis. 18, 1370–1377.

Brode, S.K., Campitelli, M.A., Kwong, J.C., Lu, H., Marchand-Austin, A., Gershon, A.S., et al., 2017. The risk of mycobacterial infections associated with inhaled corticosteroid use. Eur. Respir. J. 50 (3), 1700037–1700043.

Bryant, J.M., Grogono, D.M., Rodriguez-Rincon, D., 2016. Emergence and spread of a human-transmissible multidrug-resistant nontuberculous *Mycobacterium*. Science 354, 751–757.

Char, A., Hopkinson, N.S., Hansell, D.M., Nicholson, A.G., Shaw, E.C., Clark, S.J., et al., 2014. Evidence of mycobacterial disease in COPD patients with lung volume reduction surgery; the importance of histological assessment of specimens: a cohort study. BMC Pulm. Med. 14, 124.

Claeys, T.A., Robinson, R.T., 2018. The many lives of nontuberculous mycobacteria. J. Bacteriol., p. ii, JB.00739-17.

Cordonnier, C., Martino, R., Trabasso, P., Held, T.K., Akan, H., Ward, M.S., et al., 2004. Mycobacterial infection: a difficult and late diagnosis in stem cell transplant recipients. Clin. Infect. Dis. 38, 1229–1236.

Doucette, K., Fishman, J.A., 2004. Nontuberculous mycobacterial infection in hematopoietic stem cell and solid organ transplant recipients. Clin. Infect. Dis. 38, 1428–1439.

Esteban, J., García-Coca, M., 2018. *Mycobacterium* biofilms. Front. Microbiol. 8, 2651–2659.

Faverio, P., Stainer, A., Bonaiti, G., 2016. Characterizing non-tuberculous mycobacteria infection in bronchiectasis. Int. J. Mol. Sci. 17, 1913–1921.

Flude, L.J., Agent, P., Bilton, D., 2012. Chest physiotherapy techniques in bronchiectasis. Clin. Chest Med. 33, 351–361.

Griffith, D.E., Aksamit, T.R., 2012. Bronchiectasis and nontuberculous mycobacterial disease. Clin. Chest Med. 33 (2), 283–295.

Griffith, D.E., Aksamit, T., Brown-Elliott, B.A., 2007. An official ATS/IDSA statement: diagnosis, treatment, and prevention of nontuberculous mycobacterial diseases. Am. J. Respir. Crit. Care. Med. 175, 367–416.

Haworth, C.S., Banks, J., Capstick, T., Fisher, A.J., Gorsuch, T., Laurenson, I.F., et al., 2017. British Thoracic Society Guideline for the management of non-tuberculous mycobacterial pulmonary disease (NTM-PD). BMJ Open Respir. Res. 4 (1), e000242–e000252.

Henkle, E., Winthrop, L., 2015. Nontuberculous mycobacteria infections in immunosuppressed hosts. Clin. Chest Med. 36, 91–99.

Hirama, T., Brode, S.K., Beswick, J., Law, A.D., Lam, W., Michelis, F.V., et al., 2018. Characteristics, treatment and outcomes of nontuberculous mycobacterial pulmonary disease post allogeneic hematopoietic stem cell transplant. Eur. Respir. J. 51 (5), p. ii, 1702330.

Holland, S.M., 2001. Nontuberculous mycobacteria. Am. J. Med. Sci. 321, 49–55.

Honda, J.R., Knight, V., Chen, D., 2015. Pathogenesis and risk factors for nontuberculous mycobacterial lung disease. Clin. Chest Med. 36, 1–11.

Honda, J.R., Bai, X., Chan, E.D., 2018. Elucidating the pathogenesis of nontuberculous mycobacterial lung disease: lesson from the six blind men and the elephant. Am. J. Respir. Cell Mol. Biol. 58 (2), 142–143.

Kamata, H., Asakura, T., Suzuki, S., Namkoong, H., Yagi, K., Funatsu, Y., et al., 2017. Impact of chronic *Pseudomonas aeruginosa* infection on health-related quality of life in *Mycobacterium avium* complex lung disease. BMC Pulm. Med. 17 (1), 198–205.

Kasperbauer, S.H., De Groote, M.A., 2015. The treatment of rapidly growing mycobacterial infections. Clin. Chest Med. 36, 67–74.

Knoll, B.M., 2014. Update on nontuberculous mycobacterial infections in solid organ and hematopoietic stem cell transplant recipients. Curr. Infect. Dis. Rep. 16, 421–429.

Koh, W.J., Lee, J.H., Kwon, Y.S., Lee, K.S., Suh, G.Y., 2007. Prevalence of gastroesophageal reflux disease in patients with nontuberculous mycobacterial lung disease. Chest 131, 1825–1830.

Lee, T., Park, J.Y., Lee, H.Y., Lim, H.J., Park, J.S., Cho, Y.J., et al., 2017. Int. J. Tuberc. Lung. Dis. 21 (10), 1169–1175.

Marras, T.K., Campitelli, M.A., Kwong, J.C., 2016. Risk of nontuberculous mycobacterial pulmonary disease with obstructive lung disease. Eur. Respir. J. 48, 928–993.

Meier, E., Pennington, K., Gallo de Moraes, A., Escalante, P., 2017. Characteristics of *Mycobacterium avium* complex (MAC) pulmonary disease in previously treated lung cancer patients. Respir. Med. Case Rep. 22, 70–73.

Meije, Y., Piersimoni, C., Torre-Cisneros, J., 2014. Mycobacterial infections in solid organ transplant recipients. Clin. Microbiol. Infect. 20, 89–101.

Mirsaeidi, M., Hadid, W., Ericsoussi, B., 2013. Non-tuberculous mycobacterial disease is common in patients with non-cystic fibrosis bronchiectasis. Int. J. Infect. Dis. 17, e1000–e1004.

Mitchell, J.D., 2015. Surgical approach to pulmonary nontuberculous mycobacterial infections. Clin. Chest Med. 36, 117–125.

Mitchell, J.D., Bishop, A., Cafaro, A., Weyant, M.J., Pomerantz, M., 2008. Anatomic lung resection for nontuberculous mycobacterial disease. Ann. Thorac. Surg. 85, 1887–1892.

Nick, J.A., Pohl, K., Martiniano, S.L., 2016. Nontuberculous mycobacterial infections in cystic fibrosis: to treat or not to treat? Curr. Opin. Pulm. Med. 22 (6), 629–636.

Park, S., Jo, K.W., Lee, S.D., Kim, W.S., Shim, T.S., 2017. Clinical characteristics and treatment outcomes of pleural effusions in patients with nontuberculous mycobacterial disease. Respir. Med. 133, 36–41.

Pena, T., Klesney-Tait, J., 2017. Mycobacterial infections in solid organ and hematopoietic stem cell transplantation. Clin. Chest Med. 38 (4), 761–770.

Porvaznik, I., Solovič, I., Mokrý, J., 2017. Non-tuberculous mycobacteria: classification, diagnostics, and therapy. Adv. Exp. Med. Biol. 944, 19–25.

Prevots, D.R., Shaw, P.A., Strickland, D., Jackson, L.A., Raebel, M.A., Blosky, M.A., 2010. Nontuberculous mycobacterial lung disease prevalence at four integrated health care delivery systems. Am. J. Respir. Crit. Care. Med. 182, 970–976.

Qvist, T., 2016. Comparing the harmful effects of nontuberculous mycobacteria and Gram negative bacteria on lung function in patients with cystic fibrosis. J. Cyst. Fibros. 15 (3), 380–385.

Ringshausen, F.C., Wagner, D., de Roux, A., Diel, R., Hohmann, D., Hickstein, L., et al., 2016. Prevalence of nontuberculous mycobacterial pulmonary disease, Germany, 2009–2014. Emerg. Infect. Dis. 22 (6), 1102–1105.

Skolnik, K., Kirkpatrick, G., Quon, B.S., 2016. Nontuberculous mycobacteria in cystic fibrosis. Curr. Treat. Options Infect. Dis. 8 (4), 259–274.

van Ingen, J., Aksamit, T., Andrejak, C., Böttger, E.C., Cambau, E., Daley, C.L., et al., 2018. For NTM-NET. Treatment outcome definitions in nontuberculous mycobacterial pulmonary disease: an NTM-NET consensus statement. Eur. Respir. J. 51 (3), 1–15.

Winthrop, L., Chang, E., Yamashita, S., Iademarco, M.F., LoBue, P.A., 2009. Nontuberculous mycobacteria infections and anti-tumor necrosis. Factor-alpha therapy. Emerg. Infect. Dis. 15, 1556–1561.

Chapter 7

Clinical Presentation of Nontuberculous Mycobacteria Using Radiological and CT Scan Imagining

Payam Mehrian[1], Poopak Farnia[2], Jalaledin Ghanavi[3], Hamidreza Jamaati[4], Payam Tabarsi[5] and Ali Akbar Velayati[3]

[1]Pediatric Respiratory Diseases Research Centre (PRDRC), National Research Institute of Tuberculosis and Lung Disease (NRITLD), Shahid Beheshti University of Medical Sciences, Tehran, Iran, [2]Department of Biotechnology, School of Advanced Technologies in Medicine, Shahid Beheshti University of Medical Sciences, Tehran, Iran, [3]Mycobacteriology Research Centre (MRC), National Research Institute of Tuberculosis and Lung Disease (NRITLD), Shahid Beheshti University of Medical Sciences, Tehran, Iran, [4]Chronic Respiratory Diseases Research Center, National Research Institute of Tuberculosis and Lung Diseases (NRITLD), Shahid Beheshti University of Medical Sciences, Tehran, Iran, [5]Clinical Tuberculosis and Epidemiology Research Center, National Research Institute of Tuberculosis and Lung Diseases (NRITLD), Shahid Beheshti University of Medical Sciences, Tehran, Iran

INTRODUCTION

Chronic pulmonary infections are among the most common clinical manifestations of nontuberculous mycobacteria (NTM) disease. The clinical presentation of NTM-related pulmonary infection can be quite varied. Multiple researchers in the mid-20th century reported human infection by unusual, and previously undescribed, mycobacteria. Chronic cough is almost always universal, but fever, malaise, weight loss, dyspnea, and hemoptysis are much more variable (Griffith et al., 2007). The affected individuals can be immunologically compromised can have an underlying lung pathology, such as chronic obstructive pulmonary disease and bronchiectasis, or can be infected without any risk factors (Weiss et al., 2012). In 1868, Wolinsky and Rynearson published the first comprehensive review, stating, "Chronic pulmonary diseases" resembling tuberculosis (TB) represent the most important clinical problem associated with NTM (Wolinsky and Rynearson, 1968).

Nontuberculous Mycobacteria (NTM). DOI: https://doi.org/10.1016/B978-0-12-814692-7.00007-3

The diagnosis of NTM infection, not only requires a positive culture and identification work-up, but radiologists also play an important role in the establishment of disease presence. They can locate and define the extent of infection, determine any underlying pathology, asses the response to treatment, and outline the presence of complications (Griffith et al., 2007; Weiss et al., 2012). Additionally, since distinction between colonization and infection is difficult (Erasmus et al., 1999), the diagnostic guidelines from the American Thoracic Society (ATS) suggest to check the symptoms and radiological evidence of nodular or cavitary diseases besides identification results (Nguyen, 1997; Chung et al., 2016). Radiology provides essential information for the management and follow up of these patients and is extremely valuable for monitoring complications.

Imaging: The most useful imaging tools to evaluate patients with known or suspected pulmonary infection, to determine the location and extent of infection, and to monitor response to therapy are chest radiography and computed tomography (CT) (Chowdhury and Khandelwal, 2010). A chest radiograph, called chest X-ray or chest film, is a projection radiograph of the chest used to diagnose conditions affecting the chest, its contents, and nearby structures. Furthermore, it can detect complications like cavitation, abscess formation, pneumothorax, and pleural effusion and can sometimes guide invasive diagnostic procedures (Padley and Rubens, 2003; Mettler et al., 2008). On the other hand, CT is a computerized X-ray imaging procedure in which a narrow beam of X-ray is quickly aimed around the patient's body. This produces signals that are processed by the machine's computer to generate cross-sectional images, or "slices," of the body (Herman, 2009; Computed tomorgraphy, 2009). These slices are called tomographic images and contain more detailed information than conventional X-rays (Herman, 2009; Computed tomography, 2009). CT is particularly valuable in complicated cases or in cases with unexpectedly slow response to treatment (Hommeyer et al., 1991). Generally, CT images in pulmonary diseases are interpreted along with clinical information and the imaging findings can be grouped into a few distinct patterns. These patterns can either provide specific diagnosis or help to narrow the list of differential diagnoses (Raju et al., 2017). Overall, familiarity with the basic structure of a secondary pulmonary lobule and its radiological appearance is critical for precise interpretation of high-resolution CT (HRCT) images (Raju et al., 2017). Woodring and Vandiviere (1990) reported that NTM infection cannot be distinguished from other conditions such as TB based on chest X-ray findings (Raju et al., 2017). Although, there is an important correlation between clinical and radiological manifestations of NTM pulmonary infection, the radiological manifestation of NTM infection depends largely on host immune response (Woodring and Vandiviere, 1990; Raju et al., 2017).

In this chapter, typical chest X-ray and CT scan findings in patents infected with slow growing mycobacteria (SGM) and rapidly growing mycobacteria (RGM) are presented and discussed (Raju et al., 2017; Woodring and Vandiviere, 1990). We aim to show possible correlation between imaging findings and degrees of smear positivity and compare our results with TB infection.

IMAGING IN NONTUBERCULOUS MYCOBACTERIA GROUP

Basically, the radiological manifestations of pulmonary NTM infection are divided into five groups: (1) *Classical infection*; this is the most common reported form of pulmonary NTM infection (Hommeyer et al., 1991) and is mainly observed in elderly patients with underlying diseases such as chronic obstructive pulmonary disease or pulmonary fibrosis (Miller, 1993, 1994). The chest radiograph has almost similar features as those of postprimary TB, although progression is slower than active TB (Fraser et al., 1990). Calcified pulmonary nodules and hilar nodes are less common (Woodring et al., 1987). The most common findings are heterogeneous linear and nodular areas of increased opacity in the apical and posterior segments of the upper lobes with or without calcification (Erasmus et al., 1999; Miller, 1993, 1994; Bolursaz et al., 2013, 2014). Small, thin-walled cavities usually occur in upper lobes with mean diameters of 2.5 cm. The nodular areas have a centrilobular distribution at CT (Woodring et al., 1987; Hartman et al., 1993). In classical form, the adenopathy and pleural effusion are never detected (Miller, 1993; Bolursaz et al., 2014), but pleural thickening is common. (2) *Nonclassical infection*; this is the most common form of pulmonary NTM infection especially in elderly, Caucasian women. Underlying diseases are not very important risk factors (Hartman et al., 1993). The findings are mild to moderate cylindrical bronchiectasis, with multiple 1–5 mm diameter centrilobular nodules. Usually the disease is isolated to, or more severe, in the lingula and middle right lobe. Cavitation, ground glass areas of increased attenuation, volume loss, and adenopathy are uncommon findings (Erasmus et al., 1999). (3) *Nodules in asymptomatic patients*; sometimes NTM infection results in solitary or multiple nodules, which are incidentally detected in asymptomatic patients. The nodules are macroscopic granulomas and may represent the initial manifestation of pulmonary infection. They are usually of similar size and clustered together (Erasmus et al., 1999). (4) *Infection in patients with achalasia* usually patients with *achalasia* are prone to infection with RGM like *M. fortuitum* or *M. chelonae*. Large bilateral, confluent areas of increased opacity similar to aspiration pneumonia on chest X-rays are seen. (5) *NTM in immunocompromised patients;* the chest radiograph is often normal; Mediastinal and hilar adenopathies are the most common findings. Other pictures, like scattered alveolar and nodular areas of increased opacity, milliary nodules, and mass-like lesions, occasionally can be seen (Marinelli et al., 1986).

LUNG INFECTION WITH COMMON RAPIDLY GROWING MYCOBACTERIA

Most diseases caused by RGM in humans are due to *M. fortuitum*, *M. chelonae*, and *M. abscessus*. Cummins and Williams (1933) were the first scientists who reported the lung diseases caused by RGM. Since then, only a small number of cases of pulmonary diseases by RGM have been described that were associated with underlying conditions (Griffith et al., 1993, 2007). Griffith and his coworkers demonstrated that 18% of patients infected with RGM pulmonary diseases had previous mycobacterial lung disease, mostly TB, 6% had cystic fibrosis, and 11% had bronchiectasis. They found that 32% of patients had no underlying disorder (Cummins and Williams, 1933). Today, many cases have been reported in developed countries. Large numbers of reports have shown their presence in environmental samples, for example, seawater, wastewater, ponds, freshwater canals from hospitals, and soil (Beerwerth, 1973; Tsukamura, et al., 1974; Falkinham, 1996; Lewis and Falkinham, 2015). These mycobacteria are more likely related to traumatic, surgical wound, pulmonary and catheter-associated infections (Portaels, 1995; Lewis and Falkinham, 2015). Collins et al. (1984) documented reports of *M. fortuitum* and *M. chelonae* in drinking water. Today, their growth has been shown even in distilled water (Falkinham, 1996; Tsukamura et al., 1974). They cannot grow in temperatures more than 42°C, but they are relatively resistant to chlorine (Fischeder et al., 1991; Portaels, 1995). For these reasons, medical equipments, for example, bronchoscopes and materials used in surgery like prosthetic heart valves, have the risk of being exposed to RGM infection (Wallace, 1994; Falkinham, 1996). The advances in technology, like identification by rRNA–RFLP, could permit clinical laboratories to identify the source and route of infection (Falkinham, 1996; Wallace et al., 1989; Tsukamura et al., 1974). While *M. fortuitum* can be colonized and is frequently found in soils, *M. chelonae* and *M. abscessus* are generally present in water, especially in simulated water that contains novel compounds, such as polyhalogenated phenols and pentachlorophenol which are used for polluted soils (Falkinham, 1996; Wallace et al., 1989; Richard et al., 1990; Tsukamura et al., 1974; Wallace, 1994). Kazda et al. (1979) showed the presence of *M. chelonae* and *M. abscessus* in sphagnum vegetation in Sweden, Norway, and Germany (Richard et al., 1990). Therefore, RGM are present in the environment which we are in contact with daily and one has to take more precautions about them (Mirsaeidi et al., 2017). There are several biochemical tests to distinguish RGM types from each other in the laboratory, for example, growth on MacConkey agar media and growth in 5% NaCL. *M. fortuitum* isolates grow at 43°C on MacConkey agar media, whereas, the *M. chelonae* and *M. abscessus* isolates do not grow at 43°C or 37°C on MacConkey agar media (Falkinham, 1996; Wallace et al., 1989; Tsukamura et al., 1974). Using 16S rRNA–RFLP and DNA-to-DNA hybridization has dramatically improved identification of RGM (Kusunoki and Ezaki, 1992; Falkinham, 1996; Kazda et al.,

1979; Tsukamura et al., 1974). *M. fortuitum* is a nonpigmented RGM classified in Runyon group IV. Da Costa Cruz, in 1938, was among the first mycobacteriologists who described the characteristics and infection sources of *M. fortuitum* (Falkinham, 1996; Griffith et al., 2007; Erasmus et al., 1999). Like other NTM, they are found in soil, natural products, and processed water sources, including chlorine-treated water and hospital environments. *M. fortuitum* group consists of three subspecies: *M. fortuitum*, *M. peregrinum*, and an unnamed third species. Recently, *M. houstonense* and *M. boerickei* have been described with this group of organisms (Kusunoki and Ezaki, 1992). Other species including *M. mageritense* and *M. senegalense* have been proposed as members of this group (Griffith et al., 2007). Various species of RGM cause different types of infections, for example, *M. fortuitum* is responsible for the majority of infections following cardiac bypass surgery or augmentation mammoplasty (Wallace et al., 1989; Ho et al., 2012; Hu et al., 2012), whereas skin and soft-tissue infections are due to *M. chelonae* and *M. abscessus* (Ingram et al., 1993; Stanford and Stanford, 2012). The majority of pulmonary infections are caused by *M. abscessus*. Most patients are older than 60 years, with no predisposing conditions or underlying diseases like bronchiectasis, prior mycobacterial infection and cystic fibrosis (Griffith et al., 2007; Trovato et al., 2017; Caskey et al., 2018; Kobayashi et al., 2018). Additionally, the presence of RGM in hospital environments or laboratories is very important and is subject of many investigations (Griffith et al., 2007; Tippett et al., 2018). In a study reported by Ingram et al. (1993) out of 54 patients infected with RGM, 40 had skin involvement and 75% had a history of fever (Ingram et al., 1993). The majority of reported patients had some underlying diseases, for example, lymphoma, leukemia, or cell-mediated immune deficiency. He concluded that patients without cancer or cell-mediated immune deficiency had higher survival rates (Ingram et al., 1993). *M. abscessus* was first isolated from a knee abscess in 1952; *M. abscessus* and *M. chelonae* were originally considered as the same species, but in 1992 they were reclassified as separate species (Brown-Elliott and Wallace, 2002; Lee et al., 2015). Thereafter, they could be identified as *M. abscessus* spp. *abscessus*, *M. abscessus* spp. *massiliense*, and *M. abscessus* spp. *bolletii* (Brown-Elliott and Wallace, 2002; Lee et al., 2015). The majority of reported cases belonged to *M. abscessus* spp. *abscessus* and were identified by molecular tests. More recently, together with *M. massiliense* and *M. bolletii* they were named as the MABSC group of RGM (Adekambi et al., 2004; Adekambi et al., 2006; Santos et al., 2012). Velayati et al. (2015) reported that incidence of *M. abscessus* spp. *M. abscessus* infection was low in Iran. This can be attributed to limited laboratory equipment and low clinical suspicion. Generally, *M. abscessus* causes pulmonary disease, especially in hosts with underlying lung diseases such as cystic fibrosis, bronchiectasis, and prior TB. It has a progressive course with the decline of pulmonary function and impaired quality of life. The chest radiograph of infected patients usually shows multilobar, patchy, reticulonodular, or mixed interstitial alveolar opacities with an upper lobe predominance. Cavitation is seen in only few number of patients (Griffith et al., 1993, 2007). Overall, the usual presentation of RGM is very similar to SGM, like *M. avium* complex. HRCT of the lung frequently shows cylindrical bronchiectasis and multiple small (<5 mm) nodules (Griffith et al., 2007). Figs. 7.1–7.4 show examples of chest X-rays and CT scans of patients infected with *M. abscessus*. Generally in NTM infection, clinical manifestations, radiological findings, and the microbiological culture results are assessed together (Griffith et al., 2007). Chest X-rays and CT scans demonstrate cavitation and multilobar, small nodules. Other imaging findings include multilobar patchy, reticulonodular, or mixed interstitial and alveolar infiltrates with upper lobe predominance (Figs. 7.1–7.4) (Griffith et al., 1993; Daley and Griffith, 2002). Figs. 7.5–7.6 show examples of chest X-rays and CT scans of elderly with *M. fortuitum infection*.

LUNG INFECTION WITH COMMON SLOW GROWING MYCOBACTERIA

NTM are opportunistic pathogens that can cause lymphadenitis and infections of the lungs, skin, soft tissues, tendon sheaths, and bones. Mycobacteria which take more than 7 days to grow in a culture are called SGM and consist of more than 100 species (Katoch, 2004; Santos et al., 2012). The most common clinically important species are *M. kansasii*, *M. avium complex* (*M. avium*, *M. intracellulare*), *M. Xenopi*, *M. simiae*, *M. szulgai*, *M. malmoense*, *M. terrae-nanochromogenicum complex*, *M. hemophilum*, and *M. genavense* (Katoch, 2004; Santos et al., 2012). Due to lack of laboratory facilities, the true distribution and incidence of pulmonary diseases caused by SGM is not fully understood. *M. genavense* (Figs. 7.7–7.11) was first isolated by Hirschel et al. (1990) in a patient

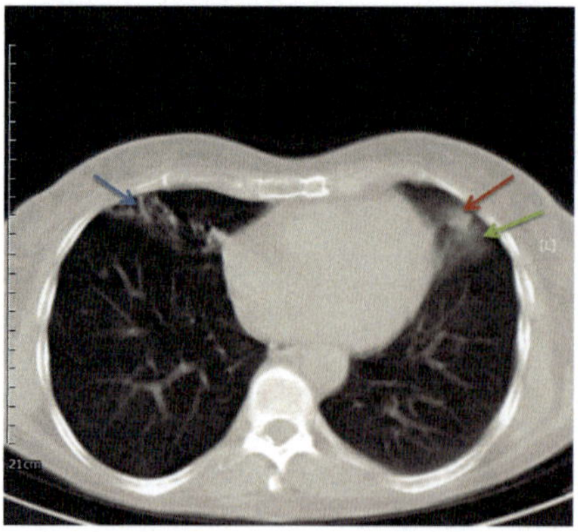

FIGURE 7.1 Bronchiectasis, ground glass opacity, and consolidation caused by *M. abscessus*. The case of a 64-year-old female infected with by *M. abscessus* (smear positivity results were 1+); bronchiectasis with ground glass opacity in right middle lobe (blue arrow), consolidation (red arrow) with surrounding ground glass opacity (green arrow) in lingula.

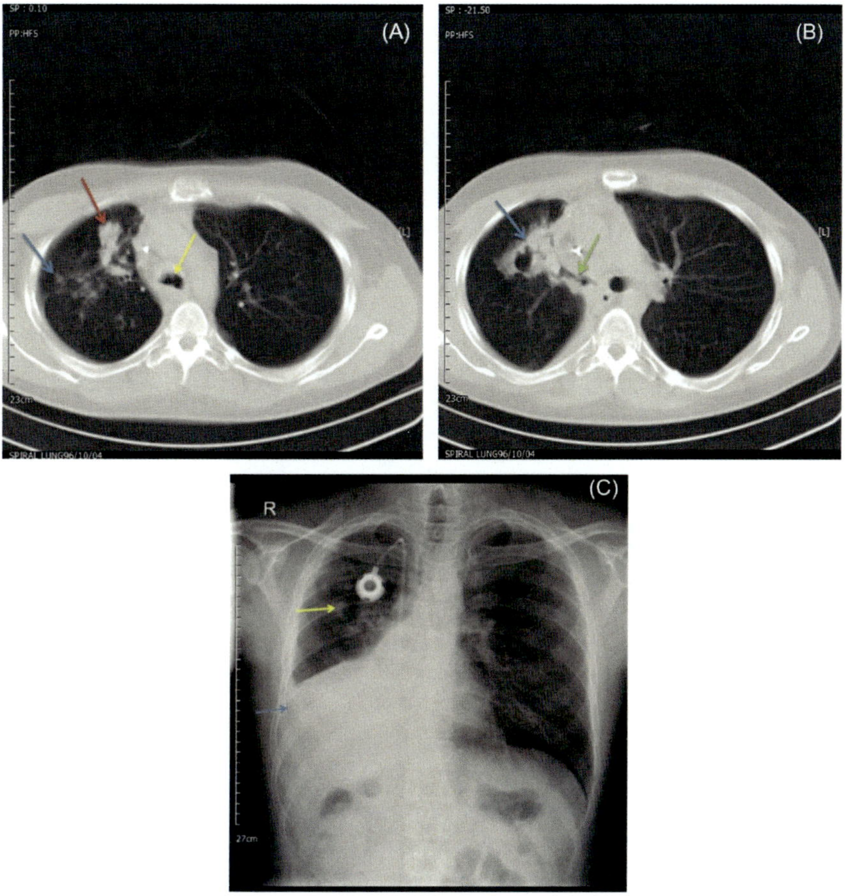

FIGURE 7.2 Nodular infiltration and consolidation with cavitation caused by *M. abscessus*. The case of a 23-year-old male infected with *M. abscessus* (smear positivity results were 3+). Nodular infiltration (blue arrow) and consolidation (red arrow) in right upper lobe; note also the nodularity in the carina indicating endotracheal lesion (yellow arrow) (A), cavitary consolidation in right upper lobe (blue arrow), and stenosis and irregularity of right main bronchus indicating endobronchial infection(green arrow) (B). There is also elevation of the right hemidiaphragm indicating subpulmonic effusion (blue arrow) and cavitary consolidation in the right upper lobe (yellow arrow) (C).

with AIDS. Later, disseminated infection with *M. genavense* has also been reported Monill et al. (2001). This mycobacterium has been isolated from tap water, pets, and the gastrointestinal tract of healthy humans (Doucette and Fishman, 2004). *M. genavense* occurs mainly in immune-compromised hosts. Tortoli et al. (1998) retrospectively reviewed all patients in Italy with *M. genavense* isolation between 1992 and 1996. All 24 patients appeared to be HIV-positive with a mean CD4 count of 23 μL. However,

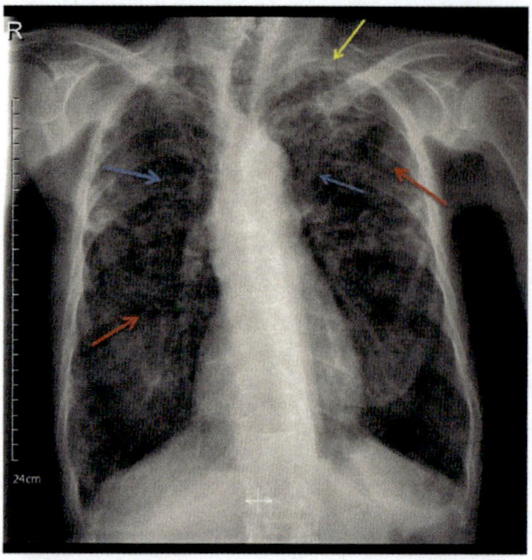

FIGURE 7.3 Bilateral bronchiectasis and nodular infiltration caused by *M. abscessus*. The case of a 74-year-old male infected with *M. abscessus* (smear positivity results were 3+). Bilateral bronchiectasis (blue arrow) and nodular infiltration (red arrow); note also the left-sided apical pleural thickening (yellow arrow).

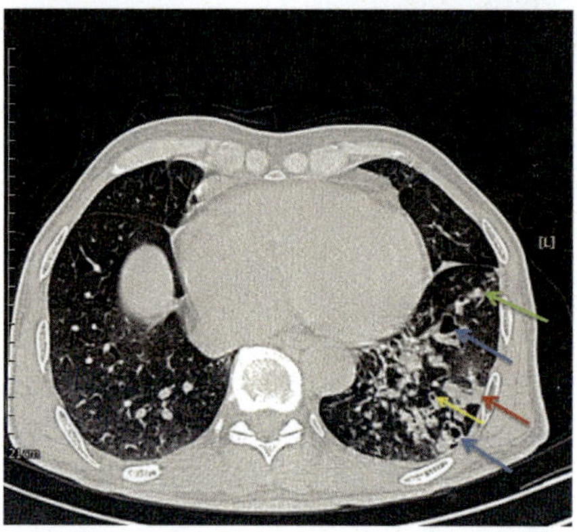

FIGURE 7.4 Cavity, consolidation, nodular infiltration, and bronchiectasis caused by *M. abscessus*. The case of a 69-year-old male infected with *M. abscessus* (smear positivity results were scanty). Cavities (blue arrow), consolidation (red arrow), nodular infiltration (green arrow) and bronchiectasis (yellow arrow) in the left, lower lobe.

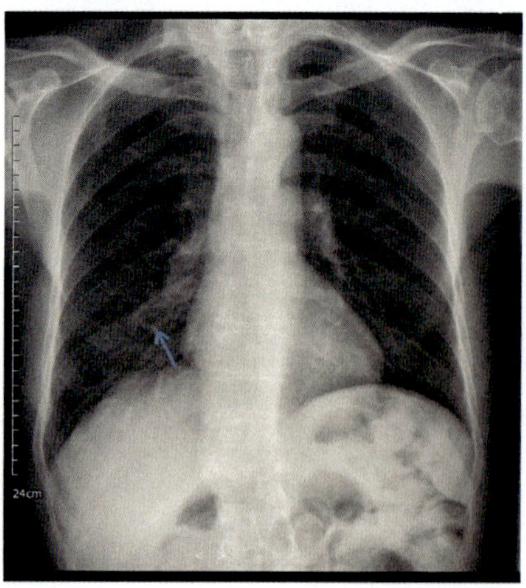

FIGURE 7.5 Cavity in right lower lung zone caused by *M. fortuitum*. The case of an 86-year-old male infected with *M. fortuitum* (smear positivity results were 3+). Cavity in right lower lung zone (blue arrow).

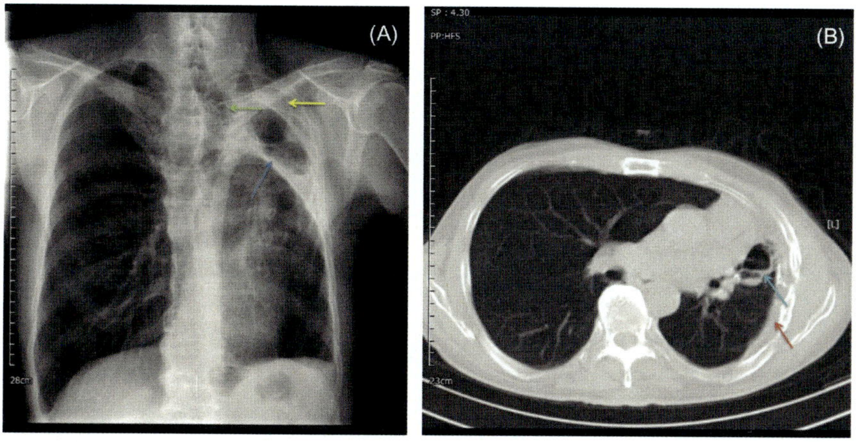

FIGURE 7.6 Cavity and pleural thickening caused by *M. fortuitum*. The case of a 70-year-old male infected with *M. fortuitum* (smear positivity results were scanty). Cavity in left upper lobe (blue arrow), pleural thickening (yellow arrow), volume loss, and tracheal shift to the left (green arrow) (A). Also note the volume loss and cavity (blue arrow) in lingula, and pleural reaction (red arrow) (B).

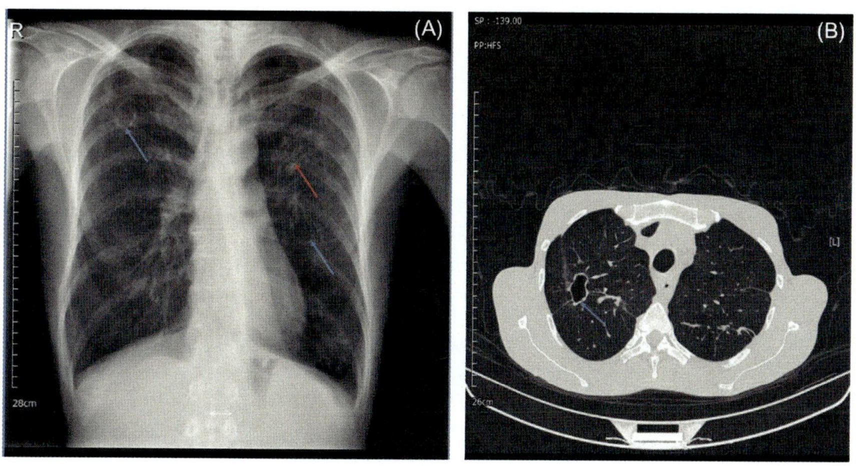

FIGURE 7.7 Bronchiectasis and cavity caused by *M. genavense*. The case of a 49-year-old male infected with *M. genavense* (smear positivity results were scanty). Bronchiectasis in left upper lobe (red arrow) and cavities in right upper lobe and left middle lung zone (blue arrow) (A). Cavity in right upper lobe (blue arrow) (B).

Hoefsloot et al. (2013) found infection with *M. genavense* among HIV-negative patients. Fig. 7.7 shows the infection of *M. genavense* in a HIV-negative case. As the isolate is usually difficult to grow, the identification of *M. genavense* is confirmed by molecular sequencing of 16s rRNA

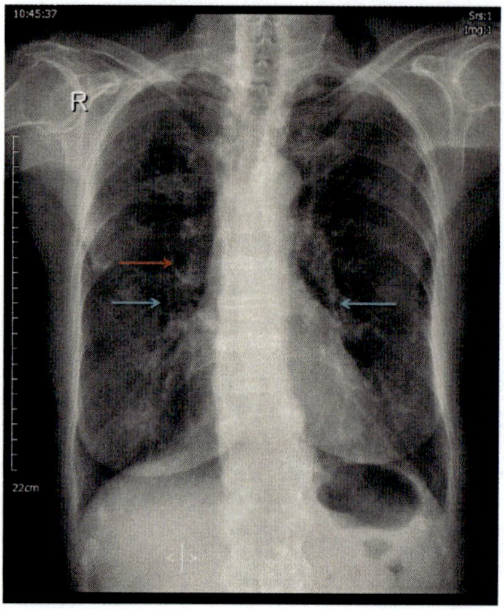

FIGURE 7.8 Bronchiectasis and cavity caused by *M. intracellular*. The case of a 76-year-old female infected with *M. intracellular* (smear positivity results were 3+). Bilateral bronchiectasis (blue arrow) and small cavity in right middle lung zone (red arrow).

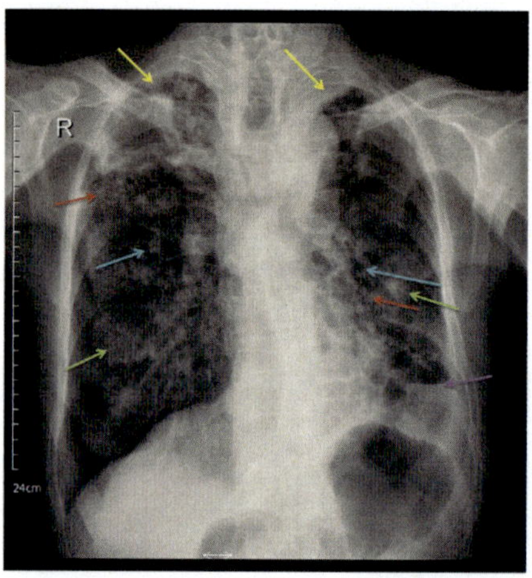

FIGURE 7.9 Apical pleural thickening, consolidation, bronchiectasis, small nodules, and cavity with air fluid level caused by *M. intracellular*. The case of an 80-year-old male infected with *M. intracellular* (smear positivity results were 2+). Bilateral apical pleural thickening (yellow arrow), bilateral consolidation (green arrow), extensive bilateral bronchiectasis (blue arrow), extensive bilateral small nodules (red arrow), and cavity with air fluid level in left lower lung zone (*empyema*) (purple arrow).

(Tortoli et al., 1998) The CT scan usually demonstrates multifocal diffuse nodular infiltrates in both lungs (Monill et al., 2001).

Regarding pulmonary NTM infection, *M. avium intracellular* complex (MAC) is the most common pathogen worldwide (Falkinham, 1996). In the immunocompetent patient, pulmonary infection likely results from the inhalation of aerosolized water droplets containing MAC. In recent years, the number of MAC pulmonary infections has been raised due to pneumoconiosis, silicosis, long-term exposure to dust, and outdoor pollution (Figs. 7.8 and 7.9).

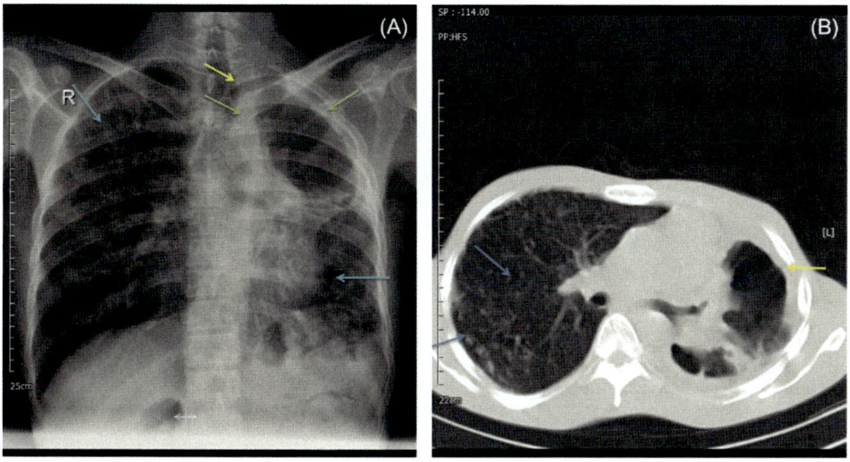

FIGURE 7.10 Small nodules, cavity, pleural thickening, and left-sided tracheal shift caused by *M. intracellular*. The case of a 44-year-old male infected with *M. intracellular* (smear positivity results were 3+). Extensive bilateral small nodules (blue arrow), cavity with marked pleural thickening (green arrow), volume loss in left upper lobe with ipsilateral tracheal shift (yellow arrow) (A). Small nodules in right lung (blue arrow), cavity with marked pleural thickening, and volume loss on the left side (yellow arrow) (B).

In a study by Christensen et al. (1981) it was reported that in the majority of patients the apical or posterior portion of the upper lobes was the site of the progenitive focus. Other less frequent locations for progenitive focus included the superior segment of the right lower lobe, lingula, middle lobe, anterior segment of the upper lobe, and basilar segments of the right lower lobe (Levin, 2002). Few of these patients had scarring with volume loss. Pleural thickening adjacent to upper lobes was also seen. Only in 16% of studied cases, the size of pleural thickening was more than 2 cm (Christensen et al., 1981). Chung et al. (2005) compared the thin section CT scan of patients that were infected with MAC and *M. abscessus* (Chung et al., 2005) They found no significant differences between MAC and *M. abscessus* with regard to small nodules, tree-in-bud patterns, and bronchiectasis. However, in MAC infection, the lobar volume decrease, airspace consolidation, thin-walled cavity, and nodule were more frequent than in patients infected with *M. abscessus* (Christensen et al., 1981; Chung et al., 2005). Overall, there was a considerable overlap in common CT findings of MAC and *M. abscessus infections* (Christensen et al., 1981; Levin, 2002; Chung et al., 2005). The CT findings of *M. tuberculosis* were compared with MAC pulmonary infection (Primack et al., 1995) in a number of studies, although similarities were high, interlobular septal thickening was more frequent in TB patients and bronchiectasis was more extensive in MAC patients. The major findings of MAC pulmonary infection are: (1) bronchiectasis and bronchial wall thickening, (2) ground glass opacities,

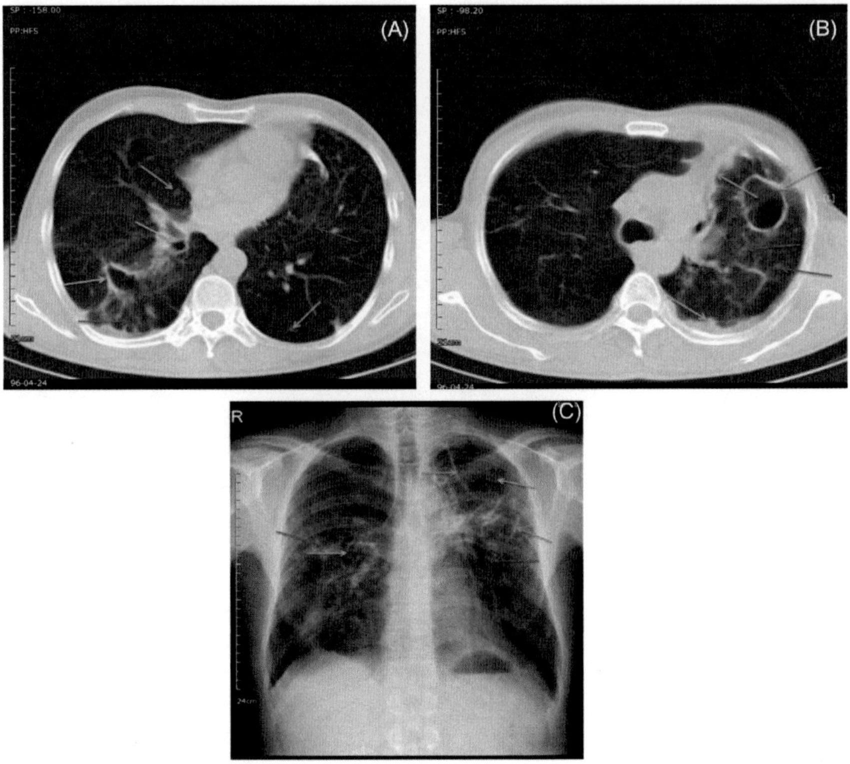

FIGURE 7.11 Bronchiectasis, small nodules, cavity, and pleural thickening caused by *M. intracellular*. The case of a 41-year-old male infected with *M. intracellular* (smear positivity results were 3+). Bilateral bronchiectasis (red arrow), bilateral small nodules (blue arrow), and cavities in right lower lobe (green arrow) (A). Left pleural thickening (blue arrow), cavity in left upper lobe (green arrow), and extensive left lung small nodules (red arrow) (B). Cavities in left upper lobe (blue arrow), bilateral patchy small nodules (red arrow), and bilateral patchy bronchiectasis (green arrow) (C).

(3) small centrilobular nodules and tree-in-bud appearance, (4) patchy consolidation, (5) predilection for the right middle lobe, (6) in lingula pleural thickening, adjacent to parenchymal change, and (7) upper lobe cavitation—as seen in TB (Field et al., 2004).

Fregnan et al. (1961) described *M. kansasii* on diabetes mellitus (DM) (basic medium containing bovine without albumin fraction V), and outlined a colony morphology with glycolipid content. *M. kansasii* is a photochromogenic SGM according to the Runyon classification of NTM (Katoch, 2004). If grown in a lighted incubator, most strains form dark red crystals of B-carotene on the surface and inside the colony (Katoch, 2004). *M. kansasii* is the second-most frequent cause of disseminated NTM diseases. Pulmonary and cutaneous manifestations have been reported in patients with *M. kansasii* infection (Falkinham, 1996; Griffith et al., 2007). Both, *M. kansasii* and *M. tuberculosis* infections have predilection for the upper lobe of lungs. In cases of severely, impaired immunity (patients with organ transplant or AIDS) *M. kansasii* can cause disseminated disease (Figs. 7.12–7.16). Although disseminated diseases are rare, patients with silicosis are also at risk. The source of *M. kansasii* is not very clear, but tap water is believed to be major reservoir associated with human diseases and usually affects the elderly. Patients are mainly smokers with one or more underlying pulmonary diseases including chronic obstructive pulmonary disease (COPD), previous

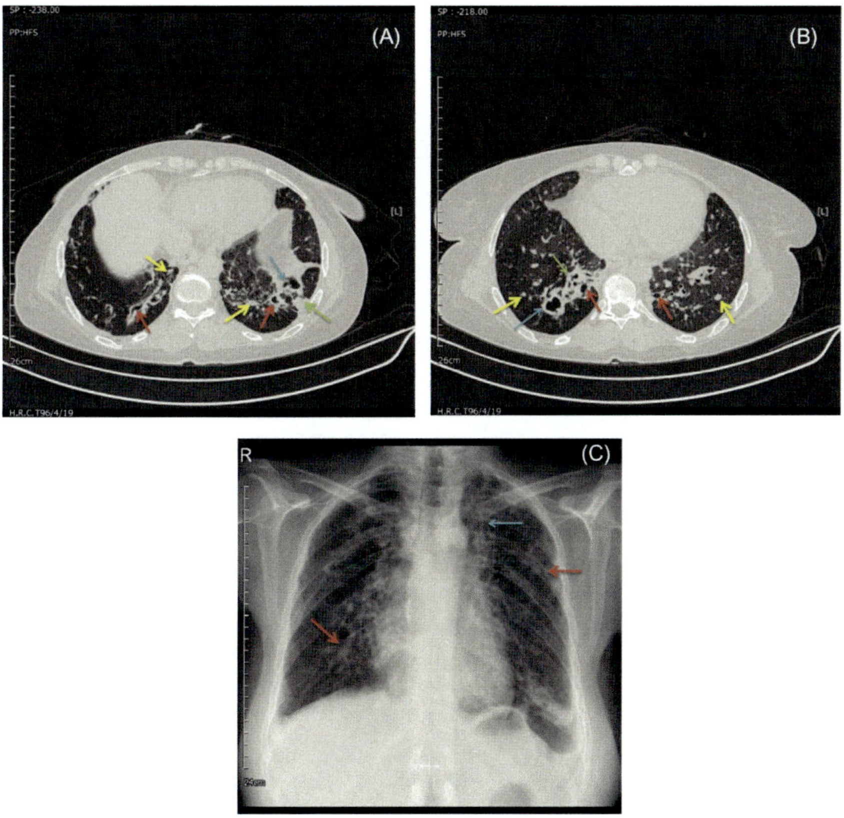

FIGURE 7.12 Tiny nodules, bronchiectasis, consolidation, and cavity caused by *M. kansasii*. The case of a 49-year-old female infected with *M. kansasii* (smear positivity results were 3+). Bilateral tiny nodules (yellow arrow), bilateral bronchiectasis (red arrow), consolidation in left lower lobe (green arrow), and cavity in left lower lobe (blue arrow) (A). Bilateral tiny nodules (yellow arrow), bilateral bronchiectasis (red arrow), and cavity in right lower lobe (blue arrow) (B). Bronchiectasis in left upper lobe (blue arrow) and diffuse bilateral nodular infiltration in lungs (red arrows) (C).

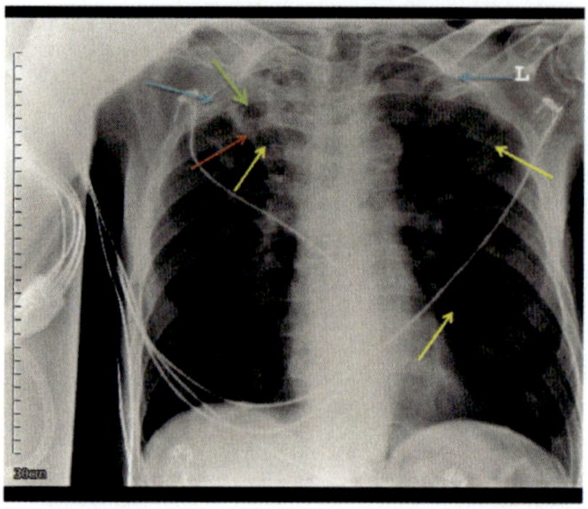

FIGURE 7.13 Pleural thickening, collapse with consolidation, cavity, and bronchiectasis caused by *M. kansasii.* The case of a 47-year-old male infected with *M. kansasii* (smear positivity results were 3+). Bilateral pleural thickening (blue arrow), collapse and consolidation (red arrow), cavity in right upper lobe (green arrow), bilateral upper lobe, and left lower zone bronchiectasis (yellow arrow).

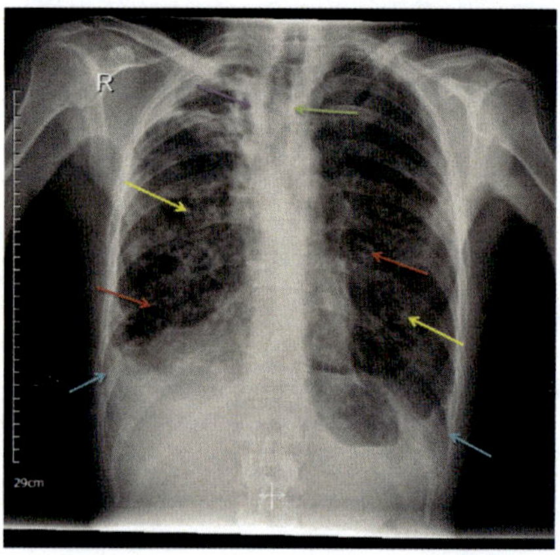

FIGURE 7.14 Pleural effusion, bronchiectasis, tiny nodules, right-sided tracheal shift, and widened right paratracheal stripe caused by *M. Kansasii.* The case of a 46-year-old male infected with *M. Kansasii* (smear positivity results were 1+). Bilateral pleural effusion (blue arrow), extensive bilateral bronchiectasis (red arrow), bilateral tiny nodules (yellow arrow), tracheal shift to right (green arrow), right paratracheal adenopathy, widened right paratracheal stripe (purple arrow), and bilateral bronchiectasis (yellow arrow).

TB, bronchiectasis, or pneumoconiosis (Evans et al., 1996; Griffith et al., 2007; Karassova et al., 1965). In imaging, it shows upper lobe predilection and cavitation is frequent; even if noncavitary, they resemble *M. avium intracellulare* lung diseases. Once a laboratory has confirmed the colonization of *M. kansasii*, immediate treatment is required.

M. simiae is a photochromogenic SGM that causes pulmonary diseases in both immune-competent and immune-compromised patients and forms rust-colored colonies when exposed to light and is the only niacin-positive NTM that is similar to *M. tuberculosis* (Griffith et al, 2007). *M. simiae* was isolated in 1965 from Macacus rhesus monkey (Karassova et al., 1965). The *M. simiae* complex group includes 18 species: *M. intermedium*, *M. interjectum*, *M. kubicae*, *M. montefiorense*, *M. florentinum*, *M. sherrisii*, *M. parmense*, *M. parascrofulaceum*, *M. lentiflavum*, *M. triplex*, *M. heidelbersengh*, *M. palustre*, *M. genavense*, *M. simiae*, *M. saskatchewanense*, *M. stomatepiae*,

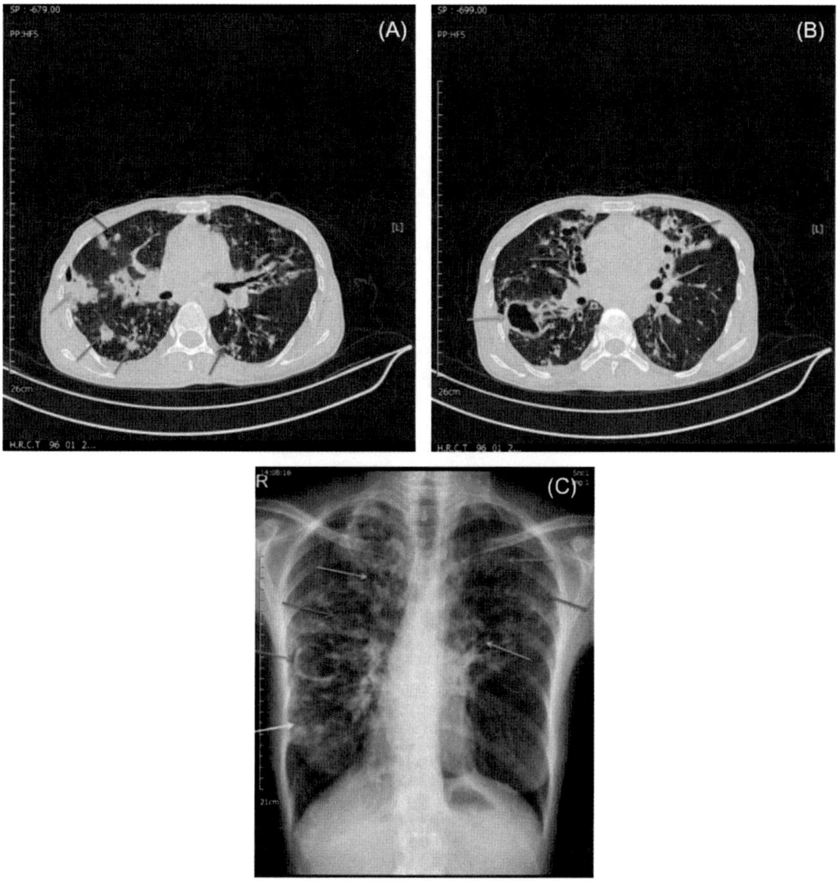

FIGURE 7.15 Consolidation, large and small nodules, cavity, and bronchiectasis caused by *M. simiae*. The case of a 42-year-old female infected with *M. simiae* (smear positivity results were 1+). Cavitary consolidation (green arrow) in right upper lobe, large nodules (blue arrow) in right lower lobe, and bilateral tiny nodules (red arrow) (A). Cavity (green arrow) in right lower lobe, bilateral bronchiectasis (blue arrow) in right middle lobe and lingula (B). Bilateral bronchiectasis (green arrow), consolidation (yellow arrow), bilateral cavities (blue arrow), and bilateral nodules (red arrow) (C).

M. eurapaeum, and *M. paraense* (Hamieh et al., 2018; Da Costa et al., 2015; Tortoli et al., 2011). *M. simiae* is the most common species in the complex group to cause human pathology (Cruz et al., 2007). It has been identified from soil and water sources, such as tap water (Velayati et al., 2014; Nasser, 2014; Hamieh et al., 2018). To date, there are no proven cases of human-to-human or animal-to-human transmission. Most patients infected with *M. simiae* have symptoms like cough, sputum production, and dyspnea (Baghizadeh et al., 2017; Hamieh et al., 2018). In the majority of cases,

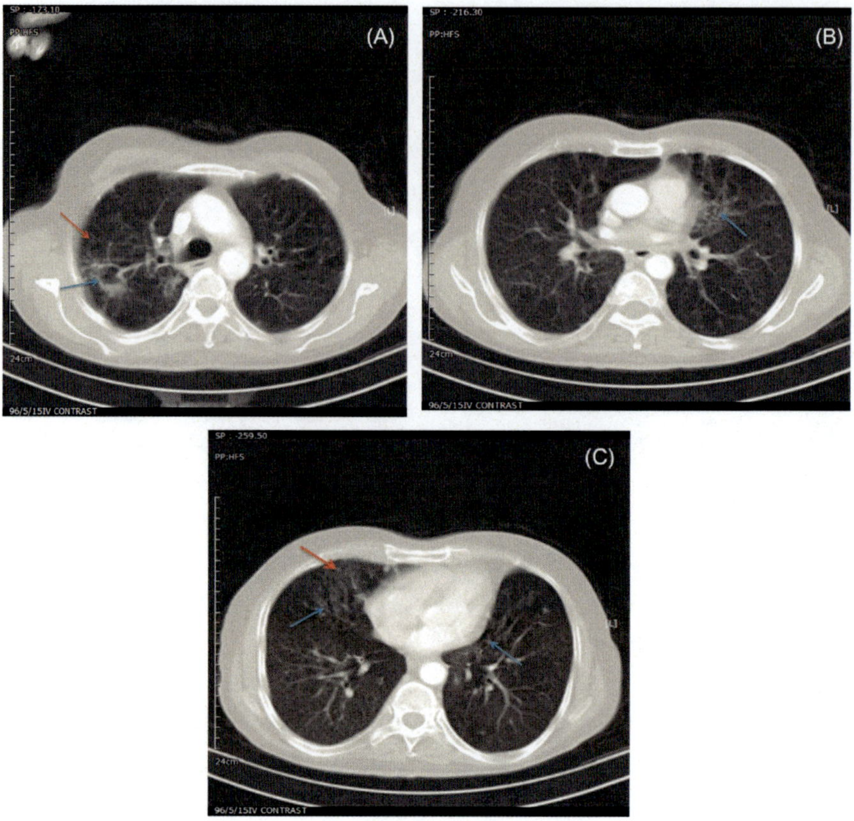

FIGURE 7.16 **Cavity, nodules, and bronchiectasis caused by *M. simiae*.** The case of a 70-year-old male infected with *M. simiae* (smear positivity results were 1+). Cavity (blue arrow) and tiny nodules (red arrow) in right lung (A). Bronchiectasis (blue arrow) in lingula (B). Bilateral bronchiectasis (blue arrow) in right middle lobe and lingula, and nodular infiltration (red arrow) in right middle lobe (C).

M. simiae is isolated mainly from the respiratory tract (sputum, bronchoalveolar lavage (BAL), or pleural fluid). Chest X-rays of infected patients show infiltrates, nodules, pleural effusion, fibrotic changes, consolidation, cavitation, pleural thickening, and emphysema—depending on the degree of smear positivity (Hamieh et al., 2018; Baghaei et al., 2011). Baghizadeh et al. (2017) found nodular lesions (100%) and bronchiectasis (85.29%) in the truly infected patients with *M. simiae* based on the ATS criteria. Christensen et al. (1978) stated that ill-defined nodules without cavity and bronchiectasis were the most common CT scan findings. Importantly, the CT manifestation of pulmonary infection with NTM might be different in immunocompetent and immunocompromised patients. Additionally, underlying

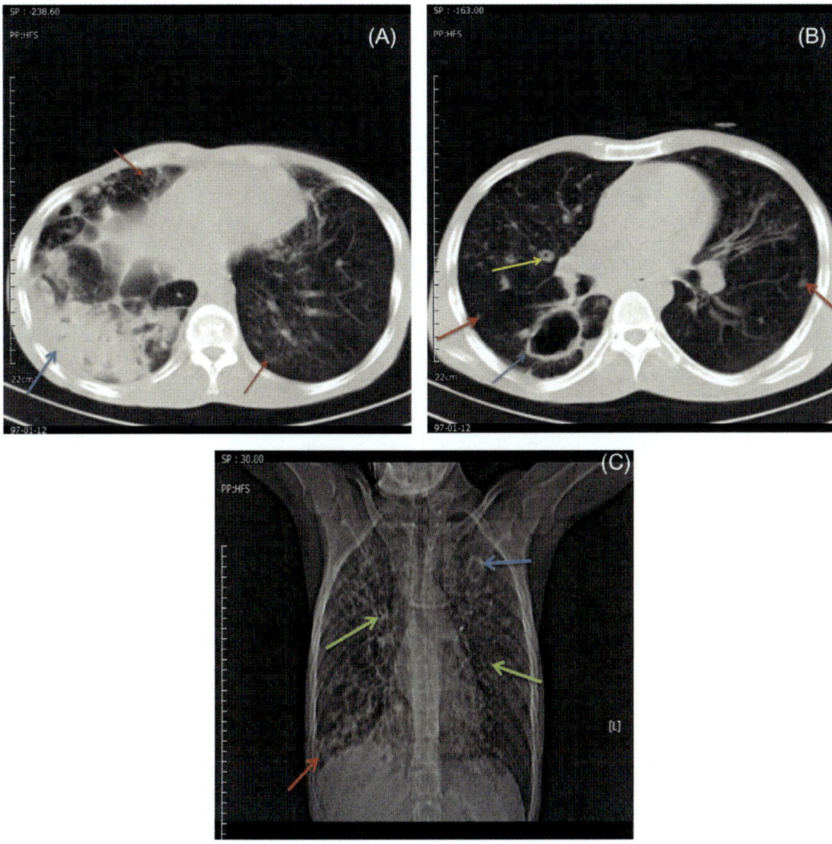

FIGURE 7.17 Consolidation, nodules, cavity, and bronchiectasis caused by *M. tuberculosis*. The case of a 30-year-old male infected with *M. tuberculosis*. Consolidation (blue arrow) in right lower lobe and bilateral nodular infiltration (red arrow) (A). Cavity (blue arrow) in right lower lobe, bronchial dilation and wall thickening (yellow arrow) in right middle lobe, and bilateral nodular infiltration (red arrow) (B). Cavity in left upper lobe (blue arrow), bilateral bronchiectasis (green arrow), consolidation (red arrow) in right lower lobe, and diffuse bilateral tiny nodular infiltration (C).

diseases or conditions such as smoking, DM, ischemic heart disease, and COPD were all associated with increased risk of *M. simiae* infection.

CONCLUSION

The clinical and imaging manifestations of pulmonary NTM infections can be divided into classical (cavitary) or nonclassical (bronchiectatic) forms. In the classical form, the most common findings in the chest radiograph include upper lobe cavitary lesions and endobronchial spread, evidenced by nodules adjacent to the foci of diseases (Christensen et al., 1978). Generally, the

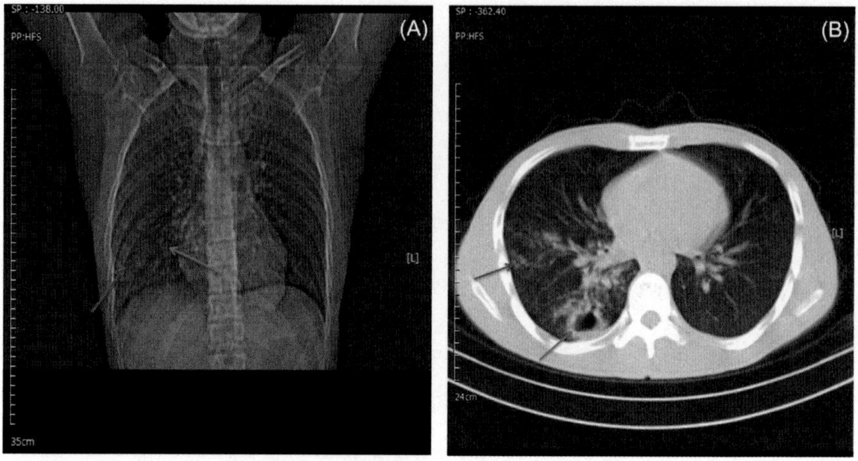

FIGURE 7.18 Bronchiectasis, nodules, and cavity caused by *M. tuberculosis*. The case of a 29-year-old male infected with *M. tuberculosis*. Bronchiectasis (blue arrow) and nodular infiltration (red arrow) in right lower lobe (A). Cavity (blue arrow) and nodular infiltration (red arrow) in right lower lobe (B).

infection is slowly progressive and cavities are smaller and have thinner walls in comparison to postprimary TB. CT scans can further characterize cavities and show associated bronchiectasis and pleural thickening. In nonclassical forms, the infection occurs mainly in elderly patients, usually without predisposing factors. Bronchiectasis can progress more rapidly and needs further investigation. CT scans characteristically show small centrilobular nodules or tree-in-bud opacities with cylindric bronchiectasis, usually in the same lobe. Figs. 7.17 and 7.18 represent pulmonary TB infection and the presence of cavities and airspace consolidation that might also be related to the degree of smear positivity. The CT manifestations are very similar in TB and NTM pulmonary infections. Thereby, for proper diagnosis and treatment, laboratory, clinical, and imaging data should be considered together.

REFERENCES

Adekambi, T., Reynaud-Gaubert, M., Greub, G., Gevaudan, M.J., La Scola, B., Raoult, D., et al., 2004. Amoebal coculture of "*Mycobacterium massiliense*" sp. nov. from the sputum of a patient with hemoptoic pneumonia. J. Clin. Microbiol. 42 (12), 5493–5501.

Adekambi, T., Berger, P., Raoult, D., Drancourt, M., 2006. rpoB gene sequence-based characterization of emerging non-tuberculous mycobacteria with descriptions of *Mycobacterium bolletii* sp. nov., *Mycobacterium phocaicum* sp. nov. and *Mycobacterium aubagnense* sp. nov. Int. J. Syst. Evol. Microbio. 56 (Pt 1), 133–143.

Baghaei, P., Tabarsi, P., Farnia, P., Marjani, M., Sheikholeslami, F.M., Chitsaz, M., et al., 2011. Pulmonary disease caused by *Mycobacterium simiae* in Iran's national referral center for tuberculosis. J. Infect. Dev. Countries 6 (01), 23–28.

Baghizadeh, A., Mehrian, P., Farnia, P., 2017. Computed tomography findings of pulmonary *Mycobacterium simiae* infection. Can. Respir. J. 2017.

Beerwerth, W., 1973. Mykobakterien in Viehtranken und Oberflachengewassern. Dtsch. Tierarztl Wochenschr. 80, 398–401.

Bolursaz, M.R., Khalilzadeh, S., Baghaie, N., Mehrian, P., Ghafaripoor, H., Aghahosseini, F., et al., 2013. Clinical and radiographic findings of pulmonary tuberculosis in infants. J. Compr. Ped. 4 (4), 170–174.

Bolursaz, M.R., Mehrian, P., Aghahosseini, F., Lotfian, F., Khalilzadeh, S., Baghaie, N., et al., 2014. Evaluation of the relationship between smear positivity and high-resolution CT findings in children with pulmonary tuberculosis. Pol. J. Radiol. 79, 120.

Brown-Elliott, B.A., Wallace Jr., R.J., 2002. Clinical and taxonomic status of pathogenic nonpigmented or late-pigmenting rapidly growing mycobacteria. Clin. Microbiol. Rev. 15 (4), 716–746.

Caskey, S., Moore, J.E., Rendall, J.C., 2018. In vitro activity of seven hospital biocides against *Mycobacterium abscessus*: implications for patients with cystic fibrosis. Int. J. Mycobacteriol. 7 (1), 45–47.

Chowdhury, V.G.A., Khandelwal, N., 2010. Diagnostic Radiology: Chest and Cardiovascular Imaging, third edition Jaypee Brothers Medical Publishers, India, p. 505.

Christensen, E.E., Dietz, G.W., Ahn, C., Chapman, J.S., Murry, R.C., Hurst, G.A., 1978. Radiographic manifestations of pulmonary *Mycobacterium kansasii* infections. Am. J. Roentgenol. 131 (6), 985–993.

Christensen, E.E., Dietz, G.W., Ahn, C.H., Chapman, J.S., Murry, R.C., Anderson, J., et al., 1981. Initial roentgenographic manifestations of pulmonary *Mycobacterium tuberculosis*, *M. kansasii*, and *M. intracellularis* infections. Chest 80 (2), 132–136.

Chung, M.J., Lee, K.S., Koh, W.-J., Lee, J.H., Kim, T.S., Kwon, O.J., et al., 2005. Thin-section CT findings of nontuberculous mycobacterial pulmonary diseases: comparison between *Mycobacterium avium-intracellulare* complex and *Mycobacterium abscessus* infection. J. Korean Med. Sci. 20 (5), 777–783.

Chung, J.H., Huitt, G., Yagihashi, K., Hobbs, S.B., Faino, A.V., Bolster Jr, B.D., et al., 2016. Proton magnetic resonance imaging for initial assessment of isolated *Mycobacterium avium* complex pneumonia. Ann. Am. Thorac. Soc. 13 (1), 49–57.

Collins, C.H., Grange, J.M., Yates, M.D., 1984. Names for mycobacteria. Mycobacteria in water. J. Appl. Bacteriol. 57 (2), 193–221.

Computed tomography—Definition from the Merriam-Webster Online Dictionary. Archived from the original on September 19, 2011. Retrieved August 18, 2009.

Cruz, A.T., Goytia, V.K., Starke, J.R., 2007. *Mycobacterium simiae* complex infection in an immunocompetent child. J. Clin. Microbiol. 45 (8), 2745–2746.

Cummins, S.L., Williams, E.M., 1933. An "acid-fast" other than Koch's bacillus cultivated from sputum. Tubercle 15, 49–53.

Da Costa, A.R.F., Fedrizzi, T., Lopes, M.L., Pecorari, M., Da Costa, W.L.O., Giacobazzi, E., et al., 2015. Characterization of 17 strains belonging to the *Mycobacterium simiae* complex and description of *Mycobacterium paraense* sp. nov. Int. J. Syst. Evol. Microbiol. 65 (2), 656–662.

Daley, C.L., Griffith, D.E., 2002. Pulmonary disease caused by rapidly growing mycobacteria. Clin. Chest Med. 23 (3), 623–632.

Doucette, K., Fishman, J.A., 2004. Nontuberculous mycobacterial infection in hematopoietic stem cell and solid organ transplant recipients. Clin. Infect. Dis. 38 (10), 1428–1439.

Erasmus, J.J., McAdams, H.P., Farrell, M.A., Patz Jr, E.F., 1999. Pulmonary nontuberculous mycobacterial infection: radiologic manifestations. Radiographics 19 (6), 1487–1503.

Evans, A.J., Crisp, A.J., Hubbard, R.B., Colville, A., Evans, S.A., Johnston, I.D., 1996. Pulmonary *Mycobacterium kansasii* infection: comparison of radiological appearances with pulmonary tuberculosis. Thorax 51 (12), 1243–1247.

Falkinham III., J.O., 1996. Epidemiology of infection by nontuberculous mycobacteria. Clin. Microbiolol. Rev. 9 (2), 177–215.

Field, S.K., Fisher, D., Cowie, R.L., 2004. *Mycobacterium avium* complex pulmonary disease in patients without HIV infection. Chest 126 (2), 566–581.

Fischeder, R., Schulze-Robbecke, R., Weber, A., 1991. Occurrence of mycobacteria in drinking water samples. Zentralbl. Hyg. Umweltmed. 192 (2), 154–158.

Fraser, R., Pare, J., Pare, P., Fraser, R., Genereux, G.E., 1990. Diagnosis of Diseases of the Chest, third edition : WB Saunders Co, Philadelphia, PA, pp. 933–941.

Fregnan, G.B., Smith, D.W., Randall, H.M., 1961. Biological and chemical studies on mycobacteria. Relationship of colony morphology to mycoside content for *Mycobacterium kansasii* and *Mycobacterium fortuitum*. J. Bacteriol. 82, 517–527.

Griffith, D.E., 2002. Management of disease due to *Mycobacterium kansasii*. Clin. Chest Med. 23 (3), 613–621.

Griffith, D.E., Girard, W.M., Wallace Jr., R.J., 1993. Clinical features of pulmonary disease caused by rapidly growing mycobacteria. An analysis of 154 patients. Am. Rev. Respir. Dis. 147 (5), 1271–1278.

Griffith, D.E., Aksamit, T., Brown-Elliott, B.A., Catanzaro, A., Daley, C., Gordin, F., et al., 2007. An official ATS/IDSA statement: diagnosis, treatment, and prevention of nontuberculous mycobacterial diseases. Am. J. Respir. Critical Care Med. 175 (4), 367–416.

Hamieh, A., Tayyar, R., Tabaja, H., Zein, S.E., Khalil, P.B., Kara, N., et al., 2018. Emergence of *Mycobacterium simiae*: a retrospective study from a tertiary care center in Lebanon. PLoS One. 13 (4), e0195390.

Hartman, T.E., Swensen, S.J., Williams, D.E., 1993. *Mycobacterium avium-intracellulare* complex: evaluation with CT. Radiology 187 (1), 23–26.

Herman, G.T., 2009. Fundamentals of Computerized Tomography: Image Reconstruction From Projections, second edition Springer Science & Business Media, London.

Hirschel, B., Chang, H.R., Mach, N., Piguet, P.-F., Cox, J., Piguet, J.-D., et al., 1990. Fatal infection with a novel, unidentified *Mycobacterium* in a man with the acquired immunodeficiency syndrome. New Engl. J. Med. 323 (2), 109–113.

Ho, Y.S., Adroub, S.A., Aleisa, F., Mahmood, H., Othoum, G., Rashid, F., et al., 2012. Complete genome sequence of *Mycobacterium fortuitum* subsp. *fortuitum* type strain DSM46621. J. Bacteriol. 194 (22), 6337–6338.

Hoefsloot, W., Van Ingen, J., Peters, E., Magis-Escurra, C., Dekhuijzen, P., Boeree, M., et al., 2013. *Mycobacterium genavense* in the Netherlands: an opportunistic pathogen in HIV and non-HIV immunocompromised patients. An observational study in 14 cases. Clin. Microbiol. Infect. 19 (5), 432–437.

Hommeyer, S.H., Godwin, D.J., Takasugi, J.F., 1991. Computed tomography of air-space disease. Radiol. Clin. North Am. 29 (5), 1065–1084.

Hu, Y.L., Bridge, B., Wang, J., Jovin, I.S., 2012. *Mycobacterium fortuitum* causing infection of a biventricular pacemaker/implantable cardioverter defibrillator. Int. J. Mycobacteriol. 1 (4), 221–223.

Ingram, C.W., Tanner, D.C., Durack, D.T., Kernodle Jr., G.W., Corey, G.R., 1993. Disseminated infection with rapidly growing mycobacteria. Clin. Infect. Dis. 16 (4), 463–471.

Karassova, V., Weiszfeiler, J., Krasznay, E., 1965. Occurrence of atypical mycobacteria in *Macacus rhesus*. Acta Microbiol. Acad. Sci. Hung. 12, 275–282.

Katoch, V.M., 2004. Infections due to non-tuberculous mycobacteria (NTM). Indian J. Med. Res. 120 (4), 290–304.

Kazda, J., Muller, K., Irgens, L.M., 1979. Cultivable mycobacteria in sphagnum vegetation of moors in South Sweden and coastal Norway. Acta Pathol. Microbiol. Scand. Sect. B Microbiol. 87b (2), 97–101.

Kobayashi, T., Tsuyuguchi, K., Yoshida, S., Kurahara, Y., Ikegami, N., Naito, M., et al., 2018. *Mycobacterium abscessus* spp. *abscessus* lung disease: drug susceptibility testing in sputum culture negative conversion. Int. J. Mycobacteriol. 7 (1), 69–75.

Kusunoki, S., Ezaki, T., 1992. Proposal of *Mycobacterium peregrinum* sp. nov., nom. rev., and elevation of *Mycobacterium chelonae* subsp. *abscessus* (Kubica et al.) to species status: *Mycobacterium abscessus* comb. nov. Int. J. Syst. Bacteriol. 42 (2), 240–245.

Lee, M.R., Sheng, W.H., Hung, C.C., Yu, C.J., Lee, L.N., Hsueh, P.R., 2015. *Mycobacterium abscessus* complex infections in humans. Emerging Infect. Dis. 21 (9), 1638–1646.

Levin, D.L., 2002. Radiology of pulmonary *Mycobacterium avium-intracellulare* complex. Clin. Chest Med. 23 (3), 603–612.

Lewis, A.H., Falkinham III., J.O., 2015. Microaerobic growth and anaerobic survival of *Mycobacterium avium*, *Mycobacterium intracellulare* and *Mycobacterium scrofulaceum*. Int. J. Mycobacteriol. 4 (1), 25–30.

Marinelli, D., Albelda, S., Williams, T., Kern, J., Iozzo, R., Miller, W., 1986. Nontuberculous mycobacterial infection in AIDS: clinical, pathologic, and radiographic features. Radiology 160 (1), 77–82.

Mettler Jr, F.A., Huda, W., Yoshizumi, T.T., Mahesh, M., 2008. Effective doses in radiology and diagnostic nuclear medicine: a catalog. Radiology 248 (1), 254–263.

Miller, W.T. (Ed.), 1993. Pulmonary infections with atypical mycobacteria in the normal host. Semin. Roentgenol. 28 (2), 139–149.

Miller Jr, W.T., 1994. Spectrum of pulmonary nontuberculous mycobacterial infection. Radiology 191 (2), 343–350.

Mirsaeidi, M., Vu, A., Leitman, P., Sharifi, A., Wisliceny, S., Leitman, A., et al., 2017. A patient-based analysis of the geographic distribution of *Mycobacterium avium* complex, *Mycobacterium abscessus*, and *Mycobacterium kansasii* infections in the United States. Chest 151 (4), 947–950.

Monill, J., Franquet, T., Sambeat, M., Martinez-Noguera, A., Villalba, J., 2001. *Mycobacterium genavense* infection in AIDS: imaging findings in eight patients. Eur. Radiol. 11 (2), 193–196.

Nasser, M., 2014. All about *Mycobacterium simiae* in brief. J. Med. Microb. Diagn. 4 (175), 2.

Nguyen, V., 1997. Diagnosis and treatment of disease caused by nontuberculous mycobacteria. This official statement of the American Thoracic Society was approved by the Board of Directors, March 1997. Medical Section of the American Lung Association. Am. J. Respir. Crit. Care Med 156 (2), S1–S25.

Padley, S.P.G., Rubens, M.B., 2003. Pulmonary infections. In: David Sutton (Ed.), Textbook of Radiology and Imaging. Churchill Livingstone, Edinburgh.

Portaels, F., 1995. Epidemiology of mycobacterial diseases. Clin. Dermatol. 13 (3), 207–222.

Primack, S.L., Logan, P.M., Hartman, T.E., Lee, K.S., Müller, N., 1995. Pulmonary tuberculosis and *Mycobacterium avium-intracellulare*: a comparison of CT findings. Radiology 194 (2), 413–417.

Raju, S., Ghosh, S., Mehta, A.C., 2017. Chest CT signs in pulmonary disease: a pictorial review. Chest 151 (6), 1356–1374.

Richard, J., Wallace, J., O'Brien, R., Glassroth, J., Raleigh, J., Dutt, A., 1990. Diagnosis and treatment of disease caused by nontuberculous mycobacteria. Am. Rev. Respir. Dis. 142 (4), 940–953.

Santos, M., Gil-Brusola, A., Morales, P., 2012. Infection by non tuberculous mycobacteria in cystic fibrosis. Cystic Fibrosis—Renewed Hopes Through Research 133–143.

Stanford, J., Stanford, C., 2012. Mycobacteria and their world. Int. J. Mycobacteriol. 1 (1), 3–12.

Tippett, E., Ellis, S., Wilson, J., Kotsimbos, T., Spelman, D., 2018. *Mycobacterium abscessus* complex: natural history and treatment outcomes at a tertiary adult cystic fibrosis center. Int. J. Mycobacteriol. 7 (2), 109–116.

Tortoli, E., Brunello, F., Cagni, A.E., Colombrita, D., Dionisio, D., Grisendi, L., et al., 1998. *Mycobacterium genavense* in AIDS patients, report of 24 cases in Italy and review of the literature. Eur. J. Epidemiol. 14 (3), 219–224.

Tortoli, E., Böttger, E.C., Fabio, A., Falsen, E., Gitti, Z., Grottola, A., et al., 2011. *Mycobacterium europaeum* sp. nov., a scotochromogenic species related to the *Mycobacterium simiae* complex. Int. J. Syst. Evol. Microbiol. 61 (7), 1606–1611.

Trovato, A., Baldan, R., Costa, D., Simonetti, T.M., Cirillo, D.M., Tortoli, E., 2017. Molecular typing of *Mycobacterium abscessus* isolated from cystic fibrosis patients. Int. J. Mycobacteriol. 6 (2), 138–141.

Tsukamura, M., Mizuno, S., Murata, H., Nemoto, H., Yugi, H., 1974. A comparative study of mycobacteria from patients' room dusts and from sputa of tuberculous patients. Source of pathogenic mycobacteria occurring in the sputa of patients as casual isolates. Jpn. J. Microbiol. 18, 271–277.

Velayati, A.A., Farnia, P., Mozafari, M., Malekshahian, D., Seif, S., Rahideh, S., et al., 2014. Molecular epidemiology of nontuberculous mycobacteria isolates from clinical and environmental sources of a metropolitan city. PLoS One 9 (12), e114428.

Velayati, A.A., Rahideh, S., Nezhad, Z.D., Farnia, P., Mirsaeidi, M., 2015. Nontuberculous mycobacteria in Middle East: current situation and future challenges. Int. J. Mycobacteriol. 4 (1), 7–17.

Wallace Jr., R.J., 1994. Recent changes in taxonomy and disease manifestations of the rapidly growing mycobacteria. Eur. J. Clin. Microbiol. Infect. Dis. 13 (11), 953–960.

Wallace Jr., R.J., Steele, L.C., Labidi, A., Silcox, V.A., 1989. Heterogeneity among isolates of rapidly growing mycobacteria responsible for infections following augmentation mammaplasty despite case clustering in Texas and other southern coastal states. J. Infect. Dis. 160 (2), 281–288.

Weiss, C.H., Glassroth, J., Stout, Subcommittee AMD, Society AT, America IDSo, et al., 2012. Pulmonary disease caused by nontuberculous mycobacteria. Expert Rev. Respir. Med. 6 (6), 597–613.

Wolinsky, E., Rynearson, T.K., 1968. Mycobacteria in soil and their relation to disease-associated strains. Am. Rev. Respir. Dis. 97 (6P1), 1032–1037.

Woodring, J.H., Vandiviere, H., 1990. Pulmonary disease caused by nontuberculous mycobacteria. J. Thorac. Imaging 5 (2), 64–76.

Woodring, J.H., Vandiviere, H., Melvin, I.G., Dillon, M.L., 1987. Roentgenographic features of pulmonary disease caused by atypical mycobacteria. South. Med. J. 80 (12), 1488–1497.

FURTHER READING

Martinez, S., McAdams, H.P., Batchu, C.S., 2007. The many faces of pulmonary nontuberculous mycobacterial infection. Am. J. Roentgenol. 189 (1), 177–186.

Chapter 8

Mapping the Footprints of Nontuberculous Mycobacteria: A Diagnostic Dilemma

Mandira Varma-Basil and Mridula Bose
Department of Microbiology, Vallabhbhai Patel Chest Institute, University of Delhi, Delhi, India

INTRODUCTION

Nontuberculous mycobacteria (NTM), also referred to at various times as atypical mycobacteria, anonymous mycobacteria, or mycobacteria other than tuberculosis (TB), have long been ignored as potential human pathogens, especially in areas endemic for TB. NTM are ubiquitous organisms and have been isolated from water, soil, dust, and plants (Tortoli, 2014). Of the more than 180 species of NTM identified so far (www.bacterio.net/mycobacterium.html), a large majority is not pathogenic to humans and animals. However, NTM are opportunistic pathogens and may cause disease in subjects with predisposing conditions or compromised immunity. Most NTM are nonpathogenic. Even the pathogenic ones demonstrate low virulence. Furthermore, since these organisms are mostly isolated from the environment, they cause easy contamination of clinical specimens. However, increasing awareness of these pathogens combined with more sophisticated diagnostic technologies has led to their recognition as causative agents in several disease presentations such as AIDS. The recent interest in NTM is due to better isolation with better technology and due to an increased predisposition of patients with AIDS to NTM disease. It is also attributed to the fact that it is now known to occur in non-AIDS patients as well.

In spite of all the advancements in recent knowledge, clinical dilemma arises due to the fact that several clinical and microbiological criteria need to be met with to establish NTM disease. The treatment is lengthy and has to be individualized since it varies with species (Primm et al., 2004). To add to this dilemma, several new species have been identified (Tortoli, 2014), although their clinical significance is not yet known, mainly because they are usually unrecognized and therefore, underreported.

Nontuberculous Mycobacteria (NTM). DOI: https://doi.org/10.1016/B978-0-12-814692-7.00008-5

This chapter summarizes some of the unique aspects of NTM infection to put in perspective the diagnosis, recognition as a causative agent, and appropriate treatment.

Potentially Pathogenic Nontuberculous Mycobacteria

NTM are environmental mycobacteria and are not obligate pathogens, unlike *M. tuberculosis* and *M. leprae*. They can be present as saprophytes, commensals, and symbionts. A few NTM are known to cause disease though they vary tremendously in their growth rate, colony characteristics, and virulence. However, even the most virulent NTM is generally less virulent than *M. tuberculosis*. Based on their growth characteristics, NTM have been classified as slow growers and rapid growers. Their generation time varies from 2 to 48 h. Slow growers take more than 7 days to form colonies. Of the slow growers, the *M. avium* complex (comprising *M. avium*, *M. intracellulare*, and other species) is the most commonly encountered NTM in pulmonary disease (Table 8.1). Other species that are slow growers, but can be pathogenic, are *M. kansasii*, *M. simiae* complex, *M. terrae* complex, *M. haemophilum*, and *M. marinum* (Jorgensen et al., 2015). Several new pathogenic species have also emerged, such as *M. asiaticum*, *M. branderi*, *M. lentiflavum*, and *M. triplex* (Tortoli, 2014). Rapid growers, as the named suggests, lead to colony formation within 7 days. Rapid growers are being increasingly seen to cause a wide spectrum of infections. Of these, *M. fortuitum*, *M. chelonae*, *M. abscessus*, and *M. mucogenicum* are the most common and have been found in cold water systems leading to nosocomial infections. It has been previously suggested that the respiratory tract can harbor NTM without it leading to a disease, especially in patients with chronic respiratory disorder. Colonization with NTM was described most commonly with *M. avium intracellulare*. However, this theory has not been tested in detail and it is possible that "colonization" is rather an indolent infection. Many a time, a patient meets the diagnostic criteria of NTM disease, but the disease is not severe enough to be treated. Such situations could arise with repeat isolations of environmental contaminants or NTM with low virulence, for example, *M. fortuitum*. Moreover, several new species are being identified due to the advances in laboratory technology, but their significance is still unknown. Investigators identify a new species if its 16S rRNA sequence, which is highly conserved, is different from a known species by 1% or more (Tortoli, 2003; McNabb et al., 2004). Therefore, it is the responsibility of clinicians and laboratories to collect a sufficient number of samples and to follow up the patients for an adequate period of time to confirm or refute the diagnosis of NTM disease.

TABLE 8.1 Potentially Pathogenic Nontuberculous Mycobacteria

Major Groups of Potentially Pathogenic NTM in Humans
Slow Growing NTM
M. avium complex
M. avium spp. *hominissuis*
M. avium spp. *paratuberculosis*
M. intracellulare
M. genavense
M. haemophilum
M. malmoense
M. shimoidei
M. xenopi
M. heidelbergense
M. branderi
M. simiae complex
M. triplex
M. conspicuum
M. szulgai
M. terrae
M. ulcerans
M. kansasii
M. asiaticum
M. marinum
M. intermedium
M. novocastrense
M. szulgai
M. scrofulaceum
M. interjectum
M. lentiflavum
M. heckeshornense
M. tusciae

(*Continued*)

TABLE 8.1 (Continued)

Major Groups of Potentially Pathogenic NTM in Humans
M. kubicae
Rapid Growers
M. abscessus spp. *abscessus*
M. abscessus spp. *bolletii*
M. fortuitum
M. chelonae
M. mucogenicum
M. phocaicum
M. boenickei
M. canariasense
M. goodii
M. neoaurum
M. senegalense
M. houstonense
M. immunogenum
M. wolinskyi

Potential Sources of Nontuberculous Mycobacterial Infection

Environmental mycobacteria, as the name suggests, have various environmental reservoirs. They can be present in municipal or natural water, soil, aerosols, animals, and humans. *M. avium* infection has been associated with municipal water supplies, including drinking water, in both humans and animals. DNA-based fingerprints of *M. avium* isolates recovered from the drinking water of AIDS patients was identical to the isolates recovered from the patients themselves (Von Reyn et al., 1994). In fact, the *M. avium* isolates obtained from macaques infected with simian immunodeficiency virus were also identical to the isolates obtained from their drinking water (Mansfield and Lackner, 1997). The presence of environmental mycobacteria in municipal water is related to the innate resistance of these organisms to chlorine. In a pilot study involving treatment of water system with ozone or chlorine, there was a dramatic shift of the resident bacterial population to actinomycetes family, which includes *Mycobacterium* sp. (Norton and LeChevallier, 2000). In addition, mycobacterial populations such as *M. fortuitum*,

M. chelonae, and *M. avium* can form biofilms and can, thus, persist in water distribution systems.

NTM are also known to withstand starvation for a long time. *M. intracellulare* has been known to survive in sterile water for more than a year. These organisms can also survive extremes of temperature and have been isolated from hot tap water and spas. *M. avium*, *M. chelonae*, *M. phlei*, and *M. xenopi* are some of the thermoresistant species. *M. kansasii*, *M. flavescence*, *M. mucogenicum*, and *M. gordonae* have been isolated from municipal water. *M. avium intracellulare*, *M. gordonae*, *M. fortuitum*, and *M. kansasii* have been found in public swimming pools and whirlpools (Primm et al., 2004).

Other sources of infection can be regions that are frequented by humans, such as soil and public bathrooms (Lipner et al., 2017). In Wardha, India, of a total of 1398 samples collected from soil and water, NTM were isolated from 170 (12.2%). Of these, 107/170 (63%) were potentially pathogenic mycobacteria. In this study, maximum isolation of mycobacteria was obtained during the rainy season, followed by the winter season. Interestingly, this study also found the highest isolation of NTM from areas with frequent community gatherings (Mishra et al., 2018).

Due to their ubiquitous nature, NTM have at times been implicated in pseudo-outbreaks. A hospital in Croatia reported *M. gordonae* in 135/150 (90%) pulmonary samples. Further investigation revealed contamination of tap water with *M. gordonae* (Zlojtro et al., 2015). In another survey of hospital water distribution systems in Tabriz, Iran, NTM were detected in 76/120 (63.3%) water samples. *Hsp65* and *16Sr RNA* sequencing revealed potentially pathogenic and saprophytic mycobacteria (Maleki et al., 2017). In a Mycobacteriology laboratory in Delhi, India, *M. fortuitum* has been recovered from six cultures performed on sputum specimens the same day. *M. fortuitum* was not obtained on repeat cultures of the same patients and hence the isolates were considered to be laboratory contaminants (Garima et al., 2012).

It is, thus, important to delineate the ecological niche of NTM in a geographical location, since the frequent isolation of potential pathogens from soil and water indicate a potential risk of infection, especially in an immune-compromised host. In contrast, their isolation may be falsely regarded as an infection. The advent of newer diagnostic tools and whole genome sequencing would make a study of the environmental sources and the relationship with human disease easier.

Transmission of Nontuberculous Mycobacteria

NTM are ubiquitous organisms and the environment is a recognized source of NTM infection. Water, soil, and dust are the most common sources of infection. Humans can acquire infection through the environment in a variety of circumstances. Source of infection can be drinking contaminated water or while swimming or bathing. Children with lymphadenitis have been hypothesized to acquire infection through ingestion of contaminated water (Primm et al., 2004). In fact, NTM isolated from patients with pulmonary disease have been traced to the plumbing systems of their homes (Falkinham, 2011). Infection with NTM can, thus, be through ingestion, inhalation, or through trauma. *M. avium* exhibits a high degree of tolerance to acid and bile salts (Primm et al., 2004). This fact is consistent with the theory that NTM infection can occur through ingestion. Infection can also occur through aerosol formation. Hypersensitivity pneumonitis due to exposure to aerosols that contain mycobacteria has also been reported. This could involve workers in certain occupations or to people using hot tubs and spas (Primm et al., 2004).

Infection has been known to spread through tap water and even through disinfectant solutions. The high hydrophobicity of these organisms favors aerosolization, on one hand, and formation of biofilms, on the other. This also accounts for the antibiotic resistance of NTM (Falkinham, 2009). Human-to-human transmission of NTM was not known to be a common route of transmission. However, recent evidence suggests that this group of organisms can be transmitted from one person to another, especially in patients of cystic fibrosis. A UK cystic fibrosis center observed similar drug susceptibility profiles among *M. abscessus* isolates from patients attending the center. Whole genome sequencing confirmed this finding, which suggests person-to-person transmission (Bryant et al., 2013). In a cystic fibrosis center in Seattle, *M. abscessus* subspecies *massiliense* was isolated from four patients who were previously acid-fast bacilli (AFB) negative. These patients visited the center on the same day as a patient with a previous history of infection with *M. abscessus* subspecies *massiliense.* Direct transmission was ruled out since the patients did not interact with each other. However, transmission through fomites could not be ruled out (Aitken et al., 2012). It has also been suggested that certain variants of *M. abscessus* may be more infectious than others (Tettelin et al., 2014).

Predisposing Conditions

NTM have for long been ignored as environmental organisms with little or no clinical relevance. It was the HIV pandemic that highlighted infections due to opportunistic pathogens like *M. avium* and *M. intracellulare* (Lake et al., 2016). Though NTM are present in the environment, at least 25 are known to be pathogenic to humans (Yeung et al., 2016), evidence of their propensity to cause disseminated infection is limited. However, there is increasing evidence of NTM causing pulmonary disease (Lake et al., 2016). This may be due to increased awareness, an aging population that is more susceptible to NTM or better diagnostic facilities. Some investigators have also reported that NTM may be present as colonizers in the respiratory tract

(Chien et al., 2014; Thanachartwet et al., 2014). However, certain conditions predispose an individual to acquire NTM infection.

Structural Pulmonary Disease

Pulmonary infections due to NTM occur most commonly due to structural lung diseases, such as chronic obstructive pulmonary disease (COPD), cystic fibrosis, bronchiectasis, previous history of TB, and pneumoconiosis (Winthrop et al., 2010). Several studies have been performed which highlight the significance of COPD in relation to NTM infection. One study examined chronic respiratory diseases and inhaled corticosteroids as risk factors for NTM infection in Denmark (Andréjak et al., 2013). In this study, chronic respiratory disease was associated with a 16.5-fold increased risk of NTM pulmonary disease. Patients with COPD on inhaled corticosteroid therapy were at a higher risk of NTM infection than patients who had never received steroids (Andréjak et al., 2013). In addition to COPD, this study provided strong evidence that the risk of NTM infection increases in patients with asthma, bronchiectasis, and previous TB (Andréjak et al., 2013). A study in the United Kingdom also concluded that a proportion of patients with severe COPD may have evidence of NTM infection even if there is no clinical suspicion, which may have an impact on long-term management of the disease in patients (Char et al., 2014).

Similarly, it has been observed that bronchiectasis and NTM infection commonly coexist. The most common NTM infection in bronchiectasis patients is *M. avium intracellulare* complex (MAC) (Wallace et al., 1998; Schweitzer et al., 2017). In addition, bronchiectasis has been seen to be more frequent in women over 65 years of age with pulmonary NTM infection (Mirsaeidi et al., 2015). Schweitzer et al. (2017), in a case control study in the United States conducted on patients with pulmonary NTM infections, reported that NTM subjects with bronchiectasis tended to be short with lower body weight. They also hypothesized that not only does bronchiectasis increase the susceptibility to NTM, NTM may also contribute to the pathogenesis of bronchiectasis (Schweitzer et al., 2017).

Another lung structural disease that predisposes to NTM infection is cystic fibrosis. NTM are found in approximately 10% patients of cystic fibrosis, however, only some of these would lead to NTM disease (Skolnik et al., 2016). The prevalence of NTM disease in cystic fibrosis, observed in various studies worldwide is 4%–14% (Fauroux et al., 1997; Roux et al., 2009; Esther et al., 2010). The most common NTM found in patients with cystic fibrosis is MAC and *M. abscessus* complex (Skolnik et al., 2016). NTM infection has also been associated with *Aspergillus fumigatus* and *Stenotrophomonas maltophilia* coinfection in patients of cystic fibrosis and with allergic bronchopulmonary aspergillosis (Viviani et al., 2016). The risk is also higher in patients who use inhaled antibiotics and in those who use bronchodilators (Viviani et al., 2016).

Age and Gender

There is a high incidence of NTM infection in elderly females. This could be due to the lack of the protective role played by estrogen in younger women. Another reason for higher susceptibility to NTM infection in elderly women may be due to low serum concentration of macrophage stimulating factor (Kamada et al., 2001). In fact, the "Lady Windermere Syndrome" has been reported in thin, postmenopausal women of European/American descent who had a tendency to suppress their cough reflex and presented with middle lobe or lingular bronchiectasis along with NTM infection. The NTM most commonly involved in this syndrome is *M. avium intracellulare* (Reich and Johnson, 1992; Schweitzer et al., 2017). Several investigators later confirmed that tall stature, low body weight, older age group, and female gender were risk factors for NTM infection (Kim et al., 2008; Okumura et al., 2008; Chan et al., 2010).

HIV and Immunodeficiency

Diseases that reduce the cell mediated immunity increase the risk of NTM infection. Mycobacteria infect macrophages and stimulate CD4 + T Helper cells which, in turn, stimulate the release of interleukin 12 (IL12) and interferon γ (IFNγ). Another proinflammatory cytokine, tumor necrosis factor $\propto$ may also be involved in inhibiting mycobacterial growth.

It has been known for a long time that HIV-1 positivity leads to enhanced susceptibility to disseminated infection, especially *M. avium intracellulare.* The risk increases manifold when the $CD4^+$ T cell count falls below 50/mm^3 (Horsburgh et al., 1991). *M. avium intracellulare* continues to be the most important NTM isolated from immunosuppressed patients. However, with better diagnostic facilities, a number of other NTM are also being recognized in the immunosuppressed state. Notable among these are the rapid growers (*M. abscessus*, *M. fortuitum*, *M. chelonae*) and certain fastidious NTM (e.g., *M. haemophilum*) (Henkle et al., 2015). The increase in immunosuppressive therapies in recent years has also led to an increase in NTM infections (Henkle et al., 2015). These could include corticosteroids, immunosuppressive therapy for autoimmune disorders including rheumatoid arthritis, psoriasis, and Crohn's disease (Henkle et al., 2015).

Several genetic disorders in humans also increase the susceptibility to NTM infection. These diseases could result from mutations in cytokine pathways that affect IL12β, IL12R, IFNγRI, and IFNγR2. It has also been reported that deficiency of nuclear factor-κB essential modulator leads to a greater susceptibility to NTM by inhibiting production of IL12 by monocytes and monocyte-derived dendritic cells (Filipe-Santos et al., 2006).

Autosomal dominant deficiency of the transcription factor GATA-2 has been shown to increase the risk of infection due to NTM. This genetic immunodeficiency manifests with aplastic anemia, fungal infections, warts, and human papilloma virus-associated squamous cell carcinomas in addition to mycobacterial infections (Hsu et al., 2011). Autoantibodies have been described against IFNγ and IL12 (Meager et al., 1997; Browne, 2014). Although, the significance of autoantibodies against IL12 in relation to NTM infection remains unclear, a cohort of patients with NTM infection have been demonstrated to have autoantibodies against IFNγ (Browne, 2014).

Clinical Manifestations of Nontuberculous Mycobacterial Infection

Pulmonary Disease

The most common clinical manifestation of NTM infection is pulmonary disease, followed by skin and soft tissue infections, lymphadenitis, or disseminated infections in severely compromised patients (Blanc et al., 2016). The symptoms of NTM pulmonary infection are nonspecific and most patients present with chronic cough with or without sputum production. The patients may also complain of constitutional symptoms, such as fatigue, malaise, dyspnea, fever, hemoptysis, and weight loss. Coexisting lung diseases like bronchiectasis, COPD, and cystic fibrosis further complicate diagnosis. Physical examination also reveals signs of the underlying pathology. Chest X-rays may reveal fibrocavitary changes similar to TB. However, the cavities are thin-walled with less surrounding parenchymal opacity, a more contiguous spread of the disease, and more involvement of the pleura. However, none of the findings are pathognomonic enough to exclude TB. Noncavitary disease will have features of the underlying structural lung disease such as bronchiectasis.

Microbiological diagnosis is necessary to commence treatment and the microbiological diagnostic criteria laid down by America Thoracic Society (ATS) are followed, especially since NTM are ubiquitously present in the environment and can, therefore, contaminate specimens. *M. gordonae* is a common lab contaminant and is often found in laboratory cultures. Other NTM that may contaminate specimens are *M. terrae* complex, *M. mucogenicum*, and *M. scrofulaceum*. Some species such as *M. simiae* are known to be present in tap water and may, thus, lead to contamination (Griffith et al., 2007). Laboratories have reported contamination of entire batches of LJ medium with *M. fortuitum* (Garima et al., 2012).

The NTM commonly found as pathogens in a patient with pulmonary symptoms are *M. avium*, *M. intracellulare*, *M. kansasii*, *M. xenopi*, *M. abscessus*, *M. fortuitum*, and *M. scrofulaceum* (Blanc et al., 2016; Garima et al., 2012; Shrivastava et al., 2017), with *M. avium intracellulare* being the most common species causing pulmonary disease (Simons et al., 2011; Blanc et al., 2016; Garima et al., 2012; Shrivastava et al., 2017) (Table 8.1).

Lymphadenitis

Cervical lymphadenitis is the most common form of NTM disease in children aged 1–5 years (Griffith et al., 2007). The disease usually affects immune-competent children. It is hypothesized that the NTM enter through the body after ingestion through mucosal lesions or through lymphatic vessels. The lymph nodes most commonly involved are jugulodigastric, submandibular, preauricular, and submental (Tortoli, 2012). Involvement of the inguinal region has also been reported (Heraud et al., 2016). The disease is asymptomatic at first, though the patient may complain of fever and malaise. The first manifestation of lymphadenitis when clinical opinion is sought is a nontender, indolent swelling which is usually unilateral. The swelling subsequently enlarges and the central part of the granuloma becomes necrosed. Regression is rare (Tortoli, 2012). Tuberculous and pyogenic lymphadenitis should be excluded, before the diagnosis of lymphadenitis due to NTM is made. Microbiological diagnosis is very important. The NTM usually reported from cases of lymphadenitis are *M. avium*, *M. scrofulaceum*, *M. simiae* complex, *M. hemophilum*, *M. malmoense*, *M. terrae*, and *M. marinum* (Tortoli, 2012; Heraud et al., 2016). Rapid growers such as *M. fortuitum*, *M. chelonae*, and *M. abscessus*, though rare, have also been reported in cases of NTM lymphadenitis (Rivron et al., 1979; Chang et al., 2006; Khalid and Ali, 2017). Recently, the first case of *M. marseillense* lymphadenitis was reported in a child in Florence, Italy (Azzali et al., 2017).

Skin, Soft Tissue, and Bone Disease

The NTM most commonly reported from skin and soft tissue infections are *M. fortuitum*, *M. abscessus*, *M. chelonae*, *M. marinum*, and *M. ulcerans*. Rapidly growing mycobacteria (RGM) (*M. fortuitum*, *M. abscessus*, *M. chelonae*) are commonly found in abscesses formed after a puncture wound or a traumatic injury. Nosocomial infections occurring after surgeries or due to invasive therapeutic interventions, such as long-term intravenous catheters, can also be infected by RGM.

Chronic granulomatous infections due to NTM have been known to develop in tendon sheaths, bursae, joints, and bones usually after accidental trauma, surgical incisions, or puncture wounds. *M. avium intracellulare* and *M. marinum* have been implicated in tenosynovitis of the hand (Griffith et al., 2007; Blanc et al., 2016). *M. massiliense* was implicated in postcesarean surgical wound infection. Failure to respond to Teicoplanin, painful cutaneous nodules, and persistent wound drainage alerted the clinicians to the possibility of NTM infection (Wu et al., 2016). NTM have also been isolated from cases of osteomyelitis. Vertebral osteomyelitis, most commonly caused by *M. avium complex* and *M. xenopi* has been reported in Korea (Kim et al., 2016), where it was suggested that NTM infection should be considered in the case of indolent vertebral osteomyelitis in immunocompromised patients if TB has been ruled out.

Disseminated Disease

Disseminated infections can occur in immunocompromised patients especially patients with advanced HIV infection, as evidenced by very low CD4 + T cell counts (Griffith et al., 2007). Studies have shown that 40% patients with less than 10 CD4 + T cells/μL developed disseminated NTM infection within 1 year. All patients with CD4 + T cell counts less than 50 CD4 + T cells/μL are at risk of developing disseminated NTM infection. The risk increases with falling CD4 + cell counts (Griffith et al., 2007). Patients with colonization of respiratory or gastrointestinal tracts with NTM have greater predisposition towards developing the disease. Most disseminated infections are caused by *M. avium intracellulare*, particularly *M. avium*, *M. kansasii*, *M. hemophilum*, and *M. chelonae* have also been implicated in disseminated disease (Henkle et al., 2015; Blanc et al., 2016). Several other NTM including, *M. scrofulaceum*, *M. gordonae*, *M. hemophilum*, *M. genavense*, *M. xenopi*, *M. fortuitum*, *M. malmoense*, *M. simiae*, and *M. celatum* have been variably implicated in disseminated disease (Griffith et al., 2007). An immunocompromised status other than AIDS, resulting from organ transplantation or prolonged corticosteroid therapy, may also predispose to disseminated NTM disease, though less commonly.

Rare Reports of Infections Due to Nontuberculous Mycobacteria

Although NTM most commonly cause pulmonary disease, certain reports with rare infections have been reported. A few examples are given in Table 8.2.

Diagnostic Criteria of Nontuberculous Mycobacterial Infection

The ATS has provided certain criteria to diagnose NTM infections. A patient suspected of pulmonary NTM disease must have nodular or cavitary opacities shown in chest X-rays (Griffith et al., 2007), or there should be evidence of multifocal bronchiectasis with multiple small nodules. Appropriate alternate diagnosis should be excluded (Griffith et al., 2007). Microbiological diagnosis is very important. The patient should have at least two positive culture results from two different sputum samples, or at least one bronchial wash or lavage must be culture positive. If these are not available, then a transbronchial or other lung biopsy must show histopathological evidence of mycobacterial infection and culture positivity for NTM; or if the culture from biopsy specimen is negative, then an additional sputum sample or bronchial wash or lavage should be culture positive (Table 8.3). Several times, a

TABLE 8.2 Rare Case Reports

Rare Case Report 1:
M. intracellulare has been reported to cause osteomyelitis and arthritis of the wrist in an immunocompetent female. The patient had chronic swelling of the wrist joint and was diagnosed as rheumatoid arthritis. Examination revealed a fistula, bone, and joint destruction was evident radiographically. Histological and microbiological examinations finally attributed the ailment to *M. intracellulare* (Yano et al., 2014).
Rare Case Report 2:
A young nulliparous female was referred to an infertility clinic. Endometrial biopsy specimen and the menstrual blood were submitted for microbiological investigations. Culture of both the tissue and menstrual blood revealed the presence of *M. intracellulare* (Shrivastava et al., unpublished results).
Rare Case Report 3:
A 66-year-old man with noninsulin dependent diabetes mellitus presented with chronic granulomatous inflammation of the left eye after multiple glaucoma surgeries. The patient was treated with systemic corticosteroids initially for panuveitis. After an initial response, the condition worsened. Culture of the vitreous fluid revealed the presence of *M. hemophilum*. The patient was treated with intraocular and systemic antimicrobials, but eventually became blind (Pinitpuwadol et al., 2018).
Rare Case Report 4:
A 67-year-old man with a history of smoking, presented with fever, night sweats, cough, dyspnea, and significant weight loss. Computed tomography of the chest revealed a cavitary lung lesion. Bronchoscopy and thoracoscopic biopsies showed necrotizing granulomatous inflammation. Culture of the biopsy specimen was negative, though microscopy showed AFB. Sputum specimens grew nontuberculous *Mycobacterium* in MGIT assay system after 13 days of incubation, which was identified as *M. branderi* by pyrosequencing of the 16S rRNA gene. The patient was started on therapy with clarithromycin, ethambutol, and moxifloxacin. Due to nonresolution of clinical signs and symptoms, the lesion was surgically resected and the patient improved symptomatically. This was a novel case of multidrug resistant *M. branderi* pulmonary infection in an immunocompetent patient, progressive despite medical therapy, and requiring surgical resection for definitive management (Turvey et al., 2017).

patient is suspected to be suffering from NTM disease, but does not meet any of the diagnostic criteria. In such cases, the patient must be followed up till the diagnosis is confirmed or excluded (Griffith et al., 2007).

The diagnosis of NTM disease remains a clinical dilemma for clinicians as NTMs exist in the environment and mere isolation from an unsterile site could be a reflection of environmental contamination. Hence, coordination between clinicians and laboratories is very important. Diagnosis requires integration of clinical, radiographic, and microbiological data. Radiological findings in NTM disease have been excellently reviewed in Chapter 5.

TABLE 8.3 Diagnostic Criteria for Pulmonary NTM Disease

Criteria	Diagnostic Signs and Symptoms[b]
Clinical symptoms[a]	Pulmonary symptoms including productive or nonproductive cough, hemoptysis, fever, fatigue, weight loss, chest pain.
Radiological	Chest X-ray may reveal nodular or cavitary opacities. High resolution computer tomography (CT) scan may show bronchiectasis with or without nodules.
Microbiological	Positive cultures for NTM from two or more expectorated sputum or positive cultures for NTM from at least one bronchial wash or Transbronchial or other lung biopsy specimen demonstrating granulomatous inflammation or AFB along with one positive sputum, bronchial wash or biopsy culture for NTM.

[a]*One or more symptoms may exist.*
[b]*Alternate diagnosis like tuberculosis or fungal infection should have been ruled out.*
Adapted from McShane, P.J., Glassroth, J., 2015. Pulmonary disease due to nontuberculous mycobacteria: current state and new insights. Chest 148, 1517–1527 (McShane and Glassroth, 2015).

In the next section, we will consider microbiological diagnosis which is of paramount importance in establishing diagnosis of NTM disease.

MICROBIOLOGICAL DIAGNOSIS OF NONTUBERCULOUS MYCOBACTERIAL DISEASE

Specimen Collection and Processing

Specimens are collected depending on the site of infection. The specimens should be collected in laboratory-approved sterile containers which should be properly labeled. Care should be taken to avoid any contamination of the specimen from environmental sources. No transport media or preservatives are required. Although mycobacteria can be recovered several days after collection, refrigeration is recommended if there is a delay of >1 h in transporting the sample to the laboratory. If the sample has to be transported to a laboratory which is located at a distance, then refrigerants such as cold packs can be used (Griffith et al., 2007). Long delays in processing the sample after collection increase the risk of bacterial overgrowth and environmental contamination. Ideally, the patient should not be on antibiotics at the time of sample collection as antibiotics may affect the yield of NTM.

Pulmonary Specimens

Pulmonary specimens can include spontaneously produced or induced sputum, gastric lavage, or bronchoscopic aspirations. Spontaneously produced sputum is the specimen of choice. Multiple early morning sputum samples improve the chances of recovery of NTM. The patient must take a deep breath and cough vigorously and deeply to produce sputum. Saliva and nasal secretions should not be collected. Sputum samples should be free of food particles and other extraneous material. Induced sputum may be used in patients who are unable to expectorate a good sputum sample, however, there are no reports on the efficiency of induced sputum for NTM detection. Bronchoscopy can be performed if indicated and if sputum cannot be obtained. Gastric lavage is used in senile patients, nonambulatory patients or children, who are not able to expectorate sputum and, hence, swallow it. However, there is limited data on the isolation of NTM in gastric lavage.

Pus, Body Fluids, and Tissue Samples

Pus from abscesses should be sent in adequate amounts. Swabs should be avoided as they often do not have enough material and are prone to become dry. In case purulent material from a site is not sufficient to be collected in a vial, the swab should be saturated well with the sample. For skin lesions, the skin should be cleansed well with alcohol before aspiration of the material into a syringe. Body fluids collected for evaluation for NTM should be in adequate amounts and must be collected aseptically by needle aspiration. The sample should not be diluted with any liquid. Though, genitourinary infections caused by NTM have rarely been reported, urine samples may be collected if indicated. Early morning, clean-catch, midstream-voided urine samples should be submitted for at least 3 days. Pooled samples are undesirable due to excessive dilution and higher contamination. Catheterized samples should be used only if voided urine sample is unavailable. Tissue biopsy specimens may be immersed in sterile saline to prevent desiccation.

Immunocompromised patients, especially patients with HIV infection can have disseminated mycobacterial infection. The NTM usually associated with such infection is *M. avium intracellulare*. A positive blood culture is always associated with clinical disease. Blood should be collected for mycobacterial culture as for routine cultures with added anticoagulants, such as sodium polyanethol sulfonate or heparin. Specialized automated systems for mycobacterial cultures such as Bactec MGIT 960 (Becton Dickinson, Sparks, MD) or the BacT/Alert 3D (Biomerieux, Durham, NC) should be used for isolation of NTM.

Identification of *Mycobacterium* spp. is laborious, costly, and time consuming. Hence specimens should be rejected and a repeat specimen requested if (1) the volume of the specimen is insufficient; (2) the patient submits saliva instead of sputum; (3) specimens such as sputum or urine are pooled; (4) dry swabs are submitted; or (5) the length of time between collection and processing is too long. It is important that laboratories are sent as much material as possible along with a brief history of the patient and a request to culture mycobacteria if NTM disease is suspected.

Processing of Specimens

Specimens from sterile sites can be inoculated directly to media or concentrated. However, most samples obtained for mycobacterial culture, such as sputum (spontaneous or induced) and gastric lavage, contain organic debris such as mucin and other proteinaceous material and may also be contaminated with bacteria. Hence, specimens from nonsterile body sites should be decontaminated. Digestion and decontamination can be done by sodium hydroxide (NaOH) method, zephiran–trisodium method, or the *N*-acetyl-L-cysteine (NALC)–2% NaOH method. The NALC–NaOH method is the most widely used and may at times be combined with 5% oxalic acid processing for specimens that are contaminated with gram-negative organisms such as *Pseudomonas aeruginosa*, as observed in patients of cystic fibrosis and bronchiectasis. However, it should be noted that NTM may be sensitive to oxalic acid treatment. Moreover, rapidly growing NTM are sensitive to prolonged exposure to greater than or equal to 2% NaOH. Hence, decontamination procedures, when one suspects NTM, should be as gentle as possible.

Briefly, after digestion and decontamination, the specimens are centrifuged and the supernatant decanted and the sediment processed further (Tille, 2014).

Large pieces of tissue should be finely minced with sterile scalpels and scissors. The material should be homogenized in a sterile tissue grinder with sterile saline and the suspension processed further.

Smear Microscopy and Culture

Smears should be stained with acid-fast stains such as fluorochrome, Ziehl Neelsen, or Kinyoun. Flourochrome staining is a more sensitive technique; however, many rapid growers may not appear fluorescent. Moreover, positive fluorochrome smears must be confirmed by Ziehl Neelsen staining. The technologist should also keep in mind that a negative smear does not necessarily mean absence of NTM in the clinical specimen.

Culture for NTM should include both solid and liquid media to improve the rate of isolation. The solid media used include Lowenstein Jensen medium, 7H10 and 7H11. Unlike broth or liquid media, a solid medium allows the observation of colony characteristics, colony counts, mixed cultures, and growth rate. The most widely used broth culture medium is the nonradiometric mycobacteria growth indicator tube (MGIT) (Becton Dickinson, Sparks, MD). The MGIT system relies on an oxygen-sensitive fluorescent compound to detect mycobacterial growth. The medium is inoculated with a sample containing mycobacteria and with subsequent growth mycobacteria utilizing the oxygen and the compound fluoresces.

Certain NTM such as *M. hemophilum*, *M. genavense*, *M. avium* spp. *paratuberculosis*, and *M. ulcerans* are fastidious organisms. *M. hemophilum* requires iron containing compounds such as ferric ammonium citrate and hemin. Hence, samples suspected to contain this species, such as skin lesions, should be inoculated in medium containing these compounds. *M. genavense* and *M. avium* spp. *paratuberculosis* need *mycobactin J* for growth, while *M. ulcerans* may require egg yolk supplementation (Griffith et al., 2007). Optimum temperature for growth for most NTM is between 28°C and 37°C.

Identification

Species identification of NTM is required since the treatment may vary for different species. However, all the isolates obtained in a laboratory need not be identified, since a number of them may be present as contaminants. The microbiologist should be aware of the clinical background of the patient to be able to discern the importance of the NTM isolated. Hence, coordination between the laboratory and clinician is of utmost importance.

Phenotypic Identification

Growth rate of NTM offers a preliminary identification of RGM and slow growing mycobacteria, since rapid growers develop colonies within 7 days of incubation. Biochemical identification includes a battery of tests and is time consuming and laborious. Moreover, many of the newly described species cannot be identified by conventional methods alone. Hence newer methods must be used for identification.

High-Performance Liquid Chromatography

High-performance liquid chromatography (HPLC) is a reliable and rapid method for identifying slow growing species of NTM. However, some new species and subspecies may fail to be recognized. Moreover, HPLC is not able to differentiate closely related species. It has been reported that HPLC could not differentiate between various species within the *M. fortuitum* group and *M. smegmatis* group (Grifith et al., 2007), or between *M. abscessus* and *M. chelonae* (Butler et al., 2001). Molecular methods of identification have become popular as they are not only rapid, sensitive, and specific, but can also differentiate between closely related species and also help in identifying new species.

Matrix Assisted Laser Desorption Ionization-Time of Flight Mass Spectrometry

MALDI-TOF is being used extensively for identification of several bacteria. Most studies performed on mycobacteria agree that MALDI-TOF can reliably differentiate between NTM and *M. tuberculosis* complex (Ceyssens et al., 2017). However, it may be unable to differentiate between closely related species. In a recent study, NTM were identified in primary MGIT cultures by MALDI-TOF. The investigators, however, recommended that an initial immunochromatographic assay be performed to screen for *M. tuberculosis* complex, since the rate of identification of *M. tuberculosis* is low in polymicrobial cultures (Huang et al., 2018). Most of the results indicate a potential role of MALDI-TOF in identification of NTM.

Molecular Methods of Identification

Several molecular methods such as probe-based assays, restriction enzyme analysis, and DNA sequencing are now being used to identify NTM. Sequencing of the 16S rRNA gene and the 16S–23S internal transcribed spacer region has often been used as a benchmark for differentiation between NTM species. Molecular methods for identification of NTM have been reviewed in detail in Chapter 7.

CONCLUSIONS

To conclude, a large number of NTMs are being identified in Mycobacteriology laboratories and application of rapid molecular assays would facilitate improved diagnosis of NTMs by obviating the use of more cumbersome and conventional diagnostic methods. It is imperative that health centers and laboratories take the initiative to increase awareness for appropriate diagnosis of NTM. Improved NTM identification through quality-controlled kits and clinico-laboratory linkages can provide necessary epidemiological data to assess the importance and relevance of these isolates that were collectively dismissed up until recently as environmental contaminants.

REFERENCES

Aitken, M.L., Limaye, A., Pottinger, P., Whimbey, E., Goss, C.H., Tonelli, M.R., et al., 2012. Respiratory outbreak of *Mycobacterium abscessus* subspecies massiliense in a lung transplant and cystic fibrosis center. Am. J. Respir. Care Med. 185 (2), 231–232.

Andréjak, C., Nielsen, R., Thomsen, V.Ø., Duhaut, P., Sørensen, H.T., Thomsen, R.W., et al., 2013. Chronic respiratory disease, inhaled corticosteroids and risk of non-tuberculous mycobacteriosis. Thorax. 68 (3), 256–262.

Azzali, A., Montagnani, C., Simonetti, M.T., Spinelli, G., de Martino, M., Galli, L., et al., 2017. First case of *Mycobacterium marseillense* lymphadenitis in a child. Ital. J. Pediatr. 43 (1), 92.

Blanc, P., Dutronc, H., Peuchant, O., Dauchy, F.A., Cazanave, C., Neau, D., et al., 2016. Nontuberculous mycobacterial infections in a French hospital: a 12-Year retrospective study. PLoS One 11 (12), e0168290.

Browne, S.K., 2014. Anticytokine autoantibody-associated immunodeficiency. Annu. Rev. Immunol. 32, 635–657.

Bryant, J.M., Grogono, D.M., Greaves, D., Foweraker, J., Roddick, I., Inns, T., et al., 2013. Whole-genome sequencing to identify transmission of *Mycobacterium abscessus* between patients with cystic fibrosis: a retrospective cohort study. Lancet 381 (9877), 1551–1560.

Butler, W.R., Guthertz, L.S., 2001. Mycolic acid analysis by high performance liquid chromatography for identification of *Mycobacterium* species. Clin. Microbiol. Rev. 14, 704–726.

Ceyssens, P.J., Soetaert, K., Timke, M., Van den Bossche, A., Sparbier, K., De Cremer, K., et al., 2017. Matrix-assisted laser desorption ionization-time of flight mass spectrometry for combined species identification and drug sensitivity testing in mycobacteria. J. Clin. Microbiol. 55 (2), 624–634.

Chan, E.D., Iseman, M.D., 2010. Slender, older women appear to be more susceptible to nontuberculous mycobacterial lung disease. Gend. Med. 7 (1), 5–18.

Chang, J.T., Huang, Y.F., Lin, Y.T., Hsieh, K.S., Liu, Y.C., Chiu, L.H., et al., 2006. *Mycobacterium abscessus* cervical lymphadenitis: an immunocompetent child. Kaohsiung J. Med. Sci. 22 (8), 415–419.

Char, A., Hopkinson, N.S., Hansell, D.M., Nicholson, A.G., Shaw, E.C., Clark, S.J., et al., 2014. Evidence of mycobacterial disease in COPD patients with lung volume reduction surgery; the importance of histological assessment of specimens: a cohort study. BMC Pulm. Med. 14 (1), 124.

Chien, J.Y., Lai, C.C., Sheng, W.H., Yu, C.J., Hsueh, P.R., et al., 2014. Pulmonary infection and colonization with nontuberculous mycobacteria, Taiwan, 2000–2012. Emerg. Infect. dis. 20 (8), 1382.

Esther Jr, C.R., Esserman, D.A., Gilligan, P., Kerr, A., Noone, P.G., et al., 2010. Chronic *Mycobacterium abscessus* infection and lung function decline in cystic fibrosis. J. Cyst. Fibros. 9, 117–123.

Falkinham III, J.O., 2009. Surrounded by mycobacteria: nontuberculous mycobacteria in the human environment. J. Appl. Microbiol. 107 (2), 356–367.

Falkinham III, J.O., 2011. Nontuberculous mycobacteria from household plumbing of patients with nontuberculous mycobacteria disease. Emerg. Infect. Dis. 17 (3), 419–424.

Fauroux, B., Delaisi, B., ClÉment, A., Saizou, C., Moissenet, D., Truffot-Pernot, C., et al., 1997. Mycobacterial lung disease in cystic fibrosis: a prospective study. Pediatr. Infect. Dis. J. 16 (4), 354–358.

Filipe-Santos, O., Bustamante, J., Haverkamp, M.H., Vinolo, E., Ku, C.-L., Puel, A., et al., 2006. X-linked susceptibility to mycobacteria is caused by mutations in NEMO impairing CD40-dependent IL-12 production. J. Exp. Med. 203, 1745–1759.

Garima, K., Varma-Basil, M., Pathak, R., Kumar, S.; Narang, A., Rawat, K.S., et al., 2012. Are we overlooking infections owing to non-tuberculous mycobacteria during routine conventional laboratory investigations? Int. J. Mycobacteriol. 1 (4), 207–211.

Griffith, D.E., Aksamit, T., Brown-Elliott, B.A., Catanzaro, A., Daley, C., Gordin, F., et al., 2007. Diseases Subcommittee; American Thoracic Society; Infectious Disease Society of

America. An official ATS/IDSA statement: diagnosis, treatment, and prevention of nontuberculous mycobacterial diseases. Am. J. Respir. Crit. Care. Med. 175 (4), 367–416.

Henkle, E., Winthrop, K.L., 2015. Nontuberculous mycobacteria infections in immunosuppressed hosts. Clin. Chest. Med. 36 (1), 91–99.

Heraud, D., Carr, R.D., McKee, J., Dehority, W., 2016. Nontuberculous mycobacterial adenitis outside of the head and neck region in children: a case report and systematic review of the literature. Int. J. Mycobacteriol. 5 (3), 351–353.

Horsburgh Jr, C.R., 1991. *Mycobacterium avium* complex infection in the acquired immunodeficiency syndrome. New Engl. J. Med. 324 (19), 1332–1338.

Hsu, A.P., Sampaio, E.P., Khan, J., Calvo, K.R., Lemieux, J.E., Patel, S.Y., et al., 2011. Mutations in GATA2 are associated with the autosomal dominant and sporadic monocytopenia and mycobacterial infection (MonoMAC) syndrome. Blood 118 (10), 2653–2655.

Huang, T.S., Lee, C.C., Tu, H.Z., Lee, S.S., 2018. Rapid identification of mycobacteria from positive MGIT broths of primary cultures by MALDI-TOF mass spectrometry. PLoS One 13 (2), e0192291.

Jorgensen, J.H.P.M., Carroll, K.C., Fnke, G., Landry, M.L., Richter, S.S., Warnock, D.W., 2015. Manual of Clinical Microbiology, 11th edition ASM Press, Washington, DC.

Kamada, M., Irahara, M., Maegawa, M., Ohmoto, Y., Takeji, T., Yasui, T., et al., 2001. Postmenopausal changes in serum cytokine levels and hormone replacement therapy. Am. J. Obstet. Gynecol. 184 (3), 309–314.

Khalid, M., Ali, S.A., 2017. *Mycobacterium abscessus* lymphadenitis in bone marrow transplant patient. J. Pak. Med. Assoc. 67 (10), 1615–1617.

Kim, R.D., Greenberg, D.E., Ehrmantraut, M.E., Guide, S.V., Ding, L., Shea, Y., et al., 2008. Pulmonary nontuberculous mycobacterial disease: prospective study of a distinct preexisting syndrome. Am. J. Respir. Crit. Care Med. 178 (10), 1066–1074.

Kim, C.J., Kim, U.J., Kim, H.B., Park, S.W., Oh, M.D., Park, K.H., et al., 2016. Vertebral osteomyelitis caused by non-tuberculous mycobacteria: predisposing conditions and clinical characteristics of six cases and a review of 63 cases in the literature. Infect. Dis. 48 (7), 509–516.

Lake, M.A., Ambrose, L.R., Lipman, M.C., Lowe, D.M., 2016. "Why me, why now?" Using clinical immunology and epidemiology to explain who gets nontuberculous mycobacterial infection. BMC Med. 14, 54.

Lipner, E.M., Knox, D., French, J., Rudman, J., Strong, M., Crooks, J.L., et al., 2017. A geospatial epidemiologic analysis of nontuberculous mycobacterial infection: an ecological study in Colorado. Ann. Am. Thorac. Soc. 14 (10), 1523–1532.

Maleki, M.R., Kafil, H.S., Harzandi, N., Moaddab, S.R., 2017. Identification of nontuberculous mycobacteria isolated from hospital water by sequence analysis of the hsp65 and 16S rRNA genes. J. Water Health 15 (5), 766–774.

Mansfield, K.G., Lackner, A.A., 1997. Simian immunodeficiency virus inoculated macaques acquire *Mycobacterium avium* from potable water during AIDS. J. Infect. Dis. 175, 184–187.

McShane, P.J., Glassroth, J., 2015. Pulmonary disease due to nontuberculous mycobacteria: current state and new insights. Chest 148, 1517–1527.

McNabb, A.D., Eisler, K., Adlie, M., Amos, M., Rodrigues, G., Stephens, W.A., et al., 2004. Assessment of partial sequencing of the 65-kilodalton heat shock protein gene (hsp65) for routine identification of mycobacterium species isolated from clinical sources. J. Clin. Microbiol. 42, 3000–3011.

Meager, A., Vincent, A., Newsom-Davis, J., Willcox, N., 1997. Spontaneous neutralising antibodies to interferon-a and interleukin-12 in thymoma-associated autoimmune disease. Lancet 350 (9091), 1596–1597.

Mirsaeidi, M., Sadikot, R.T., 2015. Gender susceptibility to mycobacterial infections in patients with non-CF bronchiectasis. Int. J. Mycobacteriol. 4 (2), 92–96.

Mishra, P.S., Narang, P., Narang, R., Goswami, B., Mendiratta, D.K., 2018. Spatio-temporal study of environmental nontuberculous mycobacteria isolated from Wardha district in Central India. Antonie Van Leeuwenhoek 111 (1), 73–87.

Norton, C.D., LeChevallier, M.W., 2000. A pilot study of bacteriological population changes through potable water treatment and distribution. Appl. Environ. Microbiol. 66, 268–276.

Okumura, M., Iwai, K., Ogata, H., Ueyama, M., Kubota, M., Aoki, M., et al., 2008. Clinical factors on cavitary and nodular bronchiectatic types in pulmonary *Mycobacterium avium* complex disease. Intern. Med. 47 (16), 1465–1472.

Pinitpuwadol, W., Sarunket, S., Boonsopon, S., Tesavibul, N., Choopong, P., 2018. Late-onset postoperative *Mycobacterium haemophilum* endophthalmitis masquerading as inflammatory uveitis: a case report. BMC Infect. Dis. 18 (1), 70.

Primm, T.P., Lucero, C.A., Falkinham III, J.O., 2004. Health impacts of environmental mycobacteria. Clin. Microbiol. Rev. 17 (1), 98–106.

Reich, J.M., Johnson, R.E., 1992. *Mycobacterium avium* complex pulmonary disease presenting as an isolated lingular or middle lobe pattern. The Lady Windermere Syndrome. Chest 101 (6), 1605–1609.

Rivron, M.J., Hughes, E.A., Sibert, J.R., Jenkins, P.A., 1979. Cervical lymphadenitis in childhood due to mycobacteria of the fortuitum group. Arch. Dis. Child 54 (4), 312–313.

Roux, A.L., Catherinot, E., Ripoll, F., Soismier, N., Macheras, E., Ravilly, S., et al., 2009. Multicenter study of prevalence of nontuberculous mycobacteria in patients with cystic fibrosis in France. J. Clin. Microbiol. 47 (12), 4124–4128.

Schweitzer, M.D., Salamo, O., Campos, M., Schraufnagel, D.E., Sadikot, R., Mirsaeidi, M., 2017. Body habitus in patients with and without bronchiectasis and non-tuberculous mycobacteria. PLoS One 12 (9), e0185095.

Shrivastava, K., Garima, K., Narang, A., Bhattacharyya, K., Vishnoi, E., Singh, R.K., et al., 2017. Rv1458c: a new diagnostic marker for identification of *Mycobacterium tuberculosis* complex in a novel duplex PCR assay. J. Med. Microbiol. 66 (3), 371–376.

Simons, S., van Ingen, J., Hsueh, P.R., Van Hung, N., Dekhuijzen, P.N., Boeree, M.J., et al., 2011. Nontuberculous mycobacteria in respiratory tract infections, eastern Asia. Emerg. Infect. Dis. 17 (3), 343–349.

Skolnik, K., Kirkpatrick, G., Quon, B.S., 2016. Nontuberculous mycobacteria in cystic fibrosis. Curr. Treat. Options Infect. Dis. 8 (4), 259–274.

Tettelin, H., Davidson, R.M., Agrawal, S., et al., 2014. High-level relatedness among *Mycobacterium abscessus* spp. massiliense strains from widely separated outbreaks. Emerg. Infect. Dis 20 (3), 364–371.

Thanachartwet, V., Desakorn, V., Duangrithi, D., Chunpongthong, P., Phojanamongkolkij, K., Jitruckthai, P., et al., 2014. Comparison of clinical and laboratory findings between those with pulmonary tuberculosis and those with nontuberculous mycobacterial lung disease. Southeast Asian J. Trop. Med. Public Health 45 (1), 85.

Tille, P.M., 2014. Bailey and Scotts Diagnostic Microbiology, 13th Edition Elsevier Mosby, St. Louis, MO.

Tortoli, E., 2003. Impact of genotypic studies on mycobacterial taxonomy: the new mycobacteria of the 1990's. Clin. Microbiol. Rev. 2, 319–354.

Tortoli, E., 2012. Epidemiology of cervico-facial pediatric lymphadenitis as a result of nontuberculous mycobacteria. Int. J. Mycobacteriol. 1 (4), 165–169.

Tortoli, E., 2014. Microbiological features and clinical relevance of new species of the genus *Mycobacterium*. Clin. Microbiol. Rev. 27 (4), 727–752.

Turvey, S.L., Tyrrell, G.J., Hernandez, C., Kabbani, D., Doucette, K., Cervera, C., 2017. *Mycobacterium branderi* infection: case report and literature review of an unusual and difficult-to-treat non-tuberculous mycobacterium. Int. J. Infect. Dis. 58, 65–67.

Viviani, L., Harrison, M.J., Zolin, A., Haworth, C.S., Floto, R.A., 2016. Epidemiology of nontuberculous mycobacteria (NTM) amongst individuals with cystic fibrosis (CF). J. Cyst. Fibros. 15 (5), 619–623.

von Reyn, C.F., Maslow, J.N., Barber, T.W., Falkinham III, J.O., Arbeit, R.D., 1994. Persistent colonisation of potable water as a source of *Mycobacterium avium* infection in AIDS. Lancet 343, 1137–1141.

Wallace Jr, R.J., Zhang, Y., Brown, B.A., Dawson, D., Murphy, D.T., et al., 1998. Polyclonal *Mycobacterium avium* complex infections in patients with nodular bronchiectasis. Am. J. Respir. Crit. Care Med. 158, 1235–1244.

Winthrop, K.L., McNelley, E., Kendall, B., Marshall-Olson, A., Morris, C., Cassidy, M., et al., 2010. Pulmonary nontuberculous mycobacterial disease prevalence and clinical features: an emerging public health disease. Am. J. Respir. Crit. Care Med. 182 (7), 977–982.

Wu, T.S., Yang, C.H., Brown-Elliott, B.A., Chao, A.S., Leu, H.S., Wu, T.L., et al., 2016. Postcesarean section wound infection caused by *Mycobacterium massiliense*. J. Microbiol. Immunol. Infect. 49 (6), 955–961.

Yano, K., Kazuki, K., Ikeda, M., Yoneda, M., 2014. Osteomyelitis and arthritis of the wrist caused by *Mycobacterium intracellulare* in an immunocompetent patient: a case report and literature review. Acta. Reumatol. Port. 39 (2).

Yeung, M.W., Khoo, E., Brode, S.K., Jamieson, F.B., Kamiya, H., Kwong, J.C., et al., 2016. Health-related quality of life, comorbidities and mortality in pulmonary nontuberculous mycobacterial infections: a systematic review. Respirology 21 (6), 1015–1025.

Zlojtro, M., Jankovic, M., Samarzija, M., Zmak, L., Jankovic, V.K., Obrovac, M., et al., 2015. Nosocomial pseudo-outbreak of *Mycobacterium gordonae* associated with a hospital's water supply contamination: a case series of 135 patients. J. Water Health 13 (1), 125–130.

Chapter 9

Nosocomial and Healthcare-Associated NTM Infections and Their Control

Sadia Shakoor[1], Maria Owais[2], Rumina Hasan[1] and Seema Irfan[1]
[1]*Department of Pathology and Laboratory Medicine, Aga Khan University, Karachi, Pakistan,*
[2]*Ziauddin Medical University, Karachi, Pakistan*

Salient features of NTM nosocomial infections are:

- NTM have specific environmental habitats in hospitals including water and water distribution systems, heater–cooler devices (HCDs), or air humidifiers.
- Resistance to disinfection and biofilm formation enable mycobacteria to flourish in these environments.
- Transmission occurs by aerosolization from aqueous habitats or contamination of critical devices.
- Direct person-to-person transmission has not been observed, but indirect *M. abscessus* transmission occurs among cystic fibrosis patients.
- Preventive measures focus on sterilizing critical equipment where possible or using mycobactericidal disinfectants, ultraviolet light (UV-C) or copper-ion generation in water systems, and filtration where applicable.
- High index of suspicion by hospital epidemiologists, microbiologists, and infectious disease physicians is critical to recognize hospital outbreaks or even sporadic iatrogenic infections.

Nontuberculous mycobacteria (NTM) are well-adapted nosocomial pathogens (Phillips and Von Reyn, 2001). Features that make NTM ideal causative agents of healthcare-associated (HCA) infections include their ubiquitous presence in water and soil, low requirement for nutrients, their resistance to a wide array of disinfectants, as well as their capacity to form and dwell in biofilms (Falkinham, 2009).

NTM transmission in the healthcare environment occurs through exposure of vulnerable sites and wounds with compromised barriers to contaminated materials or even through aerosols or fomites. The aerobiology and

Nontuberculous Mycobacteria (NTM). DOI: https://doi.org/10.1016/B978-0-12-814692-7.00009-7

hydrophobicity of NTM makes these organisms more adaptable to formation of droplets originating from the water surface of whirlpools and devices (Hruska and Kaevska, 2012). Transmission through such droplets to humans creates risk for pulmonary disease in indoor and hospital environments, but presence in dust particles in air does not pose a significant risk. Although both *M. tuberculosis* and NTM can be transmitted through the airborne route, direct transmission from patient-to-patient has not been demonstrated and, therefore, airborne isolation guidelines do not apply to patients with NTM pulmonary infections (Saiman et al., 2014). However, in view of emerging evidence of transmission within specialized populations (Bryant et al., 2016), infection prevention measures need to be contextual and tailored to the underlying illness, NTM species, and patient population being served.

Mycobacteria are more resistant to disinfectants and sterilants than other common organisms, but less resistant than prions and bacterial spores (Rutala and Weber, 2014). Therefore, sterilization practices eliminating bacterial spores also kill mycobacteria. However, disinfectants' ability to eliminate/suppress mycobacteria must be determined to avoid the risk of mycobacterial colonization and contamination of medical surfaces and equipment. Disinfectant action against mycobacteria is also relevant to water and water distribution systems from where mycobacteria can be aerosolized. Reducing mycobacterial loads in hospital water and plumbing systems and hydrotherapy pools can potentially prevent nosocomial infections.

EPIDEMIOLOGY OF HEALTHCARE-ASSOCIATED NONTUBERCULOUS MYCOBACTERIAL INFECTIONS

NTM infections in healthcare may be observed in a wide range of environments, with a very diverse epidemiology (Falkinham, 1996). Infections have been observed in dental clinics, physiotherapy and hydrotherapy suites, as well as postcosmetic surgery. Environmental sources of outbreaks often remain undetectable; however, independent studies of hospital environments may uncover possible reservoirs.

Box 9.1 presents the various heretofore definite sources of NTM in nosocomial environments and their possible routes of transmission, and risk factors in healthcare settings. These factors deviate from risks in the community/nonhealthcare settings in that immunodeficiency is not a prerequisite for NTM infection in the hospital, but may very well be an aggravating factor. However, hospitalized patients often already have compromised barriers and immune systems. Environmental exposure is, therefore, the primary determinant of NTM infection in nosocomial settings. Increasing popularity of medical tourism to avoid high healthcare costs of advanced resourced settings has also affected the epidemiology of HCA NTM infections as a high number of cosmetic surgery patients visit settings with poor infection control and higher environmental mycobacterial burden (Singh et al., 2016).

BOX 9.1 Nosocomial NTM Sources, Transmission Routes, and Risk Factors

Sources of nontuberculous mycobacteria in the healthcare environment
- Water
- Showerheads
- Heater–cooler devices
- Drinking water and Ice Machines
- Contaminated bronchoscopes
- Multidose medication vials
- Silicone implants
- Intraarticular/epidural injections
- Contaminated prosthetic devices

Transmission routes of nosocomial nontuberculous mycobacteria
- Direct inoculation
- Droplets/aerosolization from water, heater–coolers, or humidifiers

Patient risk factors for nosocomial nontuberculous Mmcobacterial infections
- Medical tourism
- Surgery/surgical wound
- Steroid use
- Prosthetic devices
- Indwelling intravascular lines catheters
- Cardiac surgery
- Acupuncture
- Rejuvenation surgery/fat grafting or liposuction

Infections may be sporadic or identified as part of an outbreak. A summary of common NTM-causing sporadic infections and outbreaks in healthcare settings, their natural ecology, possible identified sources, and associated infections is presented in Table 9.1. It is evident that of the ecological factors at play, aquatic nosocomial environments emerge as important reservoirs. Several preventive and mitigation strategies center on this fact.

In considering HCA infection due to NTM, it is important to consider two caveats of NTM isolation in culture as a marker: colonization/pseudo-infection and laboratory contamination which may lead to pseudo-outbreaks (Phillips and Von Reyn, 2001). Colonization and pseudo-infection also result from nosocomial reservoirs of NTM, but their detection and diagnosis are equally important clinically as they cause diagnostic confusion and increase healthcare costs.

Several aspects of NTM epidemiology in healthcare are unclear and warrant further investigation. Clinical significance of some NTM remains questionable. *M. xenopi* pulmonary infection has been reported in literature (Costrini et al., 1981), but it is unclear whether *M. xenopi* is a true pathogen due to its low virulence and pathogenic potential (Jiva et al., 1997).

TABLE 9.1 Nosocomial Reservoirs, Risk Factors, and Infections of Common Healthcare-Associated NTM

Ecological Niches	Nosocomial Reservoirs	Associated Risk Procedures	Infectious Manifestations	References
M. avium complex				
Natural water bodies, soil, hot tubs, swimming pools, animal feces	Recirculating/hot water systems (shower heads), potable water, reusable hemodialysis filters	Bronchoscopy, hemodialysis, transplantation, surgical trauma	Pneumonia, pneumonitis, disseminated infection	Phillips and Von Reyn (2001), Nishiuchi et al. (2017)
M. chimaera				
	Heater–cooler devices in open heart surgery, valve preserving solutions	Cardiothoracic surgery, cardiovascular implant surgery	Prosthetic Valve Endocarditis (PVE), disseminated infection (chorioretinitis, hepatitis, nephritis, discitis, wound infection)	Sommerstein et al. (2016), Tan et al. (2016), Kohler et al. (2015)
M. simiae				
Aquatic environments (possibly), animals	Tap water	Ingestion of, rinsing or showering with contaminated tap water	Pulmonary disease	Falkinham (1996), Conger et al. (2004)
M. xenopi[a]				
Rarely isolated from animals	Hot water generators, water taps, showerheads	Surgical discectomy	Spinal infections	Falkinham (1996), Astagneau et al. (2001)

M. fortuitum group				
Ground water, soil	Water distribution system, showerheads, contaminated ice machines	*Surgical*: Implant surgery, cosmetic surgery, surgical wound contamination, gastric banding, punch biopsy *Access*: use of indwelling catheters (pleural, peritoneal, central venous)	Postoperative wound infections, implant associated infections, bacteremia/sepsis, peritonitis, Postinjection abscess	Phillips and Von Reyn (2001), Jaubert et al. (2015), El Helou et al. (2013), Blair et al. (2017), Callen et al. (2011), Wright et al. (2014)
M. abscessus-chelonae[a]/*M. immunogenum*				
Water, potting soil	Tap water, water distribution systems, humidifiers, bioprosthetic valves, jet injector, contaminated gentian violet, contaminated ultrasound gel	*Surgical*: Liposuction, cosmetic procedures, rejuvenation surgery, bariatric surgery, valve replacement (predominantly bioprostheses), lung transplant, other thoracoabdominal surgery, laser assisted in situ keratomileusis (LASIK), ocular implant surgery, punch biopsy, knee arthroplasty, bronchoscopy, pulpectomy (dental) *Access*: high flux dialysis, use of indwelling catheters (peritoneal, central venous) *Others*: Steroid Injections, ultrasound	Pneumonia, bacteremia/sepsis, LASIK-related keratitis, postoperative wound infection, PVE, arteriovenous graft infections, postinjection abscess	Phillips and Von Reyn (2001), El Helou et al. (2013), Wright et al. (2014), Baker et al. (2017), John and Velotta (2005), Kheir et al. (2015), Bouchiat et al. (2015), Jung et al. (2015), Peralta (2016), Cheng et al. (2016)

(*Continued*)

TABLE 9.1 (Continued)

Ecological Niches	Nosocomial Reservoirs	Associated Risk Procedures	Infectious Manifestations	References
Others: *M. conceptionense, M. phlei, M. wolinskyi,* and *M. mucogenicum*				
Rivers, brook waters, lakes	Tap water, automatic faucets, possibly heater–cooler devices (*M. wolinskyi*)	Implant surgery, cardiac surgery, use of indwelling central venous catheters, ultrasound	Postoperative wound infections, implant infections, bacteremia/sepsis	Thibeaut et al. (2010), Karnam et al. (2011), Nagpal et al. (2014), Livni et al. (2008), Tagashira et al. (2015)

[a]*Pseudo-outbreaks have been associated with contaminated bronchoscopes.*

In many instances, NTM have been recovered from highly invasive devices which can potentially pose serious infection risk, but no identifiable infections have been reported in exposed patients. An example is *M. chimaera* associated with extracorporeal membrane oxygenation (ECMO), with the reservoir being ECMO-associated thermoregulatory devices, but no associated infections have been reported in patients undergoing ECMO (Trudzinski et al., 2016).

Another emerging concept is that of patient-to-patient transmission of *M. abscessus* in cystic fibrosis (CF) populations. There is no evidence of direct human-to-human transmission, but a high level of strain relatedness and clonality most likely stem from indirect transmission (Bryant et al., 2013). Near-identical *M. abscessus* isolates have been observed among CF patients visiting the same centers (Davidson et al., 2014). Prevention and control guidelines recommend segregation of infected patients from each other and other infected patients to minimize the risk of cross transmission (Saiman et al., 2014; Group TUCFTMaICW, 2013). Use of single, well-ventilated patient rooms with negative pressure in CF centers has been suggested (Group TUCFTMaICW, 2013); however, there is little evidence to support this recommendation, and effectiveness has also not yet been determined (Haworth et al., 2017).

CHALLENGES OF IDENTIFYING AND REPORTING HEALTHCARE-ASSOCIATED NONTUBERCULOUS MYCOBACTERIAL INFECTIONS AND OUTBREAKS

Appropriate management of HCA NTM infections requires timely diagnosis and assessment of transmission potential. However, NTM infections are underdiagnosed due to difficulties of culture, unavailability of culture-independent methods for direct detection from clinical samples, and insidious clinical presentation. The disease manifests months after initial exposure, making associations with nosocomial exposure difficult to determine. Incubation periods are often long even for HCA caused by rapidly growing mycobacteria (RGM) (Jung et al., 2015). Changes in taxonomy and emerging evolutionary phylogenetic linkages leading to species and subspecies reassignments mean that historical controls cannot reliably be used to determine associations (Falkinham, 1996).

Molecular strain typing is essential to outbreak confirmation and investigations. Conventional typing methods such as antibiograms, serotyping, and biotyping have an unacceptably low level of discrimination for genetic relatedness (Falkinham, 1996). Newer methods including variable number tandem repeat markers and 16S rRNA sequencing and single nucleotide polymorphism genotyping (Chen et al., 2017), plasmid typing, and restriction fragment length polymorphism (Falkinham, 1996), which also require technical expertise and method validation, but these can be performed in reference-level laboratories. However, since mycobacterial typing methods are only available in reference-level diagnostic centers, this aspect may be overlooked especially in low resource settings where NTM outbreaks are likely to occur due to lack of compliance with infection control measures.

Even after all measures have been taken to confirm and detect outbreaks, reporting may be suppressed to avoid liability (Jung et al., 2015; National Research Council US Committee on Achieving Sustainable Global Capacity for Surveillance and Response to Emerging Diseases of Zoonotic Origin et al., 2009). Iatrogenic cases may also be underreported due to lack of surveillance definitions, audit standards related to NTM infections, and lack of standardized control measures.

PREVENTIVE AND CONTROL MEASURES

The three common drivers of NTM HCA infections are: (1) nosocomial exposure through compromised barriers; (2) contaminated/colonized critical devices and instruments; and (3) presence of NTM in hospital potable water and water distribution systems. Prevention, therefore, centers on these drivers.

1. *Minimizing exposure*: Standard precautions and basic principles of infection prevention are essential to avoid exposure of compromised cutaneous barriers with water or aqueous solutions. Patients should be advised to avoid bathing and showering for 48 h after surgery (Kannaiyan et al., 2015). Use of transparent occlusive dressings around catheter insertion sites is also recommended to avoid contamination with water. Hydrotherapy tubs and pools should also be disinfected with high concentration of chlorine compounds (1000 parts per million of free chlorine) (Phillips and Von Reyn, 2001; Rutala and Weber, 2014) to overcome colonization and transmission potential. It is also important to avoid decorative water fountains in facilities housing or serving immunocompromised patient populations (Kanamori et al., 2016). More research is needed into optimal methods for skin antisepsis as chlorhexidine is not active and povidone-iodine, although mycobactericidal, is irritant and sometimes not acceptable as a first-line antiseptic (Cheng et al., 2018).
2. *Device disinfection and reprocessing*: FDA guidelines recommend that a high-level disinfectant should have mycobactericidal activity to achieve 6-$\log_{10}$ kill of an adequate reporter mycobacterial strain (Rutala and Weber, 2014; Steinhauer et al., 2010). Table 9.2 summarizes currently available high-level disinfectants active against mycobacteria. Quaternary ammonium compounds are poorly mycobactericidal and not used for critical or semicritical item disinfection (Chatterjee et al., 2016). Applications for use should be tailored to efficacy for intended use and compatibility with material of critical/semicritical items, while ensuring healthcare worker and environmental safety (Ling et al., 2018).

TABLE 9.2 Summary of Mycobactericidal High-Level Disinfectants for Critical and Semicritical Items

Disinfectant Active Ingredient	Concentration Recommended for Mycobactericidal Action	Exposure Time	Comment	References
Glutaraldehyde 2.4%–3.2% products	>2%	20 min at 20°C	Slow mycobactericidal activity; can be used for critical and semicritical devices, flexible endoscopes, endocavitary probes, transducers; not recommended for bronchoscopes *M. abscessus* subspecies massiliense is aldehyde resistant, some strains of *M. avium* and *M. gordonae* may be resistant	Rutala and Weber (2014), Du Rand et al. (2013), Fisher et al. (2012)
OPA (*o*-phthalaldehyde)	0.55%	12 min at 20°C		
Peracetic acid	0.2%	12 min at 50°C	Peracetic aid can be used in automated endoscope reprocessors (AERs) Combinations of peracetic acid and hydrogen peroxide are also available at mycobactericidal concentrations of either chemical	Phillips and Von Reyn (2001), Rutala and Weber (2014), Rutala and Weber (2008), Rutala et al. (2012)
Hydrogen peroxide	7% standard and >2% improved	10–25 min at 20°C		

Guidelines have been proposed for disinfection and reprocessing of endoscopes and bronchoscopes. Methods are available for manual and automated high-level disinfection of reusable endoscopes. Both guidelines recommend use of automated endoscope reprocessors (AERs), prescribe using manufacturer's recommendations for disinfectant compatibility, efficacy, exposure times, and monitoring and maintaining logs for each procedure and patient for traceability in the event of a contamination event (Du Rand et al., 2013; Petersen et al., 2017). Guidelines for bronchoscope use also recommend using disposable flexible bronchoscope accessories where possible, and advise against use of manual disinfection, aldehyde disinfectants, and use of tap water in AERs (Du Rand et al., 2013).

Facilities using HCDs should follow manufacturer guidelines for cleaning and disinfection of devices, use only water passed through a 0.22 μm filter in HCDs, prevent aerosolized water from entering the sterile operative field by directing the exhaust vent away from it. Moreover, a quality management program for device maintenance, cleaning and disinfection should be implemented in facilities employing HCDs (Food and Drug Administration. Nontuberculous Mycobacterium Infections Associated with Heater–Cooler Devices: FDA Safety Communication, 2015). Recommended frequency of disinfection for ECMO devices has been debated (Garvey et al., 2017), but since no infections have been reported to date, manufacturer-recommended monthly disinfection is feasible (Sommerstein et al., 2018).

3. *Source control and disinfection of reservoir*: The use of chloramine instead of chlorine has affected an increase in NTM colonization of potable water (Falkinham, 2016). NTM in hospital water can be reduced by use of short wavelength UV-C. In a recent study, significant reductions in NTM counts were observed suggesting its potential in use for water disinfection, however, *M. abscessus* count reductions were less than those for other bacteria (Yang et al., 2017). Point-of-use membrane filters and copper–silver ion generation in water distribution systems have also been effective against RGM (Williams et al., 2011; Kusnetsov et al., 2001).

There are no recommendations for routine surveillance of hospital water NTM counts. Water and biofilm sampling procedures are also not standardized. However, in view of emerging rile of NTM in nosocomial infections and new technologies to prevent NTM growth in hospital water, guidance on evaluation of such methods is needed.

A risk analysis in hospital environments has been suggested, but is difficult to undertake owing to the many different species with varying pathogenic potentials causing various infections, and lack of information on virulence determinants (Falkinham, 2013).

In summary, while NTM nosocomial and HCA infections are not uncommon, they are underreported. To institute timely and effective prevention measures against the vast array of NTM HCA infections, epidemiology of these infections needs to be elicited through appropriately designed studies. Heretofore well-known common reservoirs and sources of NTM in the hospital environment are water storage and distribution systems, and HCDs for cardiac surgery which should be subjected to stringent maintenance and quality control measures to minimize the risk of HCA infections.

REFERENCES

Astagneau, P., Desplaces, N., Vincent, V., Chicheportiche, V., Botherel, A.H., Maugat, S., et al., 2001. *Mycobacterium xenopi* spinal infections after discovertebral surgery: investigation and screening of a large outbreak. Lancet 358 (9283), 747–751.

Baker, A.W., Lewis, S.S., Alexander, B.D., Chen, L.F., Wallace Jr, R.J., Brown-Elliott, B.A., et al., 2017. Two-phase hospital-associated outbreak of *Mycobacterium abscessus*: investigation and mitigation. Clin. Infect. Dis. 64 (7), 902–911.

Blair, P., Moshgriz, M., Siegel, M., 2017. *Mycobacterium fortuitum* empyema associated with an indwelling pleural catheter: case report and review of the literature. J. Infect. Chemother. 23 (3), 177–179.

Bouchiat, C., Saison, J., Boisset, S., Flandrois, J.P., Issartel, B., Dauwalder, O., et al., 2015. Nontuberculous mycobacteria: an underestimated cause of bioprosthetic valve infective endocarditis. Open Forum Infect Dis 2 (2), ofv047. Available from: https://doi.org/10.1093/ofid/ofv047.

Bryant, J.M., Grogono, D.M., Greaves, D., Foweraker, J., Roddick, I., Inns, T., et al., 2013. Whole-genome sequencing to identify transmission of *Mycobacterium abscessus* between patients with cystic fibrosis: a retrospective cohort study. Lancet 381 (9877), 1551–1560.

Bryant, J.M., Grogono, D.M., Rodriguez-Rincon, D., Everall, I., Brown, K.P., Moreno, P., et al., 2016. Emergence and spread of a human-transmissible multidrug-resistant nontuberculous mycobacterium. Science 354 (6313), 751–757.

Callen, E.C., Kessler, T.L., 2011. *Mycobacterium fortuitum* infections associated with laparoscopic gastric banding. Obes. Surg. 21 (3), 404–406.

Chatterjee, S.S., Chumber, S.K., Khanduri, U., 2016. Commercial disinfectants during disinfection process validation: more failures than success. J. Clin. Diagn. Res. 10 (8), DM01.

Chen, K., Zhang, Y., Peng, Y., 2017. Variable-number tandem repeat markers for *Mycobacterium intracellulare* genotyping: comparison to the 16S rRNA gene sequencing. J. Infect. Dev. Countries 11 (02), 158–165.

Cheng, A., Sheng, W.H., Huang, Y.C., Sun, H.Y., Tsai, Y.T., Chen, M.L., et al., 2016. Prolonged postprocedural outbreak of *Mycobacterium massiliense* infections associated with ultrasound transmission gel. Clin. Microbiol. Infect. 22 (4), 382-e1.

Cheng, A., Sun, H.Y., Tsai, Y.T., Wu, U.I., Chuang, Y.C., Wang, J.T., et al., 2018. In vitro evaluation of povidone-iodine and chlorhexidine against outbreak and nonoutbreak strains of *Mycobacterium abscessus* using standard quantitative suspension and carrier testing. Antimicrob. Agents Chemother. 62 (1), e01364–17.

Conger, N.G., O'Connell, R.J., Laurel, V.L., Olivier, K.N., Graviss, E.A., Williams-Bouyer, N., et al., 2004. *Mycobacterium simiae* outbreak associated with a hospital water supply. Infect. Control. Hosp. Epidemiol. 25 (12), 1050–1055.

Costrini, A.M., Mahler, D.A., Gross, W.M., Hawkins, J.E., Yesner, R., D'Esopo, N.D., 1981. Clinical and roentgenographic features of nosocomial pulmonary disease due to *Mycobacterium xenopi*. Am. Rev. Respir. Dis. 123 (1), 104–109.

Davidson, R.M., Hasan, N.A., Reynolds, P.R., Totten, S., Garcia, B., Levin, A., et al., 2014. Genome sequencing of *Mycobacterium abscessus* isolates from patients in the United States and comparisons to globally diverse clinical strains. J. Clin. Microbiol. 52 (10), 3573–3582.

Du Rand, I.A., Blaikley, J., Booton, R., Chaudhuri, N., Gupta, V., Khalid, S., et al., 2013. British Thoracic Society guideline for diagnostic flexible bronchoscopy in adults: accredited by NICE. Thorax 68 (Suppl. 1), i1–i44.

El Helou, G., Viola, G.M., Hachem, R., Han, X.Y., Raad, I.I., 2013. Rapidly growing mycobacterial bloodstream infections. Lancet. Infect. Dis. 13 (2), 166–174.

Falkinham III, J.O., 1996. Epidemiology of infection by nontuberculous mycobacteria. Clin. Microbiol. Rev. 9 (2), 177.

Falkinham III, J.O., 2009. Surrounded by mycobacteria: nontuberculous mycobacteria in the human environment. J. Appl. Microbiol. 107 (2), 356–367.

Falkinham III, J.O., 2013. Ecology of nontuberculous mycobacteria—where do human infections come from? Semin. Respir. Crit. Care. Med. 34 (1), 95–102. Available from: https://doi.org/10.1055/s-0033-1333568.

Falkinham, J.O., 2016. Current epidemiologic trends of the nontuberculous mycobacteria (NTM). Curr. Environ. Health Rep. 3 (2), 161–167.

Food and Drug Administration. Nontuberculous Mycobacterium Infections Associated With Heater-Cooler Devices: FDA Safety Communication, 2015. US Food and Drug Administration. <http://wayback.archiveit.org/7993/20170722215713/; https://www.fda.gov/MedicalDevices/Safety/AlertsandNotices/ucm466963.htm>.

Fisher, C.W., Fiorello, A., Shaffer, D., Jackson, M., McDonnell, G.E., 2012. Aldehyde-resistant mycobacteria bacteria associated with the use of endoscope reprocessing systems. Am. J. Infect. Control 40 (9), 880–882.

Garvey, M.I., Phillips, N., Bradley, C.W., Holden, E., 2017. Decontamination of an extracorporeal membrane oxygenator contaminated with *Mycobacterium chimaera*. Infect. Control. Hosp. Epidemiol. 38 (10), 1244–1246.

Group TUCFTMaICW, 2013. *Mycobacterium abscessus* Suggestions for Infection Prevention and Control (Interim Guidance—October 2013). Cystic Fibrosis Trust.

Haworth, C.S., Banks, J., Capstick, T., Fisher, A.J., Gorsuch, T., Laurenson, I.F., et al., 2017. British Thoracic Society guidelines for the management of non-tuberculous mycobacterial pulmonary disease (NTM-PD). Thorax 72 (Suppl. 2), ii1–ii64.

Hruska, K., Kaevska, M., 2012. Mycobacteria in water, soil, plants and air: a review. Vet. Med. (Praha) 57 (12), 623–679.

Jaubert, J., Mougari, F., Picot, S., Boukerrou, M., Barau, G., Ahmed, S.A.A., et al., 2015. A case of postoperative breast infection by *Mycobacterium fortuitum* associated with the hospital water supply. Am. J. Infect. Control 43 (4), 406–408.

Jiva, T.M., Jacoby, H.M., Weymouth, L.A., Kaminski, D.A., Portmore, A.C., 1997. *Mycobacterium xenopi*: innocent bystander or emerging pathogen? Clin. Infect. Dis. 24 (2), 225–232.

John, T., Velotta, E., 2005. Nontuberculous (atypical) mycobacterial keratitis after LASIK: current status and clinical implications. Cornea 24 (3), 245–255.

Jung, S.Y., Kim, B.G., Kwon, D., Park, J.H., Youn, S.K., Jeon, S., et al., 2015. An outbreak of joint and cutaneous infections caused by non-tuberculous mycobacteria after corticosteroid injection. Int. J. Infect. Dis. 36, 62–69.

Kanamori, H., Weber, D.J., Rutala, W.A., 2016. Healthcare outbreaks associated with a water reservoir and infection prevention strategies. Clin. Infect. Dis. 62 (11), 1423–1435.

Kannaiyan, K., Ragunathan, L., Sakthivel, S., Sasidar, A.R., 2015. Surgical site infections due to rapidly growing mycobacteria in Puducherry, India. J. Clin. Diagn. Res. 9 (3), DC05.

Karnam, S., Alla, V.M., Kwon, J., Harbert, T., Sharma, A., Airey, K., et al., 2011. *Mycobacterium phlei*, a previously unreported cause of pacemaker infection: thinking outside the box in cardiac device infections. J. Cardiol. 18 (6), 687–690.

Kheir, W.J., Sheheitli, H., Abdul Fattah, M., Hamam, R.N., 2015. Nontuberculous mycobacterial ocular infections: a systematic review of the literature. Biomed. Res. Int. 2015.

Kohler, P., Kuster, S.P., Bloemberg, G., Schulthess, B., Frank, M., Tanner, F.C., et al., 2015. Healthcare-associated prosthetic heart valve, aortic vascular graft, and disseminated *Mycobacterium chimaera* infections subsequent to open heart surgery. Eur. Heart J. 36 (40), 2745–2753.

Kusnetsov, J., Iivanainen, E., Elomaa, N., Zacheus, O., Martikainen, P.J., 2001. Copper and silver ions more effective against legionellae than against mycobacteria in a hospital warm water system. Water Res. 35 (17), 4217–4225.

Nagpal, A., Wentink, J.E., Berbari, E.F., Aronhalt, K.C., Wright, A.J., Krageschmidt, D.A., et al., 2014. A cluster of *Mycobacterium wolinskyi* surgical site infections at an academic medical center. Infect. Control. Hosp. Epidemiol. 35 (9), 1169–1175.

Ling, M.L., Ching, P., Widitaputra, A., Stewart, A., Sirijindadirat, N., 2018. APSIC guidelines for disinfection and sterilization of instruments in health care facilities. Antimicrob. Resist. Infect. Control. 7 (1), 25.

Livni, G., Yaniv, I., Samra, Z., Kaufman, L., Solter, E., Ashkenazi, S., et al., 2008. Outbreak of *Mycobacterium mucogenicum* bacteraemia due to contaminated water supply in a paediatric haematology–oncology department. J. Hosp. Infect. 70 (3), 253–258.

National Research Council (US) Committee on Achieving Sustainable Global Capacity for Surveillance and Response to Emerging Diseases of Zoonotic Origin, Keusch, G.T., Pappaioanou, M., Gonzalez, M.C., et al., (Eds.), 2009. Sustaining Global Surveillance and Response to Emerging Zoonotic Diseases. National Academies Press, Washington, DC, Incentives for Disease Surveillance, Reporting, and Response. Available from: https://www.ncbi.nlm.nih.gov/books/NBK215309/.

Nishiuchi, Y., Iwamoto, T., Maruyama, F., 2017. Infection sources of a common non-tuberculous mycobacterial pathogen, *Mycobacterium avium* complex. Front. Med. 4, 27.

Peralta, G., 2016. Notes from the field: *Mycobacterium abscessus* infections among patients of a pediatric dentistry practice—Georgia, 2015. MMWR. Morb. Mortal. Wkly. Rep. 65, 355–356.

Petersen, B.T., Cohen, J., Hambrick, R.D., Buttar, N., Greenwald, D.A., Buscaglia, J.M., et al., 2017. Multisociety guideline on reprocessing flexible GI endoscopes: 2016 update. Gastrointest. Endosc. 85 (2), 282–294.

Phillips, M.S., Von Reyn, C.F., 2001. Nosocomial infections due to nontuberculous mycobacteria. Clin. Infect. Dis. 33 (8), 1363–1374.

Rutala, W.A., Weber, D.J., Healthcare Infection Control Practices Advisory Committee. Guideline for disinfection and sterilization in healthcare facilities, 2008. Available from <https://www.cdc.gov/infectioncontrol/pdf/guidelines/disinfection-guidelines.pdf>.

Rutala, W.A., Weber, D.J., 2014. Selection of the ideal disinfectant. Infect. Control. Hosp. Epidemiol. 35 (7), 855–865.

Rutala, W.A., Weber, D.J., 2014. Disinfection, sterilization, and control of hospital waste, Mandell, Douglas, and Bennett's Principles and Practice of Infectious Diseases, vol. 2. Elsevier Inc, pp. 3294–3309, e1. https:\\doi.org\10.1016\B978-1-4557-4801-3.00301-5.

Rutala, W.A., Gergen, M.F., Weber, D.J., 2012. Efficacy of improved hydrogen peroxide against important healthcare-associated pathogens. Infect. Control. Hosp. Epidemiol. 33 (11), 1159–1161.

Saiman, L., Siegel, J.D., LiPuma, J.J., Brown, R.F., Bryson, E.A., Chambers, M.J., et al., 2014. Infection prevention and control guideline for cystic fibrosis: 2013 update. Infect. Control. Hosp. Epidemiol. 35 (S1), s1–s67.

Singh, M., Dugdale, C.M., Solomon, I.H., Huang, A., Montgomery, M.W., Pomahac, B., et al., 2016. Rapid-growing mycobacteria infections in medical tourists: our experience and literature review. Aesthet. Surg. J. 36 (8), NP246–NP253.

Sommerstein, R., Merz, T.M., Jakob, S.M., Takala, J., Marschall, J., 2018. Decontamination of extracorporeal membrane oxygenator devices with an intensified disinfection protocol: how strict is too strict? Infect. Control. Hosp. Epidemiol. 39 (3), 366–367.

Sommerstein, R., Rüegg, C., Kohler, P., Bloemberg, G., Kuster, S.P., Sax, H., 2016. Transmission of *Mycobacterium chimaera* from heater–cooler units during cardiac surgery despite an ultraclean air ventilation system. Emerg. Infect. Dis. 22 (6), 1008.

Steinhauer, K., Eschenbacher, I., Radischat, N., Detsch, C., Niederweis, M., Goroncy-Bermes, P., 2010. Rapid evaluation of the mycobactericidal efficacy of disinfectants in the quantitative carrier test EN 14563 by using fluorescent *Mycobacterium terrae*. Appl. Environ. Microbiol. 76 (2), 546–554.

Tagashira, Y., Kozai, Y., Yamasa, H., Sakurada, M., Kashiyama, T., Honda, H., 2015. A cluster of central line–associated bloodstream infections due to rapidly growing nontuberculous mycobacteria in patients with hematologic disorders at a Japanese tertiary care center: an outbreak investigation and review of the literature. Infect. Control. Hosp. Epidemiol. 36 (1), 76–80.

Tan, N., Sampath, R., Abu Saleh, O.M., Tweet, M.S., Jevremovic, D., Alniemi, S., et al., 2016. Disseminated Mycobacterium chimaera infection after cardiothoracic surgery. Open Forum Infect Dis 3 (3), ofw131.

Thibeaut, S., Levy, P.Y., Pelletier, M.L., Drancourt, M., 2010. *Mycobacterium conceptionense* infection after breast implant surgery, France. Emerg. Infect. Dis. 16 (7), 1180.

Trudzinski, F.C., Schlotthauer, U., Kamp, A., Hennemann, K., Muellenbach, R.M., Reischl, U., et al., 2016. Clinical implications of *Mycobacterium chimaera* detection in thermoregulatory devices used for extracorporeal membrane oxygenation (ECMO), Germany, 2015 to 2016. Euro Surveillance 21 (46).

Williams, M.M., Chen, T.H., Keane, T., Toney, N., Toney, S., Armbruster, C.R., et al., 2011. Point-of-use membrane filtration and hyperchlorination to prevent patient exposure to rapidly growing mycobacteria in the potable water supply of a skilled nursing facility. Infect. Control. Hosp. Epidemiol. 32 (9), 837–844.

Wright, H.L., Thomson, R.M., Reid, A.B., Carter, R., Bartley, P.B., Newton, P., et al., 2014. Rapidly growing mycobacteria associated with laparoscopic gastric banding, Australia, 2005–2011. Emerg. Infect. Dis. 20 (10), 1612.

Yang, J.H., Wu, U.I., Tai, H.M., Sheng, W.H., 2017. Effectiveness of an ultraviolet-C disinfection system for reduction of healthcare-associated pathogens. J. Microbiol. Immunol. Infect. Sep 18. pii: S1684-1182(17)30200-1.

Chapter 10

Epidemiological Distribution of Nontuberculous Mycobacteria Using Geographical Information System

Parissa Farnia[1], Poopak Farnia[2], Jalaledin Ghanavi[1] and Ali Akbar Velayati[1]

[1]*Mycobacteriology Research Centre (MRC), National Research Institute of Tuberculosis and Lung Disease (NRITLD), Shahid Beheshti University of Medical Sciences, Tehran, Iran,*
[2]*Department of Biotechnology, School of Advanced Technologies in Medicine, Shahid Beheshti University of Medical Sciences, Tehran, Iran*

INTRODUCTION

The nontuberculous mycobacteria (NTM) include those *Mycobacterium* species that are not members of the *Mycobacterium tuberculosis* complex (Falkinham, 1996). They have been identified as early as 1885; but it has taken 60–70 years for these pathogens to be considered as playing a significant role in human infections. This is due to molecular advances and the emergence of HIV/AIDS (Falkinham, 1996; Benson and Ellner, 1993; Portaels, 1995; Devana et al., 2018; Tortoli, 2003; Maurya et al., 2017). In fact, between the early 1950s and 1980s there were increasing reports of NTM infection, although this number has increased considerably with the onset of the AIDS epidemic (Benson and Ellner, 1993). At first, the appearance of AIDS resulted in the recurrence of tuberculosis throughout the world, particularly in developing countries (Portaels, 1995) Subsequently, AIDS also led to the emergence of opportunistic infections (Portaels, 1995). The main species of NTM before the advent of HIV, was pulmonary, confined to cervical lymph nodes, skin, and in rare disseminated cases (Mortaz et al., 2015). The basis of clinical criteria for mycobacterial diseases is divided into three groups: (1) strict pathogens in humans and animals that are members of the *M. tuberculosis* complex; these strict pathogens are generally not found in the environment (Portaels, 1995); (2) consisting of pathogens in

Nontuberculous Mycobacteria (NTM). DOI: https://doi.org/10.1016/B978-0-12-814692-7.00010-3

humans and animals that are available everywhere in nature, also called opportunistic mycobacteria, for example, *M. malmoense* and *M. ulcerans*; and (3) normally sapro*phytic* species that are nonpathogenic or only exceptionally pathogenic, usually called NTM, paratuberculous, "anonymous" mycobacterial other than tuberculosis or "atypical" mycobacteria (Portaels, 1995). Generally, these species are aerobic, nonmotile organisms that have a rigid and thick cell walls. The thickness of NTM cell-walls function as a natural protective shield against disinfectants (Maurya et al., 2017; Halstrom et al., 2015; Griffith et al., 1993, 2007). These characteristics enable NTM to thrive in most natural resources surrounding humans (Maurya et al., 2017; Halstrom et al., 2015; Griffith et al., 1993, 2007). The reservoir of NTM is mainly environmental and the majority of these mycobacteria are naturally resistant to antimycobacterial drugs, making eradication unfeasible (Halstrom et al., 2015; Mirsaeidi et al., 2007). Among the most reported NTM, *M. avium*, *M. intracellular*, *M. kansasi*, *M. chelonae*, and *M. fortuitum* were the major isolates that cause clinical symptoms (Griffith et al., 1993, 2007; WHO, 2014; Velayati et al., 2014, 2015). However, the clinical picture of NTM has changed with the emergence of AIDS. About 25%–50% of patients with AIDS were infected with various kinds of NTM. According to the latest report released by WHO (2014) about 35.5 million people in the world are living with HIV/AIDS, one of the most vulnerable populations to NTM infection (Velayati et al., 2015; Weiss and Glassroth, 2012). The other competent individuals at risk are those receiving immunosuppressive therapy secondary to organ transplantation, cancer, and autoimmune diseases and those with diabetes mellitus. From 1980–2000, most reported NTM cases were mainly from developed countries, but advances in molecular technology could show a rise of NTM infection worldwide (Hoefsloot et al., 2013). A classical view of NTM was written by Wolinsky (Tsang et al., 1985), but the definition of the condition was updated by Wayne and Sramek (1992). Since then, several new NTM species have been characterized and published NTM species have reached more than 180 (Tortoli, 2003; Petrini, 2006). Based on new technologies, such as high-performance liquid chromatography (Tortoli and Bartoloni, 1996), as well as amplification of specific gene fragments by polymerase chain reaction techniques followed by restriction analysis have been described for the analysis of gene regions like 65KD and rRNA (Petrini, 2006; Tortoli and Bartoloni, 1996). Among the most important and specific approaches mainly applicable at reference level involves the amplification of 16S rRNA followed by sequencing. DNA fingerprinting was used to identify the subtypes/strains associated with outbreaks in hospitals and communities (Joao et al., 2014; Edirisinghe et al., 2014). Among fingerprinting techniques, pulse field gel electrophoresis and random amplified polymorphic DNA have helped to identify new species (Portaels, 1995; Griffith et al., 1993, 2007). Gene probes, which are commercially available, make the growth from solid slants/liquid cultures (e.g., BACTEC) more rapid and reliable. As a consequence, countries from Asia to Africa reported different species of *Mycobacteria* that cause infection in humans. Therefore, it has become vital to understand the treatment protocol and diagnostic methods for NTM (Velayati et al., 2014, 2015; Joao et al., 2014; Edirisinghe et al., 2014). However, in many countries, due to limited laboratory set-up and improper identification procedures, the issue of NTM is lacking (Velayati et al., 2014, 2015; Martin-Casabona et al., 2004). During 1991–96, the Working Group of the Bacteriology and Immunology Section of International Union Against Tuberculosis and Lung Diseases contacted 50 laboratories in different countries. They identified 36,099 patients with NTM from 14 countries (Martin-Casabona et al., 2004). The first five major isolates were *M. avium* complex, *M. gordonae*, *M. xenopi*, *M. kansasii*, and *M. fortuitum*. Additionally, they showed an increase from their previous data in 1976 and 1989 (Martin-Casabona et al., 2004). Today, with the availability of new and rapid communication systems, it is easier to collect this kind of data from different regions. In this review, after collecting the information from previous publications, the geographical information system (GIS) for mapping and/or visualizing of disease distribution was used. This could ultimately improve our understanding of NTM diversities and their spatial patterns. The value of GIS information depends on variables including its timelines, the context in which it is applied, and cost of collection, storage, manipulation, and presentation (Aghajani et al., 2017; Rajaei et al., 2018). Two types of information systems can be identified: transaction processing systems and decision support systems. Overall, comparison of the epidemiological distribution of NTM in environmental and clinical isolates by GIS helps the health authorities to analyzed and decided about diseases distribution in specific populations.

Geographical Information System on Clinical and Environmental Isolates of Nontuberculous Mycobacteria

Europe Geography

Europe can be divided into four major physical regions, running from north to south: The ten countries in *Eastern Europe* are Belarus, Bulgaria, Czech Republic, Hungary, Poland, Republic of Moldova, Romania, Russian Federation, Slovakia, and Ukraine. Poland and Finland, which border the Baltic Sea with countries like Estonia, Latvia, and Lithuania are the Baltic state. *Northern Europe* consists of the countries in the Scandinavia Peninsula; Norway, Sweden, Finland, Iceland, and the France Islands. The Aland Islands; Greenland and Denmark constitute Scandinavia. *Southern European* countries lying on the Balkan Peninsula are often referred to as the Balkan States and include Albania, Bosnia, Herzegovina, Bulgaria, Croatia, Greece, Macedonia, Romania, Slovenia and Yugoslavia. The Balkan Peninsula also houses a small region of Turkey that lies northwest of the Sea of Marmara. *Western Europe* consists of countries like Austria, Belgium, France, Germany, Liechtenstein, Luxembourg, Monaco, the Netherlands, and Switzerland.

Distribution of Nontuberculous Mycobacteria Based on Clinical and Environmental Specimens From Europe

The present report is based on data compiled from information published on the distribution of NTM from 24 countries in different regions of Europe between 1967 and 2016 (Tables 10.1 and 10.2). Tables 10.1 and 10.2 shows the number of NTM detected and at least three species with the highest frequency and type of sample in each country. In general, 55,976 and 4305 NTMs have been reported in clinical and environmental specimens from different regions of the

TABLE 10.1 Overview of Studies on Nontuberculous Mycobacteria From Clinical Specimens in Europe

Countries (Period)	Types of Samples	Common Species (*n*)	Total NTM
Austria (Hoefsloot et al., 2013) (2008)	Pulmonary samples	*M. gordonae* (25), *M. intracellulare* (25), *M. avium* (22)	117
Belgium (Martin-Casabona et al., 2004) (1991–96)	Clinical specimens	*M. xenopi* (418), *M. gordonae* (281), *M. avium* complex (140)	1180
Belgium (Hoefsloot et al., 2013) (2008)	Pulmonary samples	*M. xenopi* (112), *M. avium* (88), *M. intracellulare* (80)	440
Croatia (Jankovic et al., 2013) (2006–10)	Respiratory specimen	*M. gordonae* (509), *M. xenopi* (184), *M. fortuitum* (136)	1027
Croatia (Hoefsloot et al., 2013) (2008)	Pulmonary samples	*M. gordonae* (61), *M. xenopi* (33), *M. fortuitum* (31)	197
Croatia (Van der Werf et al., 2014) (2008–10)	Pulmonary and nonpulmonary	*M. gordonae* (377), *M. xenopi* (122), *M. fortuitum* (98)	848
Czech Republic (Martin-Casabona et al., 2004) (1991–96)	Clinical specimens	*M. gordonae* (857), *M. kansasii* (744), *M. fortuitum* (517)	2955
Denmark (Martin-Casabona et al., 2004) (1991–96)	Clinical specimens	*M. avium* complex (512), *M. fortuitum* (38), *M. xenopi* (28)	713
Denmark (Hermansen et al., 2017) (1991–2015)	Pulmonary and nonpulmonary	*M. avium* (1520), *M. gordonae* (527), *M. intracellulare* (233)	3462

(*Continued*)

TABLE 10.1 (Continued)

Countries (Period)	Types of Samples	Common Species (*n*)	Total NTM
Denmark (Hoefsloot et al., 2013) (2008)	Pulmonary samples	*M. avium* (55), *M. gordonae* (24), *M. intracellulare* (13)	135
Denmark (Van der Werf et al., 2014; Andréjak et al., 2010) (2011)	Pulmonary and nonpulmonary	*M. avium* (36), *M. intracellulare* (15), *M. malmoense* (12)	182
Estonia (Van der Werf et al., 2014) (2004–11)	Pulmonary and nonpulmonary	*M. avium* (308), *M. gordonae* (96), *M. fortuitum* (65), *M. intracellulare* (52), *M. chelonae* (20)	698
Finland (Martin-Casabona et al., 2004) (1991–96)	Clinical specimens	*M. avium* complex (163), *M. gordonae* (37), *M. fortuitum* (17)	251
Finland (Van der Werf et al., 2014) (1995–2011)	Pulmonary and nonpulmonary	*M. avium* (1568), *M. gordonae* (1286), *M. intracellulare* (636)	5778
Finland (Hoefsloot et al., 2013) (2008)	Pulmonary samples	*M. avium* (92), *M. gordonae* (50), *M. intracellulare* (40)	344
France (Martin-Casabona et al., 2004) (1991–96)	Clinical specimens	*M. avium* complex (1296), *M. xenopi* (682), *M. kansasii* (365)	3200
France (Hoefsloot et al., 2013) (2008)	Pulmonary samples	*M. gordonae* (175), *M. intracellulare* (135), *M. avium* (79)	597
French Hospital (Blanc et al., 2016) (2002–13)	Pulmonary and nonpulmonary	*M. avium* (54), *M. intracellulare* (34), *M. marinum* (23)	170
Germany (Martin-Casabona et al., 2004) (1991–96)	Clinical specimens	*M. gordonae* (492), *M. avium* complex (359), *M. xenopi* (181)	1390
Germany (Hoefsloot et al., 2013) (2008)	Pulmonary samples	*M. avium* (44), *M. intracellulare* (32), *M. gordonae* (29)	197
Germany (Van der Werf et al., 2014) (2011)	Pulmonary and nonpulmonary	*M. intracellulare* (272), *M. avium* (176), *M. gordonae* (176)	1068

(Continued)

TABLE 10.1 (Continued)

Countries (Period)	Types of Samples	Common Species (*n*)	Total NTM
Greece (Van der Werf et al., 2014) (2005–11)	Pulmonary and nonpulmonary	*M. avium* (101), *M. fortuitum* (66), *M. intracellulare* (49)	426
Greece (Hoefsloot et al., 2013) (2008)	Pulmonary samples	*M. avium* (13), *M. fortuitum* (13), *M. chelonae* (11)	63
Greece, Athens (Panagiotou et al., 2014) (2007–13)	Respiratory specimen	*M. gordonae* (17), *M. avium* (16), *M. fortuitum* (15)	122
Hungary (Hoefsloot et al., 2013) (2008)	Pulmonary samples	*M. xenopi* (503), *M. gordonae* (127), *M. avium* (102)	1021
Italy (Martin-Casabona et al., 2004) (1991–96)	Clinical specimens	*M. xenopi* (165), *M. avium* complex (154), *M. gordonae* (37)	550
Italy (Van der Werf et al., 2014) (2001–10)	Pulmonary and nonpulmonary	*M. xenopi* (687), *M. avium* (350), *M. gordonae* (253)	2498
Italy (Hoefsloot et al., 2013) (2008)	Pulmonary samples	*M. gordonae* (112), *M. xenopi* (102), *M. intracellulare* (69)	468
Italy, Tuscany (Rindi and Garzelli, 2015) (2004–14)	Clinical specimens	*M. avium* (61), *M. intracellulare* (21), *M. gordonae* (17)	147
Italy, Florence (Mencarini et al., 2017) (2011–12)	Sputum, bronchoalveolar lavage, bronchoalveolar aspirated, Pleural fluid	*M. xenopi* (219), *M. chimaera* (100), *M. avium* (78)	554
Luxembourg (Van der Werf et al., 2014) (2009–11)	Pulmonary and nonpulmonary	*M. gordonae* (26), *M. avium* (23), *M. xenopi* (13)	123
The Netherlands (Van der Werf et al., 2014) (2006–11)	Pulmonary and nonpulmonary	*M. avium* (1274), *M. gordonae* (623), *M. intracellulare* (359)	3996
The Netherlands (Hoefsloot et al., 2013) (2008)	Pulmonary samples	*M. avium* (152), *M. gordonae* (90), *M. chimaera* (44)	577
Norway (Hoefsloot et al., 2013) (2008)	Pulmonary samples	*M. avium* complex (76), *M. avium* (32), *M. gordonae* (31)	212

(*Continued*)

TABLE 10.1 (Continued)

Countries (Period)	Types of Samples	Common Species (*n*)	Total NTM
Poland (Hoefsloot et al., 2013) (2008)	Pulmonary samples	*M. kansasii* (17), *M. xenopi* (11), *M. avium* (9)	49
Portugal (Martin-Casabona et al., 2004) (1991–96)	Clinical specimens	*M. avium* complex (275), *M. gordonae* (226), *M. fortuitum* (105)	639
Portugal (Oliveira et al., 2017) (2002–12)	Clinical specimens	*M. avium* complex (355), *M. kansasii* (72), *M. gordonae* (41)	632
Portugal (Hoefsloot et al., 2013) (2008)	Pulmonary samples	*M. avium* complex (34), *M. gordonae* (21), *M. intracellulare* (18)	149
Slovakia (Hoefsloot et al., 2013) (2008)	Pulmonary samples	*M. kansasii* (16), *M. chelonae* (10), *M. gordonae* (5)	45
Slovenia (Van der Werf et al., 2014) (2000–10)	Pulmonary and nonpulmonary	*M. xenopi* (615), *M. gordonae* (365), *M. avium* (349)	2240
Spain (Hoefsloot et al., 2013) (2008)	Pulmonary samples	*M. avium* (199), *M. gordonae* (78), *M. avium* complex (77)	805
Spain, Catalonia (Miguel et al., 2018) (1994–2014)	Respiratory specimen	*M. kansasii* (194), *M. avium* complex (139), *M. xenopi* (96)	680
Sweden (Hoefsloot et al., 2013) (2008)	Pulmonary samples	*M. avium* (59), *M. intracellulare* (26), *M. abscessus* (14)	126
Switzerland (Martin-Casabona et al., 2004) (1991–96)	Clinical specimens	*M. avium* complex (597), *M. gordonae* (534), *M. xenopi* (323)	2341
United Kingdom (Martin-Casabona et al., 2004) (1991–96)	Clinical specimens	*M. avium* complex (371), *M. gordonae* (81), *M. fortuitum* (53)	875
United Kingdom (Moore et al., 2010) (1995–2006)	Pulmonary and nonpulmonary	*M. malmoense* (1510), *M. kansasii* (1380), *M. xenopi* (827)	11,039
United Kingdom-London (Hoefsloot et al., 2013) (2008)	Pulmonary samples	*M. fortuitum* (36), *M. avium* (34), *M. kansasii* (19)	180

TABLE 10.2 Overview of Studies on Nontuberculous Mycobacteria From Environmental Specimens in Europe

Country (Period)	Type of Sample	Most Common Species	Total NTM
Belgium (Sajduda et al., 2012) (2001–10)	Water	*M. chelonae* 15 (34.88%), *M. abscessus* 15 (34.88%), *M. kansasii* 13 (30.23%)	43
Czech Republic (Kubalek and Mysak, 1996) (1984–89)	Running water	*M. gordonae* 73 (54.08%), *M. lentiflavum* 47 (34.81%), *M. scrofulaceum* 4 (2.96%)	135
Czech Republic (Kubalek and Mysak, 1996) (1984–89)	Swab or scraping	*M. gordonae* 61 (43.27%), *M. flavescens* 44 (31.20%), *M. fortuitum* 2 (1.41%)	141
Czech Republic (Trcka et al., 2006) (2002–05)	Wild boar	*M. avium* complex 9 (9.68%), *Mycobacterium* sp. 65 (69.89%), *M. fortuitum* 8 (8.60%), *M. phlei* 2 (2.15%)	93
Czech Republic (Beran et al., 2006) (2003–04)	Farm facilities and peat	*M. avium* 12 (6.48%), *M. fortuitum* 6 (3.24%), *M. flavescens* 6 (3.24%)	185
Czech Republic (Beran et al., 2006) (2003–04)	Aquariums and breed tanks	*M. fortuitum* 28 (13.93%), *M. marinum* 16 (36.37%)	44
Czech Republic (Sebakova et al., 2008) (2006)	Water	*M. xenopi* 51 (59.31%), *M. gordonae* 17 (19.76%), M. kansasii 15 (17.44%)	86
Czech Republic (Slany et al., 2010) (2010)	Animal	*M. marinum* 9, *M. thermoresistable* 9 (28.58%), *M. xenopi* 4 (7.54%)	53
Czech Republic (Slany et al., 2010) (2010)	Environment	*M. peregrinum* 32 (17.59%), *M. gordonae* 28 (34.14%), *M. chelonae* 18 (21.95%)	82
Czech Republic (Makovcova et al., 2014) (2014)	Fish ponds	*M. montefiorense* 3 (18.75%), *M. fortuitum* 2 (12.55), *M. gordonae* 2 (12.5%)	16
Czech Republic (Makovcova et al., 2014) (2014)	Drinking water reservoirs	*M. triplex* 5 (17.85%), *M. lentiflavum* 5 (17.85%), *M. avium* complex 4 (14.28%)	28
Czech Republic (Makovcova et al., 2014) (2014)	Storage ponds	*M. peregrinum* 2 (33.34%), *M. intracellulare,* (33.34%), *M. gordonae* 1 (16.67%)	6

(*Continued*)

TABLE 10.2 (Continued)

Country (Period)	Type of Sample	Most Common Species	Total NTM
Czech Republic (Makovcova et al., 2014) (2014)	Experimental recirculation system	*M. peregrinum* 14 (41.18%), *M. gordonae* 4 (14.28%), *M. szulgai* 4 (14.28%)	28
Finland (Torvinen et al., 2004) (2000–04)	Water	*M. lentiflavum* 192 (58.89%), *M. tusciae* 22 (6.74%), *M. gordonae* 11 (3.37%)	326
Finland (Torvinen et al., 2014) (2012)	Water	*M. gordonae* 8 (2.69%), *M. chelonae* 7 (3.55%), *M. mucogenicum* 6 (3.04%)	197
France (Thorel et al., 2004) (2004)	Soil	*M. chelonae* 42 (56.75%), *M. malmoense* 12 (16.21%), *M. fortuitum* 5 (6.75%)	74
France (Thorel et al., 2004) (2004)	Sphagnum	*M. chelonae* 5 (33.34%), *M. gordonae* 2 (3.03%), *M. fortuitum* 1 (1.51%)	20
Germany (Fischeder et al., 1991) (1991)	Dirking water	*M. gordonae* 82 (95.35%), *M. kansasii* 42 (37.16%), *M. chelonae* 8 (7.07%)	113
Germany (Peters et al., 1995) (1991–92)	Water (hot)	*M. gordonae* 12 (57.15%), *M. gastri* 4 (8.33%), *M. chelonae* 1 (2.08%)	48
Germany (Peters et al., 1995) (1991–92)	Water (cold)	*M. gordonae* 21 (48.84%), *M. chelonae* 8 (11.42%), *M. gastri* 2 (2.85%)	70
Germany (Peters et al., 1995) (1991–92)	Taps water hospital	*M. gordonae* 5 (5.68%), *M. fortuitum* 5 (5.68%), *M. avium* 4 (4.54%)	88
Germany (Sajduda et al., 2012) (2001–10)	Water	*M. chelonae* 13 (21.66%), *M. abscessus* 13 (21.66%), *M. immunogenum* 11 (18.33%)	60
Germany (Lahiri, 2014) (2014)	Surface water	*M. gordonae* 20 (36.37%), *M. peregrinum* 8 (14.54%), *M. chelonae* 6 (10.90%)	55
Germany (Lahiri, 2014) (2014)	Drinking water	*M. gordonae* 35 (35%), *M. chelonae* 31 (31%), *M. kansasii* 8 (8%)	100
Hungary (Rónai et al., 2016) (2006–13)	Wild Animal	*M. nonchromogenicum* 26 (11.55%), *M. kansasii* 5 (2.22%), *M. intermedium* 6 (2.66%)	225

(*Continued*)

TABLE 10.2 (Continued)

Country (Period)	Type of Sample	Most Common Species	Total NTM
Islands (Honda et al., 2016) (2016)	Soil	*M. chimaera* 60(80%), *M. chelonae* 22 (16.17%), *M. porcinum* 16 (11.76%)	136
Italy (Zanoni et al., 2008) (2002–05)	Freshwater fish	*M. peregrinum* 6 (20%), *M. chelonae* 17 (56.66%), *M. abscessus* 9 (30%)	30
Italy (Zanoni et al., 2008) (2002–05)	Marine fish	*M. fortuitum* 4 (13.79%), *M. marinum*; 3 (37.5%), *M. chelonae* 4 (13.79%)	29
Italy (Galassi et al., 2003) (2003)	Water	*M. gordonae* 69(64.49%), *M. fortuitum* 21 (19.62%), *M. smegmatis* 8 (7.47%)	107
Italy (Ebani et al., 2012) (2004–06)	Feces samples of reptiles	*M. fortuitum* 31 (86.12%), *M. peregrinum* 4 (2.91%), *M. chelonae* 1 (0.72%)	137
Italy (Briancesco et al., 2014) (2013)	Water	*M. llatzerense* 3 (27.28%), *M. mucogenicum* 4 (14.28%), *M. terrae* 4 (14.28%)	28
London (McSwiggan and Collins, 1974) (1967–71)	Taps water	*M. kansasii* 27 (47.37%), *M. xenopi* 8 (14.03%), *M. Scoto-*Nonchromogens 17 (29.82%)	57
Netherlands (van Ingen et al., 2010) (2010)	Water	*M. salmoniphilum* 144 (34.45%), *M. chelonae* 32 (7.65%), *M. llatzerense* 83 (19.85%)	418
North Spain (Lecuona et al., 2016) (2016)	Water supplies	*M. fortuitum* 17 (26.57%), *M. mucogenicum* 2 (3.12%), *M. canariasense* 2 (3.12)%	64
Norway (Agdestein et al., 2014) (2014)	Pig	*M. avium* complex 159 (61.39%), *M. malmoense* 23 (8.88%), *M. avium* 32 (12.35%)	259
Norwegian (Zerihun et al., 2011) (2006–10)	Fish	*M. salmoniphilum* 8 (44.45%), *M. marinum* 4 (15.38%), *M. fortuitum* 2 (7.69%)	26
Paris (France) (Le Dantec et al., 2002) (2002)	Groundwater	*M. nonchromogenicum* 8 (19.52%), *M. gordonae* 4 (9.75%), *M. fortuitum* 3 (7.31%)	41

(*Continued*)

TABLE 10.2 (Continued)

Country (Period)	Type of Sample	Most Common Species	Total NTM
Paris (France) (Le Dantec et al., 2002) (2002)	Treated surface water	*M. gordonae* 16 (48.49%), *M. chelonae* 3 (3.57%), *M. fortuitum* 1 (1.19%)	84
Paris (France) (Le Dantec et al., 2002) (2002)	Ground or treated surface water	*M. gordonae* 10 (33.34%), *M. nonchromogenicum* 4 (4.93%), *M. chelonae* 3 (3.70%)	81
France (Käser et al., 2009) (2008)	Aquatic insects	*M. ulcerans* 41 (78.85%), *M. gordonae* 10 (19.23%), *M. chelonae* 45 (86.53%)	52
Poland (Beran et al., 2006) (2001–10)	Tap water	*M. chelonae* 9 (30%), *M. immunogenum* 2 (6.66%), *M. abscessus* 15 (50%)	30
Poland (Ledwon et al., 2008) (2004–08)	Parrot feces	*M. fortuitum* 36 (47.36%), *M. avium* complex 16 (21.05%), *M. neoaurum* 8 (10.52%)	76
Portugal (Santos et al., 2005) (2003)	Water	*M. gordonae* 24 (63.16%), *M. chelonae* 2 (5.26%), *M. kansasii* 2 (5.26%)	38
Slovakia (Fischer et al., 2001) (1993–95)	Cattle and pigs	*M. fortuitum* 4 (28.58%), *M. paratuberculosis, M. avium* complex 3 (21.42%)	14
Slovenia (Kušar et al., 2017) (2009–11)	Fish	*M. fortuitum* 56 (61.54%), *M. marinum* 9 (9.78%), *M. chelonae* 7 (7.60%)	92
Switzerland (Bono et al., 1995) (1994)	Pet birds, animal and environmental sources	*M. genavense* 5 (7.82%), *M. avium* complex 31 (44.28%), *M. fortuitum* 14 (20%)	70
United Kingdom (Matthews et al., 1976) (1975)	Sludge	*M. fortuitum* 13 (40.63%), *M. peregrinum* 6 (14.28%), *M. gordonae* 5 (11.90%)	42
United Kingdom (Pourahmad et al., 2009) (2009)	Water	*M. fortuitum* 32 (25.40%), *M. marinum* 18 (37.5%), *M. conceptionence* 17 (35.41%)	48

Europe. Among European countries, United Kingdom (12,094/55,976; 21.60%) reported major NTM from pulmonary and nonpulmonary specimens, followed by Finland (6373/55,976; 11.38%), Denmark (4492/55,976; 8.02%), Netherlands (4573/55,976; 8.16%) and France (4359/55,976; 7.78%). Also, the Czech Republic (897/4275; 20.98%) reported major NTM from environmental specimens, followed by Germany (534/4275; 12.49%), Finland (523/4275; 12.23%), and France (352/4275; 8.23%), and Italy (331/4275; 7.74%).

Overall, 99 NTM species were identified from clinical and environmental specimens. *Mycobacterium avium*, *Mycobacterium avium* complex, *Mycobacterium gordonae*, *Mycobacterium xenopi*, *Mycobacterium kansasii*, *Mycobacterium intracellulare*, and *Mycobacterium fortuitum* were identified as the most frequent species.

Clinical Nontuberculous Mycobacteria Reported From Northern Europe

The Northern region includes data collected from seven countries: Finland, Sweden, Estonia, Denmark, United Kingdom, Sweden, Norway, and Ireland. The geographical distribution of the northern part of Europe is shown in Fig. 10.1 by a higher frequency of species (Fig. 10.1). In general, in northern

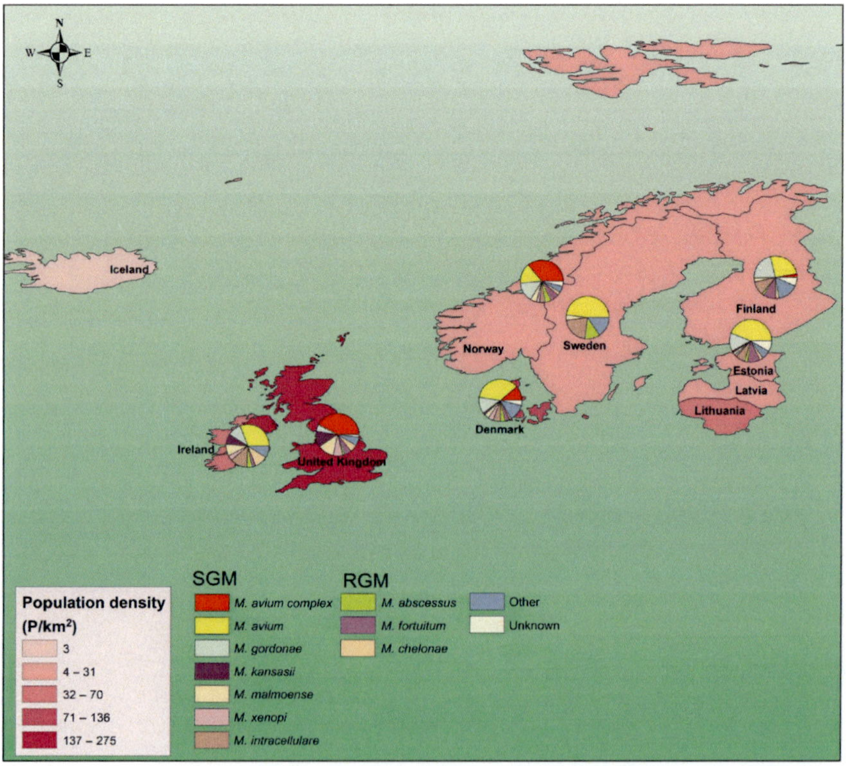

FIGURE 10.1 The distribution of nontuberculous mycobacteria species from clinical samples identified in Northern Europe. *Mycobacterium avium* complex (5859/24,073; 24.33%), *Mycobacterium avium* (3728/24,073; 15.48%), and *Mycobacterium gordonae* (2710/24,073; 11.25%) were the most frequent species in northern Europe (Table 10.3). Table 10.3 shows the data recorded for Northern Europe by country, period of study, and the frequency of different species.

TABLE 10.3 Species Distribution of Nontuberculous Mycobacteria From Clinical Samples in Northern Europe

Species	Country (Reference) (Period)							
	Finland (Hoefsloot et al., 2013; Martin-Casabona et al., 2004; Van der Werf et al., 2014) (1991–2011)	Estonia (Van der Werf et al., 2014) (2004–11)	Denmark (Hoefsloot et al., 2013; Martin-Casabona et al., 2004; Van der Werf et al., 2014; Hermansen et al., 2017; Andréjak et al., 2010) (1991–2015)	United Kingdom (Hoefsloot et al., 2013; Martin-Casabona et al., 2004; Moore et al., 2010) (1991–2008)	Sweden (Hoefsloot et al., 2013) (2008)	Norway (Hoefsloot et al., 2013) (2008)	Ireland (Hoefsloot et al., 2013) (2008)	Total Northern Europe
M. avium complex	163		517	5103		76		(5859/24,073; 24.33%)
M. avium	1660	308	1611	34	59	32	24	(3728/24,073; 15.48%)
M. gordonae	1373	96	586	615	1	31	8	(2710/24,073; 11.25%)
M. xenopi		12	179	884	1	5	3	(1084/24,073; 4.50%)
M. intracellulare	676	52	261	6	26	9	10	(1040/24,073; 4.32%)
M. fortuitum	569	65	147	607		13	2	(1403/24,073; 5.82%)
M. kansasii	57	11	25	1448		2	6	(1549/24,073; 6.43%)

(*Continued*)

TABLE 10.3 (Continued)

Species	Country (Reference) (Period)							
	Finland (Hoefsloot et al., 2013; Martin-Casabona et al., 2004; Van der Werf et al., 2014) (1991–2011)	Estonia (Van der Werf et al., 2014) (2004–11)	Denmark (Hoefsloot et al., 2013; Martin-Casabona et al., 2004; Van der Werf et al., 2014; Hermansen et al., 2017; Andréjak et al., 2010) (1991–2015)	United Kingdom (Hoefsloot et al., 2013; Martin-Casabona et al., 2004; Moore et al., 2010) (1991–2008)	Sweden (Hoefsloot et al., 2013) (2008)	Norway (Hoefsloot et al., 2013) (2008)	Ireland (Hoefsloot et al., 2013) (2008)	Total Northern Europe
M. chelonae	187	20	70	719	1	6	8	(1011/24,073; 4.19%)
M. abscessus	104	17	149	177	14	10	3	(474/24,073; 1.96%)
M. chelonae/M. abscessus				12				(12/24,073; 0.049%)
M. malmoense	194	14	156	1512	5	10	7	(1898/24,073; 7.88%)
M. lentiflavum	266	5	9					(280/24,073; 1.16%)
M. marinum	51	3	107	274				(435/24,073; 1.80%)
M. terrae	163		5	33	1			(202/24,073; 0.83%)

M. peregrinum	85	15	22	79		4		(205/24,073; 0.85%)
M. simiae	76		23	2	3	4		(108/24,073; 0.44%)
M. bohemicum	117		6					(123/24,073; 0.51%)
M. chimaera			4					(4/24,073; 0.016%)
M. interjectum	100	3	18		1	1	2	(125/24,073; 0.519%)
M. nonchromogenicum	45		5					(50/24,073; 0.20%)
M. arupense			1					(1/24,073; 0.004%)
M. celatum		1	81					(82/24,073; 0.34%)
M. mucogenicum			8	6				(14/24,073; 0.05%)
M. scrofulaceum		16	20	15	3			(54/24,073; 0.22%)
M. fortuitum/chelonae group				23				(23/24,073; 0.095%)
M. szulgai	1	3	28	30	2	1	2	(67/24,073; 0.24%)
M. phlei	1							(1/24,073; 0.004%)

(Continued)

TABLE 10.3 (Continued)

Species	Country (Reference) (Period)							
	Finland (Hoefsloot et al., 2013; Martin-Casabona et al., 2004; Van der Werf et al., 2014) (1991–2011)	Estonia (Van der Werf et al., 2014) (2004–11)	Denmark (Hoefsloot et al., 2013; Martin-Casabona et al., 2004; Van der Werf et al., 2014; Hermansen et al., 2017; Andréjak et al., 2010) (1991–2015)	United Kingdom (Hoefsloot et al., 2013; Martin-Casabona et al., 2004; Moore et al., 2010) (1991–2008)	Sweden (Hoefsloot et al., 2013) (2008)	Norway (Hoefsloot et al., 2013) (2008)	Ireland (Hoefsloot et al., 2013) (2008)	Total Northern Europe
M. genavense			11	1				(12/24,073; 0.049%)
M. haemophilum			1					(1/24,073; 0.004%)
M. triviale			1					(1/24,073; 0.004%)
M. triplex					1			(1/24,073; 0.004%)
M. arosiense			4					(4/24,073; 0.016%)
M. neoaurum	2		3					(5/24,073; 0.020%)
M. vaccae			2					(2/24,073; 0.008%)

M. parascrofulaceum			3					(3/24,073; 0.012%)
M. europaeum			1					(1/24,073; 0.004%)
M. flavescens			4	1				(5/24,073; 0.020%)
M. shimoidei	2		5		4			(11/24,073; 0.045%)
M. gilvum	3		2					(5/24,073; 0.020%)
M. goodii	1							(1/24,073; 0.004%)
M. aichiense	1							(1/24,073; 0.004%)
M. branderi			6		1			(7/24,073; 0.029%)
M. smegmatis			4					(4/24,073; 0.016%)
M. pulveris	2							(2/24,073; 0.008%)
M. duvalii	1							(1/24,073; 0.004%)
M. cosmeticum			2					(2/24,073; 0.008%)

(Continued)

TABLE 10.3 (Continued)

Species	Country (Reference) (Period)							
	Finland (Hoefsloot et al., 2013; Martin-Casabona et al., 2004; Van der Werf et al., 2014) (1991–2011)	Estonia (Van der Werf et al., 2014) (2004–11)	Denmark (Hoefsloot et al., 2013; Martin-Casabona et al., 2004; Van der Werf et al., 2014; Hermansen et al., 2017; Andréjak et al., 2010) (1991–2015)	United Kingdom (Hoefsloot et al., 2013; Martin-Casabona et al., 2004; Moore et al., 2010) (1991–2008)	Sweden (Hoefsloot et al., 2013) (2008)	Norway (Hoefsloot et al., 2013) (2008)	Ireland (Hoefsloot et al., 2013) (2008)	Total Northern Europe
M. heckeshornense			5					(5/24,073; 0.020%)
M. agri			5					(5/24,073; 0.020%)
M. frederiksbergense			1					(1/24,073; 0.004%)
M. heidelbergense			3					(3/24,073; 0.012%)
M. palustre			3					(3/24,073; 0.012%)
M. gastri			2					(2/24,073; 0.008%)
M. novocastrense			2					(2/24,073; 0.008%)

M. phocaicum			1					(1/24,073; 0.004%)
M. colombiense			3					(3/24,073; 0.012%)
M. conceptionense			1					(1/24,073; 0.004%)
M. brisbanense			1					(1/24,073; 0.004%)
M. chitae			2					(2/24,073; 0.008%)
M. septicum	1		3					(4/24,073; 0.016%)
M. ulcerans			2					(2/24,073; 0.008%)
Other	43		96	282	3		3	(427/24,073; 1.77%)
Unknown	429	57	175	231		8		(900/24,073; 3.73%)
Total NTM	6373	698	4492	12,094	126	212	78	(24,073/ 24,073; 100%)

Europe (24,073/55,976) 43%) NTM has been reported from clinical samples. The highest prevalence of NTM is in the United Kingdom with (12,094/24,073; 50.23%), followed by Finland (6373/24,073; 26.47%), Denmark (4492/24,073; 18.65%).

Distribution of Nontuberculous Mycobacteria Based on Environmental Specimens From Northern Europe

The frequency of NTM in environmental samples includes data collected from Iceland, Norway, England, and Finland. The geographical distribution of NTM based on environmental specimens is shown in Fig. 10.2 by the higher frequency of species (Fig. 10.2). In general, in northern Europe (1091/4275; 25.52%) NTM has been reported from environmental specimens. The highest prevalence of NTM is in Finland with (523/1091; 47.93%), followed by Norway (285/1091; 26.12%), United Kingdom (147/1091; 13.47%).

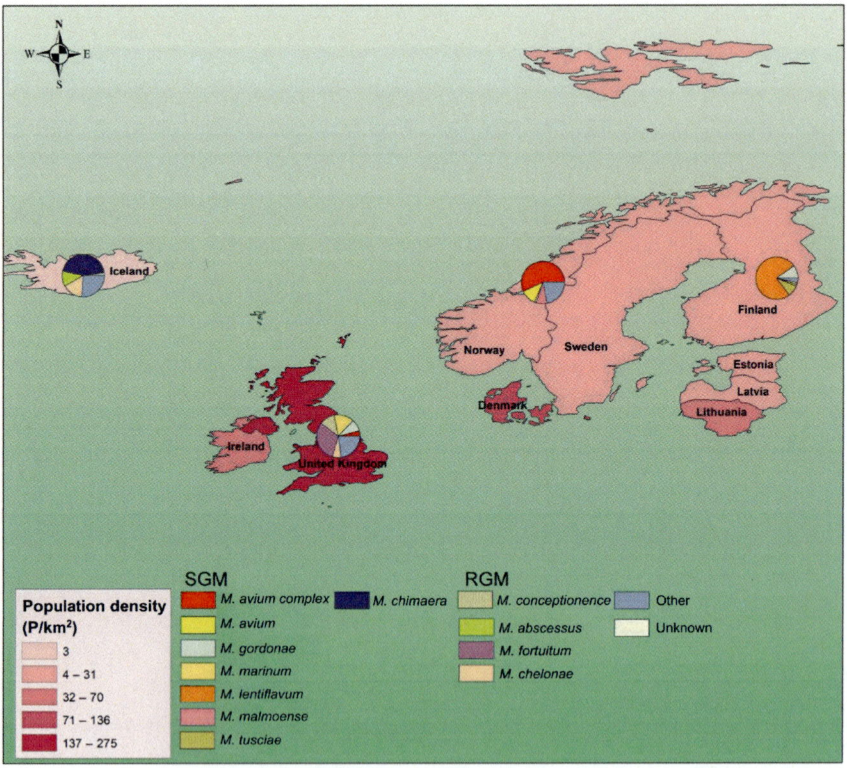

FIGURE 10.2 The distribution of nontuberculous mycobacteria species from environmental samples identified in northern Europe. *M. avium* complex (167/1091; 15.30%), *M. lentiflavum* (387/1091; 35.47%), *M. gordonae* (69/1091; 6.32%), and *M. chimaera* (60/1091; 5.49%) were the most frequent species among different regions in Northern Europe (Table 10.4).

TABLE 10.4 Species Distribution of Nontuberculous Mycobacteria From Environmental Samples in Northern Europe

Species	Country (Reference) (Period)				
	Islands (Honda et al., 2016) (2016)	**Norwegian (Agdestein et al., 2014; Zerihun et al., 2011) (2006–10, 2009, 2014)**	**United Kingdom (McSwiggan and Collins, 1974; Matthews et al., 1976; Pourahmad et al., 2009; Bullin et al., 1970) (1975, 2009)**	**Finland (Torvinen et al., 2004) (2000–04, 2012)**	**Total Northern Europe**
M. fortuitum		2 (0.71%)	45 (22.73%)	2 (0.39%)	(49/1091; 4.49%)
M. avium complex		159 (55.79%)	6 (4.08%)	2 (0.39%)	(167/1091; 15.30%)
M. gordonae	3 (2.20%)	2 (0.71%)	13 (8.84%)	51 (9.76%)	(69/1091; 6.32%)
M. intracellulare				2 (0.39%)	(2/1091; 0.18%)
M. hominissuis			2 (1.36%)		(2/1091; 0.18%)
M. lentiflavum				387 (73.99%)	(387/1091; 35.47%)
M. flavescens	2 (1.47%)				(2/1091; 0.18%)
M. gilvum				4 (0.77%)	(4/1091; 0.36%)
M. paraffinicum	2 (1.47%)				(2/1091; 0.18%)
M. intermedium			1 (0.68%)		(1/1091; 0.09%)
M. peregrinum			10 (6.80%)	2 (0.39%)	(12/1091; 1.09%)
M. kansasii		1 (0.35%)			(1/1091; 0.09%)
M. poriferae			1 (0.68%)		(1/1091; 0.09%)

(*Continued*)

TABLE 10.4 (Continued)

Species	Country (Reference) (Period)				
	Islands (Honda et al., 2016) (2016)	Norwegian (Agdestein et al., 2014; Zerihun et al., 2011) (2006–10, 2009, 2014)	United Kingdom (McSwiggan and Collins, 1974; Matthews et al., 1976; Pourahmad et al., 2009; Bullin et al., 1970) (1975, 2009)	Finland (Torvinen et al., 2004) (2000–04, 2012)	Total Northern Europe
M. scrofulaceum			4 (2.73%)		(4/1091; 0.36%)
M. marinum		4 (1.41%)	22(14.97%)		(26/1091; 2.38%)
M. senegalense			2 (1.36%)		(2/1091; 0.18%)
M. abscessus	16 (11.75%)			7	(23/1091; 2.10%)
M. conceptionence			17 (11.57%)		(17/1091; 1.55%)
M. chelonae	22 (16.18%)	2 (0.71%)	10 (6.80%)	7 (1.34%)	(41/1091; 3.75%)
M. gadium	3 (2.20%)				(3/1091; 0.27%)
M. alvei	2 (1.47%)				(2/1091; 0.18%)
M. porcinum	16 (11.75%)				(16/1091; 1.46%)
M. phocaicum	4 (2.94%)				(4/1091; 0.36%)
M. triviale		14 (4.92%)			(14/1091; 1.28%)
M. colombiense	2 (1.47%)				(2/1091; 0.18%)
M. frederiksbergense			7 (4.77%)	4 (0.77%)	(11/1091; 1%)

M. marseillense	4 (2.94%)				(4/1091; 0.36%)
M. chimaera	60 (44.12%)				(60/1091; 5.49%)
M. bohemicum		21 (7.37%)			(21/1091; 1.92%)
M. celatum		12 (4.21%)			(12/1091; 1.09%)
M. malmoense		23 (8.07%)			(23/1091; 2.10%)
M. palustre		4 (1.41%)			(4/1091; 0.36%)
M. avium		32 (11.23%)	3 (2.04%)		(35/1091; 3.20%)
M. salmoniphilum		8 (2.81%)	2 (1.37%)		(10/1091; 0.91%)
M. branderi		1 (0.35%)			(1/1091; 0.09%)
M. mucogenicum			2 (1.37%)	6 (1.15%)	(8/1091; 0.73%)
M. tusciae				44 (8.42%)	(44/1091; 4.03%)
Unidentified				5 (0.96%)	(5/1091; 0.45%)
Total NTM	136	285	147	523	(1091/1091; 100%)

Clinical Nontuberculous Mycobacteria Reported From Southern Europe

The Sothern region includes data collected from six countries: Portugal, Spain, Italy, Croatia, Slovenia, and Greece. The geographical distribution of the southern part of Europe is shown in Fig. 10.3 by the higher frequency of species. In general, in south Europe (12,045/55,976; 21.51%) NTM has been reported from clinical samples. The highest prevalence of NTM is in Italy with (4217/12,045; 35.01%), followed by Slovenia (2240/12,045; 18.59%), Croatia (2072/12,045; 17.20%).

Distribution of Nontuberculous Mycobacteria Based on Environmental Specimens From Southern Europe

The frequency of NTM in environmental samples includes data collected from Spain, Slovenia, Italy, Portugal, and Croatia. The geographical

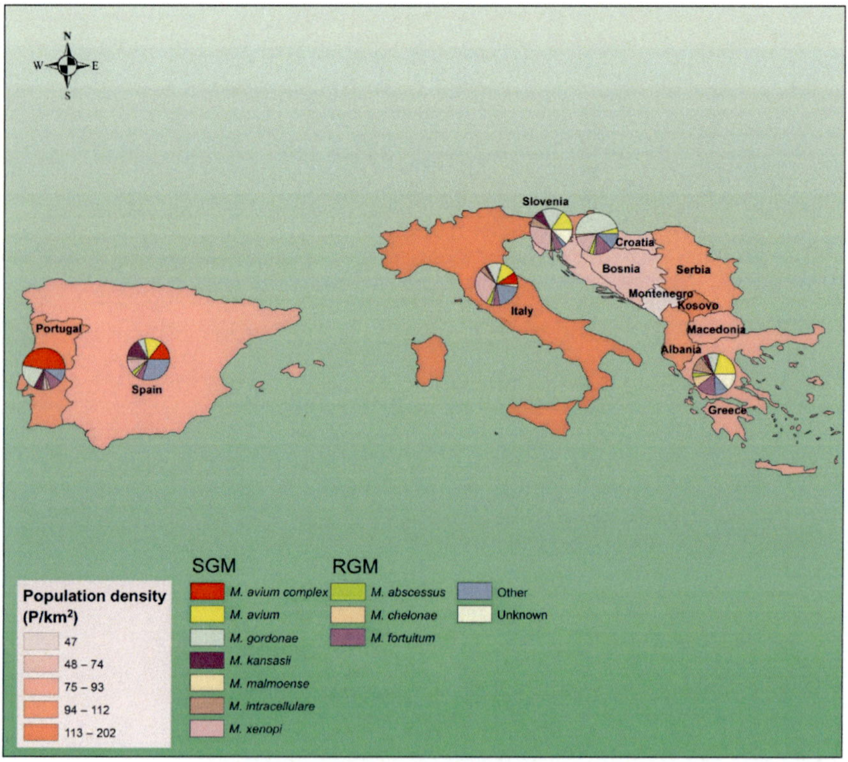

FIGURE 10.3 The distribution of nontuberculous mycobacteria species from clinical samples identified in southern Europe. *M. xenopi* (2355/12,045; 19.55%), *M. gordonae* (2203/12,045; 18.28%) and *M. avium* (1332/12,045; 11.05%), and *M. avium* complex (1052/12,045; 8.73%) were the most frequent species among different regions in Southern Europe (Table 10.5).

TABLE 10.5 Species Distribution of Nontuberculous Mycobacteria From Clinical Samples in Southern Europe

Species	Country (Reference) (Period)						
	Portugal (Hoefsloot et al., 2013; Martin-Casabona et al., 2004; Oliveira et al., 2017) (1991–2012)	Spain (Hoefsloot et al., 2013; Santin et al., 2018) (1991–2014)	Italy (Hoefsloot et al., 2013; Martin-Casabona et al., 2004; Van der Werf et al., 2014; Rindi and Garzelli, 2015; Mencarini et al., 2017) (1991–2014)	Croatia (Hoefsloot et al., 2013; Jankovic et al., 2013; Van der Werf et al., 2014) (2006–10)	Slovenia (Van der Werf et al., 2014) (2000–10)	Greece (Hoefsloot et al., 2013; Van der Werf et al., 2014; Panagiotou et al., 2014) (2005–11)	Total South Europe
M. avium complex	664	216	172				(1052/ 12,045; 8.73%)
M. avium	7	199	557	90	349	130	(1332/ 12,045; 11.05%)
M. gordonae	288	78	471	947	365	54	(2203/ 12,045; 18.28%)
M. xenopi	43	158	1187	339	615	13	(2355/ 12,045; 19.55%)
M. intracellulare	18	37	231	51	156	71	(564/ 12,045; 4.68%)
M. fortuitum mageritense						1	(1/12,045; 0.008%)

(Continued)

TABLE 10.5 (Continued)

Species	Country (Reference) (Period)						
	Portugal (Hoefsloot et al., 2013; Martin-Casabona et al., 2004; Oliveira et al., 2017) (1991–2012)	Spain (Hoefsloot et al., 2013; Santin et al., 2018) (1991–2014)	Italy (Hoefsloot et al., 2013; Martin-Casabona et al., 2004; Van der Werf et al., 2014; Rindi and Garzelli, 2015; Mencarini et al., 2017) (1991–2014)	Croatia (Hoefsloot et al., 2013; Jankovic et al., 2013; Van der Werf et al., 2014) (2006–10)	Slovenia (Van der Werf et al., 2014) (2000–10)	Greece (Hoefsloot et al., 2013; Van der Werf et al., 2014; Panagiotou et al., 2014) (2005–11)	Total South Europe
M. fortuitum	139	72	201	265	150	94	(921/ 12,045; 7.64%)
M. kansasii	83	211	83	19	181	23	(600/ 12,045; 4.98%)
M. chelonae	30	29	71	36	32	51	(249/ 12,045; 2.06%)
M. abscessus	3	42	151	65	24	21	(306/ 12,045; 2.54%)
M. malmoense			17		22	8	(47/12,045; 0.39%)
M. lentiflavum	9	21	53	9		13	(105/ 12,045; 0.87%)

M. marinum			17		6	5	(28/12,045; 0.23%)
M. terrae		4	17	157			(178/ 12,045; 1.47%)
M. peregrinum	11	2	34	9		26	(82/12,045; 0.68%)
M. peregrinum/alvei/ septicum						3	(3/12,045; 0.024%)
M. simiae	1	3	23				(27/12,045; 0.22%)
M. bohemicum			4				(4/12,045; 0.033%)
M. chimaera			152				(152/ 12,045; 1.26%)
M. interjectum	1	1	10	1			(13/12,045; 0.10%)
M. nonchromogenicum			42	22	21		(85/12,045; 0.70%)
M. arupense		1	92				(93/12,045; 0.77%)
M. celatum			17	4	21	9	(51/12,045; 0.42%)
M. mucogenicum	4	10	45	3		1	(63/12,045; 0.52%)

(*Continued*)

TABLE 10.5 (Continued)

Species	Country (Reference) (Period)						
	Portugal (Hoefsloot et al., 2013; Martin-Casabona et al., 2004; Oliveira et al., 2017) (1991–2012)	Spain (Hoefsloot et al., 2013; Santin et al., 2018) (1991–2014)	Italy (Hoefsloot et al., 2013; Martin-Casabona et al., 2004; Van der Werf et al., 2014; Rindi and Garzelli, 2015; Mencarini et al., 2017) (1991–2014)	Croatia (Hoefsloot et al., 2013; Jankovic et al., 2013; Van der Werf et al., 2014) (2006–10)	Slovenia (Van der Werf et al., 2014) (2000–10)	Greece (Hoefsloot et al., 2013; Van der Werf et al., 2014; Panagiotou et al., 2014) (2005–11)	Total South Europe
M. scrofulaceum	1	7	14	4	12	1	(39/12,045; 0.32%)
M. szulgai	2	2	12		6		(22/12,045; 0.18%)
M. kumamotonense		3	46				(49/12,045; 0.40%)
M. heraklionense			46				(46/12,045; 0.38%)
M. genavense			5				(5/12,045; 0.041%)
M. phlei			3	5			(8/12,045; 0.066%)
M. vulneris			1				(1/12,045; 0.008%)

M. haemophilum			1				(1/12,045; 0.008%)
M. triviale			2	20			(22/12,045; 0.18%)
M. arosiense			17				(17/12,045; 0.14%)
M. florentinum			6				(6/12,045; 0.049%)
M. nebraskense			1				(1/12,045; 0.008%)
M. neoaurum			5				(5/12,045; 0.041%)
M. vaccae			1	14			(15/12,045; 0.12%)
M. elephantis		1	11				(12/12,045; 0.09%)
M. insubricum			9				(9/12,045; 0.07%)
M. parascrofulaceum			5				(5/12,045; 0.041%)
M. sherrisii			14				(14/12,045; 0.11%)
M. alvei			3				(3/12,045; 0.024%)
M. porcinum			5				(5/12,045; 0.041%)

(Continued)

TABLE 10.5 (Continued)

Species	Country (Reference) (Period)						
	Portugal (Hoefsloot et al., 2013; Martin-Casabona et al., 2004; Oliveira et al., 2017) (1991–2012)	Spain (Hoefsloot et al., 2013; Santin et al., 2018) (1991–2014)	Italy (Hoefsloot et al., 2013; Martin-Casabona et al., 2004; Van der Werf et al., 2014; Rindi and Garzelli, 2015; Mencarini et al., 2017) (1991–2014)	Croatia (Hoefsloot et al., 2013; Jankovic et al., 2013; Van der Werf et al., 2014) (2006–10)	Slovenia (Van der Werf et al., 2014) (2000–10)	Greece (Hoefsloot et al., 2013; Van der Werf et al., 2014; Panagiotou et al., 2014) (2005–11)	Total South Europe
M. europaeum			4				(4/12,045; 0.033%)
M. flavescens			1	10			(11/12,045; 0.09%)
M. shimoidei			5				(5/12,045; 0.041%)
M. gilvum			6				(6/12,045; 0.049%)
M. longobardum			6				(6/12,045; 0.049%)
M. paraffinicum			4				(4/12,045; 0.033%)
M. triplex			7				(7/12,045; 0.058%)
M. goodie			1				(1/12,045; 0.008%)

M. holsaticum			2				(2/12,045; 0.016%)
M. smegmatis			4			2	(6/12,045; 0.049%)
M. thermoresistibile		1	5	1			(7/12,045; 0.058%)
M. cosmeticum			1				(1/12,045; 0.008%)
M. heckeshornense			1				(1/12,045; 0.008%)
M. brumae		1	2				(3/12,045; 0.024%)
M. frederiksbergense			3				(3/12,045; 0.024%)
M. mantenii			2				(2/12,045; 0.016%)
M. chlorophenolicum				1			(1/12,045; 0.008%)
M. noviomagense			1				(1/12,045; 0.008%)
M. palustre			3				(3/12,045; 0.024%)
M. setense			3				(3/12,045; 0.024%)
M. aichiense			2				(2/12,045; 0.016%)
M. engbaekii			2				(2/12,045; 0.016%)

(Continued)

TABLE 10.5 (Continued)

Species	Country (Reference) (Period)						
	Portugal (Hoefsloot et al., 2013; Martin-Casabona et al., 2004; Oliveira et al., 2017) (1991–2012)	Spain (Hoefsloot et al., 2013; Santin et al., 2018) (1991–2014)	Italy (Hoefsloot et al., 2013; Martin-Casabona et al., 2004; Van der Werf et al., 2014; Rindi and Garzelli, 2015; Mencarini et al., 2017) (1991–2014)	Croatia (Hoefsloot et al., 2013; Jankovic et al., 2013; Van der Werf et al., 2014) (2006–10)	Slovenia (Van der Werf et al., 2014) (2000–10)	Greece (Hoefsloot et al., 2013; Van der Werf et al., 2014; Panagiotou et al., 2014) (2005–11)	Total South Europe
M. bolletii			4				(4/12,045; 0.033%)
M. gastri			2				(2/12,045; 0.016%)
M. immunogenum		1	2				(3/12,045; 0.024%)
M. intermedium			2				(2/12,045; 0.016%)
M. llatzerense			2				(2/12,045; 0.016%)
M. mageritense		9					(9/12,045; 0.074%)
M. monacense			2				(2/12,045; 0.016%)
M. novocastrense			1				(1/12,045; 0.008%)
M. phocaicum			2				(2/12,045; 0.016%)

M. simulans			1				(1/12,045; 0.008%)
M. pulveris			2				(2/12,045; 0.016%)
M. rhodesiae			2				(2/12,045; 0.016%)
M. asiaticum		1					(1/12,045; 0.008%)
M. riyadhense			1				(1/12,045; 0.008%)
M. colombiense			1				(1/12,045; 0.008%)
M. conceptionense					1		(1/12,045; 0.008%)
M. duvalii			1				(1/12,045; 0.008%)
M. gadium			1				(1/12,045; 0.008%)
M. hiberniae			1				(1/12,045; 0.008%)
M. massiliense			3				(3/12,045; 0.024%)
M. parmense		1	1				(2/12,045; 0.016%)
M. poriferae			1				(1/12,045; 0.008%)
M. senegalense			1				(1/12,045; 0.008%)

(Continued)

TABLE 10.5 (Continued)

Species	Country (Reference) (Period)						
	Portugal (Hoefsloot et al., 2013; Martin-Casabona et al., 2004; Oliveira et al., 2017) (1991–2012)	Spain (Hoefsloot et al., 2013; Santin et al., 2018) (1991–2014)	Italy (Hoefsloot et al., 2013; Martin-Casabona et al., 2004; Van der Werf et al., 2014; Rindi and Garzelli, 2015; Mencarini et al., 2017) (1991–2014)	Croatia (Hoefsloot et al., 2013; Jankovic et al., 2013; Van der Werf et al., 2014) (2006–10)	Slovenia (Van der Werf et al., 2014) (2000–10)	Greece (Hoefsloot et al., 2013; Van der Werf et al., 2014; Panagiotou et al., 2014) (2005–11)	Total South Europe
M. senuense			1				(1/12,045; 0.008%)
M. tokaiense			1				(1/12,045; 0.008%)
M. tusciae			1				(1/12,045; 0.008%)
M. bouchedurhonense			1				(1/12,045; 0.008%)
Other	116	369	164				(649/ 12,045; 5.38%)
Unknown		2	97		279	85	(463/ 12,045; 3.84%)
Total NTM	1420	1485	4217	2072	2240	611	(12,045/ 12,045; 100%)

distribution of NTM based on environmental specimens is shown in Fig. 10.4 by the higher frequency of species. In general, in southern Europe (525/4275; 12.28%) NTM has been reported from environmental specimens. The highest prevalence of NTM is in Italy with (331/525; 63.04%), followed by Slovenia (92/525; 17.52%), Spain (64/525; 12.19%) and Portugal (38/525; 7.23%).

Clinical Nontuberculous Mycobacteria Reported From Eastern Europe

The eastern region includes data collected from four countries: Poland, Slovakia, Hungary, and Czech Republic. The geographical distribution of the

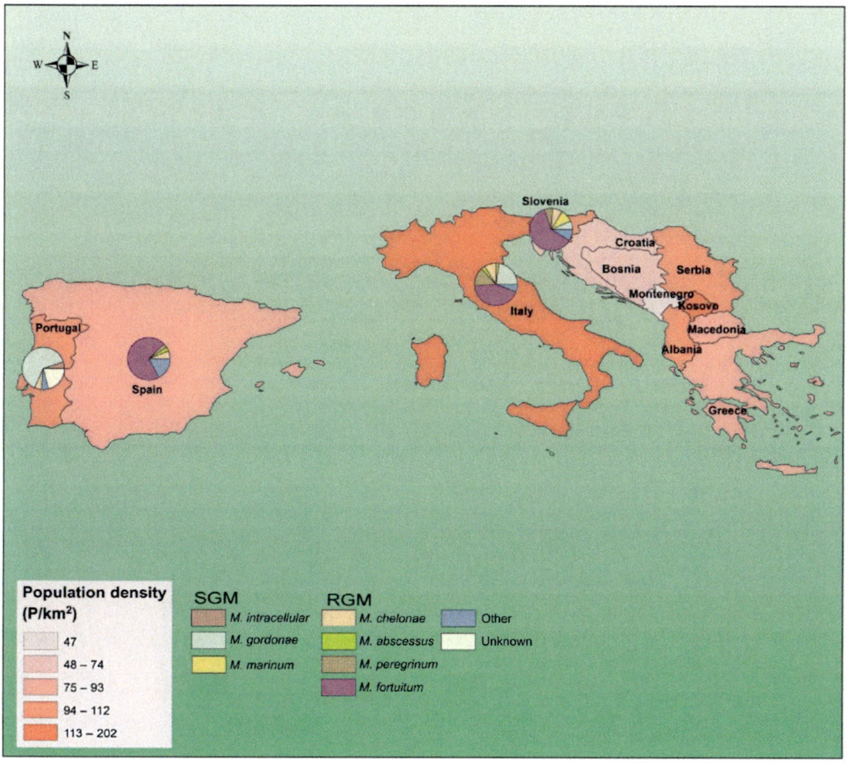

FIGURE 10.4 The distribution of nontuberculous mycobacteria species from environmental samples identified in Southern Europe. *M. fortuitum* (242/525; 46.09%), *M. gordonae* (106/525; 20.19%), *M. peregrinum* (47/525; 8.95%), and *M. chelonae* (42/525; 8%) were the most frequent species among different regions in Southern Europe (Table 10.6).

TABLE 10.6 Species Distribution of Nontuberculous Mycobacteria From Environmental Samples in Southern Europe

Kind of Species	Country (Reference) (Period)				
	Spain (Lecuona et al., 2016) (2013–16)	Slovenia (Kušar et al., 2017; Pavlik et al., 2005) (2000–11)	Italy (Zanoni et al., 2008; Galassi et al., 2003; Ebani et al., 2012; Briancesco et al., 2014) (2002–06, 2013)	Portugal (Santos et al., 2005) (2003, 2005–07)	Total/Southern Europe
M. fortuitum	45 (70.32%)	56 (60.87%)	141 (42.60%)		(242/555; 43.60%)
M. chelonae	4 (6.25%)	7 (7.61%)	29 (8.76%)	2 (5.27%)	(42/555; 7.56%)
M. mucogenicum	2 (3.13%)		1 (0.30%)		(3/555; 0.54%)
M. gordonae		6 (6.52%)	76 (22.96%)	24 (63.16%)	(106/555; 19.09%)
M. intracellulare				2 (5.27%)	(2/555; 0.36%)
M. arupense		3 (3.26%)			(3/555; 0.54%)
M. nonchromogenicum			5 (1.51%)		(5/555; 0.90%)
M. peregrinum			47 (14.20%)		(47/555; 8.46%)
M. kansasii		2 (2.17%)		2 (5.27%)	(4/555; 0.72%)
M. scrofulaceum			3 (0.90%)		(3/555; 0.54%)
M. marinum		9 (9.79%)	8 (2.42%)		(17/555; 3.06%)
M. abscessus	2 (3.13%)		9 (2.72%)		(11/555; 1.98%)
M. porcinum	1 (1.57%)				(1/555; 0.18%)
M. conceptionence	2 (3.13%)				(2/555; 0.36%)

M. canariasense	4 (6.25%)				(4/555; 0.72%)
M. peregrinum	2 (3.13%)	6 (6.52%)			(8/555; 1.44%)
M. septicum	1 (1.57%)				(1/555; 0.18%)
M. phocaicum	1 (1.57%)				(1/555; 0.18%)
M. africanum		1 (1.09%)			(2/555; 0.36%)
M. smegmatis			8 (2.42%)		(8/555; 1.44%)
M. setense		1 (1.09%)			(1/555; 0.18%)
Other		1 (1.09%)	1 (0.30%)		(31/555; 5.58%)
Unidentified			3 (0.90%)	8 (21.06%)	(11/555; 1.98%)
Total NTM	64	92	331	38	(555/555; 100%)

Eastern part of Europe is shown in Fig. 10.5 by a higher frequency of species. In general, in Eastern Europe (4070/55,976; 7.27%) NTM has been reported from clinical samples. The highest prevalence of NTM is in Czech Republic with (2955/4070; 72.60%), followed by Hungary (1021/4070; 25.08%), and Poland (49/4070; 1.20%).

Distribution of Nontuberculous Mycobacteria Based on Environmental Specimens From Eastern Europe

The frequency of NTM in environmental samples includes data collected from Hungary, Slovakia, Czech Republic and Poland. The geographical

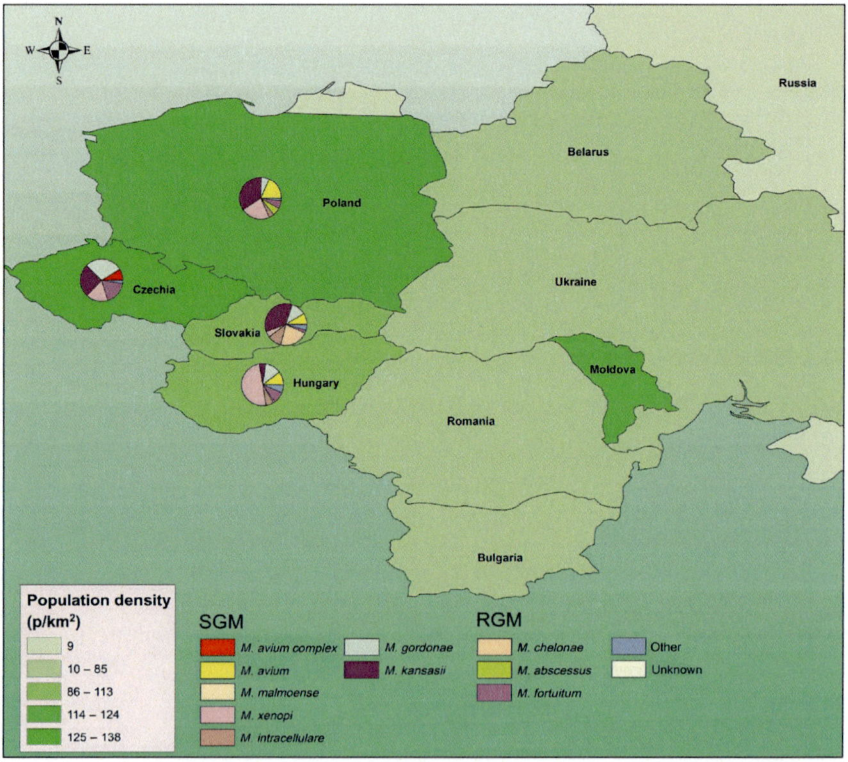

FIGURE 10.5 The distribution of nontuberculous mycobacteria species from clinical samples identified in Eastern Europe. *M. xenopi* (2355/12,045; 19.55%), *M. gordonae* (2203/12,045; 18.28%), *M. avium* (1332/12,045; 11.05%), and *M. avium* complex (1052/12,045; 8.73%) were the most frequent species among different regions in Southern Europe (Table 10.7).

TABLE 10.7 Species Distribution of Nontuberculous Mycobacteria From Clinical Samples in Eastern Europe

Species	Country (Reference) (Period)				
	Poland (Hoefsloot et al., 2013) (2008)	Slovakia (Hoefsloot et al., 2013) (2008)	Hungary (Hoefsloot et al., 2013) (2008)	Czech Republic (Martin-Casabona et al., 2004) (1991–96)	Total Eastern Europe
M. avium complex				254	(254/4070; 6.24%)
M. avium	9	4	102		(115/4070; 2.82%)
M. gordonae	3	5	127	857	(992/4070; 24.37%)
M. xenopi	11	2	503	505	(1021/4070; 25.8%)
M. intracellulare	2	5	62		(69/4070; 1.69%)
M. fortuitum	3	1	90	517	(611/4070; 15.01%)
M. kansasii	17	16	52	744	(829/4070; 20.36%)
M. chelonae		10	12		(22/4070; 0.54%)
M. abscessus	3		7		(10/4070; 0.24%)
M. malmoense			3		(3/4070; 0.073%)

(*Continued*)

TABLE 10.7 (Continued)

Species	Country (Reference) (Period)				
	Poland (Hoefsloot et al., 2013) (2008)	Slovakia (Hoefsloot et al., 2013) (2008)	Hungary (Hoefsloot et al., 2013) (2008)	Czech Republic (Martin-Casabona et al., 2004) (1991–96)	Total Eastern Europe
M. peregrinum		2	19		(21/4070; 0.51%)
M. simiae			11		(11/4070; 0.27%)
M. interjectum			1		(1/4070; 0.024%)
M. scrofulaceum			1		(1/4070; 0.024%)
M. aurum			1		(1/4070; 0.024%)
M. flavescens	1		1		(2/4070; 0.049%)
M. intermedium			2		(2/4070; 0.049%)
Other			27	78	(105/4070; 2.57%)
Total NTM	49	45	1021	2955	(4070/4070; 100%)

distribution of NTM based on environmental specimens is shown in Fig. 10.6 by a higher frequency of species. In general, in Eastern Europe (1242/4275; 29.05%) NTM has been reported from environmental specimens. The highest prevalence of NTM is in Czech Republic with (897/1242; 72.22%), followed by Hungary (225/1242; 18.11%), Poland (106/1242; 8.53%), and Slovakia (14/1242; 1.12%).

Clinical Nontuberculous Mycobacteria Reported From Western Europe

The western region includes data collected from seven countries: The Netherlands, France, Germany, Luxemburg, Belgium, Austria, and

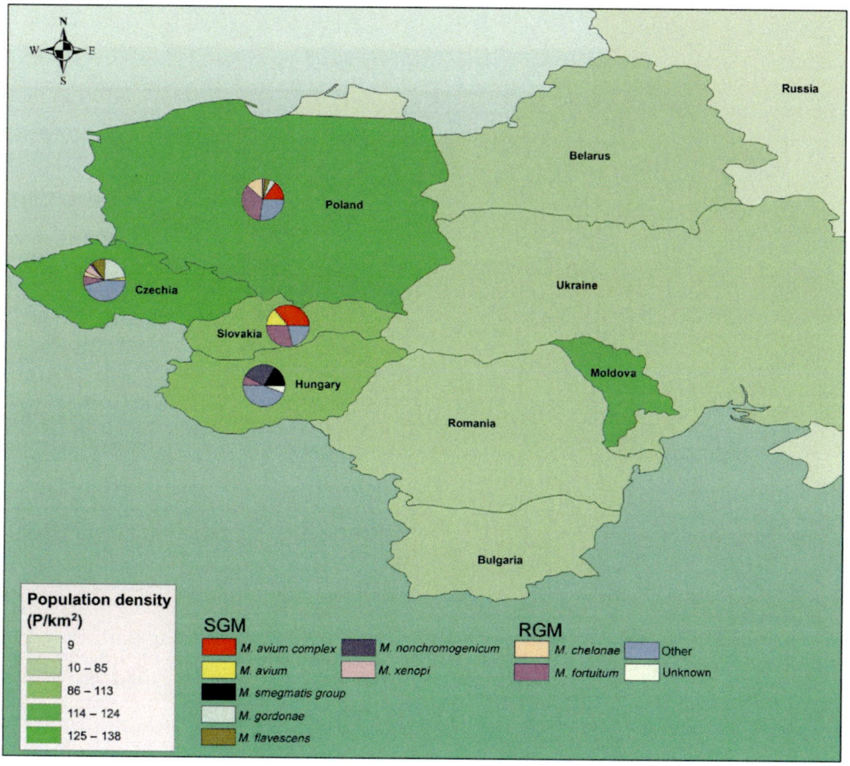

FIGURE 10.6 The distribution of nontuberculous mycobacteria species from environmental samples identified in Eastern Europe. *M. gordonae* (205/1242; 16.51%), *M. fortuitum* (123/1242; 9.90%), and *M. flavescens* (96/1242; 7.73%) were the most frequent species among different regions in Eastern Europe (Table 10.8).

TABLE 10.8 Species Distribution of Nontuberculous Mycobacteria From Environmental Samples in Eastern Europe

Species	Country (Reference) (Period)				
	Hungary (Rónai et al., 2016) (2000–13)	Slovakia (Fischer et al., 2001) (1997–2000)	Czech Republic (Kubalek and Mysak, 1996; Trcka et al., 2006; Sebakova et al., 2008; Makovcova et al., 2014; Beran et al., 2006) (1984–2014)	Poland (Sajduda et al., 2012; Ledwon et al., 2008) (2000–10)	Total/Eastern European
M. fortuitum	16 (7.12%)	4 (28.58%)	67 (7.47%)	36 (33.97%)	(123/1242; 9.90%)
M. avium complex		5 (35.72%)	7 (0.78%)	16 (15.09%)	(28/1242; 2.26%)
M. avium		2 (14.29%)	21 (2.35%)		(23/1242; 1.86%)
M. simiae			2 (0.23%)		(2/1242; 0.17%)
M. gordonae	1 (0.45%)		199 (22.19%0	5 (4.72%)	(205/1242; 16.51%)
M. terrae			12 (1.34%)		(12/1242; 0.97%)
M. intracellulare	6 (2.67%)		13 (1.45%)		(19/1242; 1.53%)
M. arupense	10 (4.45%)		8 (1.45%)		(18/1242; 1.44%)

M. nonchromogenicum	58 (25.78%)		16 (1.79%)		(74/1242; 5.96%)
M. lentiflavum			11 (1.23%)		(11/1242; 0.88%)
M. thermoresistable	9 (4%)				(9/1242; 0.72%)
M. kansasii	11 (4.89%)		23 (2.57%)	12 (11.32%)	(46/1242; 3.70%)
M. africanum				6 (5.66%)	(6/1242; 0.48%)
M. smegmatis	37 (16.45%)		1 (0.12%)		(38/1242; 3.06%)
M. intermedium	15 (6.67%)				(15/1242; 1.21%)
M. flavescens			92 (10.26%)	4 (3.78%)	(96/1242; 7.73%)
M. bohemicum			1 (0.12%)		(1/1242; 0.08%)
M. vaccae	4 (1.78%)		1 (0.12%)		(5/1242; 0.40%)
M. paraffinicum			2 (0.23%)		(2/1242; 0.16%)
M. parafortuitum	1 (0.45%)				(1/1242; 0.08%)

(*Continued*)

TABLE 10.8 (Continued)

Species	Country (Reference) (Period)				
	Hungary (Rónai et al., 2016) (2000–13)	Slovakia (Fischer et al., 2001) (1997–2000)	Czech Republic (Kubalek and Mysak, 1996; Trcka et al., 2006; Sebakova et al., 2008; Makovcova et al., 2014; Beran et al., 2006) (1984–2014)	Poland (Sajduda et al., 2012; Ledwon et al., 2008) (2000–10)	Total/Eastern European
M. confluentis			1 (0.12%)		(1/1242; 0.08%)
M. peregrinum	3 (1.34%)		51 (5.69%)		(54/1242; 4.35%)
M. palustre	2 (0.89%)				(2/1242; 0.16%)
M. sinense	12 (5.34%)				(12/1242; 0.96%)
M. alvei			1 (0.12%)		(1/1242; 0.08%)
M. chelonae	1 (0.45%)		33 (3.68%)	14 (13.20%)	(48/1242; 3.87%)
M. scrofulaceum	2 (0.89%)	1 (7.15%)	10 (1.12%)		(13/1242; 1.04%)
M. marinum			25 (2.79%)		(25/1242; 2.02%)
M. mucogenicum			1 (0.12%)		(1/1242; 0.08%)

M. phlei	2 (0.89%)	1 (7.15%)	2 (0.23%)		(5/1242; 0.40%)
M. triviale			10 (1.12%)	1 (0.94%)	(11/1242; 0.88%)
M. xenopi			56 (6.25%)	2 (1.89%)	(58/1242; 4.67%)
M. abscessus	4 (1.78%)		5 (0.56%)		(9/1242; 0.73%)
M. gastri			1 (0.12%)		(1/1242; 0.08%)
M. interjectum			5 (0.56%)		(5/1242; 0.40%)
M. triplex			6 (0.67%)		(6/1242; 0.48%)
M. celatum			1 (0.12%)		(1/1242; 0.08%)
M. chitae	1 (0.45%)		3 (0.34%)		(4/1242; 0.32%)
M. engbaekii			3 (0.34%)		(3/1242; 0.24%)
M. hiberniae			2 (0.23%)		(2/1242; 0.16%)
M. kumamotonense			5 (0.56%)		(5/1242; 0.40%)

(*Continued*)

TABLE 10.8 (Continued)

Species	Country (Reference) (Period)				
	Hungary (Rónai et al., 2016) (2000–13)	Slovakia (Fischer et al., 2001) (1997–2000)	Czech Republic (Kubalek and Mysak, 1996; Trcka et al., 2006; Sebakova et al., 2008; Makovcova et al., 2014; Beran et al., 2006) (1984–2014)	Poland (Sajduda et al., 2012; Ledwon et al., 2008) (2000–10)	Total/Eastern European
M. nebraskense	1 (0.45%)		4 (0.45%)		(5/1242; 0.40%)
M. szulgai			9 (1%)		(9/1242; 0.72%)
M. europaeum	2 (0.89%)				(2/1242; 0.16%)
M. porcinum			2 (0.23%)		(2/1242; 0.16%)
M. malmoense	1 (0.45%)				(1/1242; 0.08%)
M. asiaticum			1 (0.12%)		(1/1242; 0.08%)
M. chimaera			1 (0.12%)		(1/1242; 0.08%)
M. llatzerense	1 (0.45%)				(1/1242; 0.08%)
M. montefiorense			5 (0.56%)		(5/1242; 0.40%)

M. hominissuis			1 (0.12%)		(1/1242; 0.08%)
M. neoaurum	1 (0.45%)			8 (7.55%)	(9/1242; 0.72%)
M. bourgelatii	5 (2.23%)				(5/1242; 0.40%)
M. immunogenum				2 (1.89%)	(2/1242; 0.16%)
M. arosiense	1 (0.45%)				(1/1242; 0.08%)
M. shimoidei	3 (1.34%)				(3/1242; 0.24%)
Other	2 (0.89%)	1 (7.15%)	177 (19.74%)		(180/1242; 14.49%)
Unidentified	13 (5.78%)				(13/1242; 1.04%)
Total NTM	225	14	897	106	(1242/1242; 100%)

Switzerland. The geographical distribution of the western part of Europe is shown in Fig. 10.7 by the higher frequency of species. In general, in Western Europe (15,788/55,976; 28.20%) NTM has been reported from clinical samples. The highest prevalence of NTM is in the Netherlands with (4573/15,788; 28.96%), followed by France (4359/15,788; 27.60%), and Germany (2655/15,788; 16.81%).

Distribution of Nontuberculous Mycobacteria Based on Environmental Specimens From Western Europe

The frequency of NTM in environmental samples includes data collected from France, The Netherlands, Germany, Belgium, and Switzerland. The

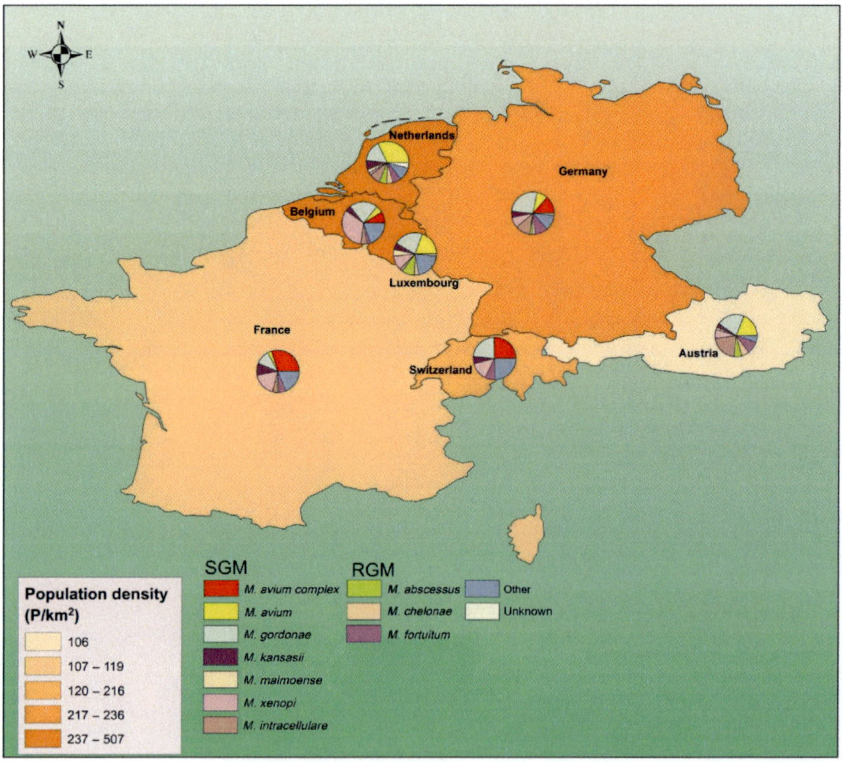

FIGURE 10.7 The distribution of nontuberculous mycobacteria species from clinical samples identified in Western Europe. *M. gordonae* (2810/15,788; 17.79%), *M. avium* complex (2417/15,788; 15.30%), *M. xenopi* (1941/15,788; 12.29%), and *M. avium* (1912/15,788; 12.11%) were the most frequent species among different regions in Western Europe (Table 10.9).

TABLE 10.9 Species Distribution of Nontuberculous Mycobacteria From Clinical Samples in Western Europe

Species	Country (Reference) (Period)							
	The Netherlands (Hoefsloot et al., 2013; Van der Werf et al., 2014) (2006–11)	France (Hoefsloot et al., 2013; Martin-Casabona et al., 2004; Blanc et al., 2016) (1991–2013)	Germany (Hoefsloot et al., 2013; Martin-Casabona et al., 2004; Van der Werf et al., 2014) (1991–2011)	Luxembourg (Van der Werf et al., 2014) (2009–11)	Belgium (Hoefsloot et al., 2013; Martin-Casabona et al., 2004) (1991–2008)	Austria (Hoefsloot et al., 2013) (2008)	Switzerland (Martin-Casabona et al., 2004) (2008)	Total Western Europe
M. avium complex	25	1296	359		140		597	(2417/15,788; 15.30%)
M. avium	1426	133	220	23	88	22		(1912/15,788; 12.11%)
M. gordonae	713	467	697	26	348	25	534	(2810/15,788; 17.79%)
M. xenopi	92	765	210	13	530	8	323	(1941/15,788; 12.29%)
M. intracellulare	381	169	304	3	80	25		(962/15,788; 6.09%)
M. fortuitum	280	233	253		37	11	176	(990/15,788; 6.27%)

(*Continued*)

TABLE 10.9 (Continued)

Species	Country (Reference) (Period)							
	The Netherlands (Hoefsloot et al., 2013; Van der Werf et al., 2014) (2006–11)	France (Hoefsloot et al., 2013; Martin-Casabona et al., 2004; Blanc et al., 2016) (1991–2013)	Germany (Hoefsloot et al., 2013; Martin-Casabona et al., 2004; Van der Werf et al., 2014) (1991–2011)	Luxembourg (Van der Werf et al., 2014) (2009–11)	Belgium (Hoefsloot et al., 2013; Martin-Casabona et al., 2004) (1991–2008)	Austria (Hoefsloot et al., 2013) (2008)	Switzerland (Martin-Casabona et al., 2004) (2008)	Total Western Europe
M. kansasii	324	405	130	6	71	4	138	(1078/15,788; 6.82%)
M. chelonae	202	22	65	4		7		(300/15,788; 1.90%)
M. chelonae-abscessus					16			(16/15,788; 0.10%)
M. abscessus	253	44	43	11		7		(358/15,788; 2.26%)
M. malmoense	122	8	16	6	6	2		(160/15,788; 1.01%)
M. lentiflavum	5	8	5					(18/15,788; 0.11%)
M. marinum	115	23	29	4				(117/15,788; 0.74%)

M. terrae	6	1	1					(8/15,788; 0.05%)
M. terrae complex	2				2			(4/15,788; 0.02%)
M. peregrinum	6	5	8	1		1		(21/15,788; 0.13%)
M. simiae	68	7	5	3	8			(91/15,788; 0.57%)
M. bohemicum	12		2	3				(17/15,788; 0.10%)
M. chimaera	112			5				(117/15,788; 0.74%)
M. interjectum	17	1	7		2	1		(28/15,788; 0.17%)
M. intermedium					1			(1/15,788; 0.006%)
M. nonchromogenicum	2		1					(3/15,788; 0.019%)
M. arupense	8		7	3				(18/15,788; 0.11%)
M. celatum	15		5	1				(21/15,788; 0.13%)
M. mucogenicum	17	1	6					(24/15,788; 0.15%)
M. scrofulaceum	12	4						(16/15,788; 0.10%)

(Continued)

TABLE 10.9 (Continued)

Species	Country (Reference) (Period)							
	The Netherlands (Hoefsloot et al., 2013; Van der Werf et al., 2014) (2006–11)	France (Hoefsloot et al., 2013; Martin-Casabona et al., 2004; Blanc et al., 2016) (1991–2013)	Germany (Hoefsloot et al., 2013; Martin-Casabona et al., 2004; Van der Werf et al., 2014) (1991–2011)	Luxembourg (Van der Werf et al., 2014) (2009–11)	Belgium (Hoefsloot et al., 2013; Martin-Casabona et al., 2004) (1991–2008)	Austria (Hoefsloot et al., 2013) (2008)	Switzerland (Martin-Casabona et al., 2004) (2008)	Total Western Europe
M. scrofulaceum-parascrofulaceum						1		(1/15,788; 0.006%)
M. szulgai	19	1	13		3			(36/15,788; 0.22%)
M. kumamotonense	1		1					(2/15,788; 0.012%)
M. genavense	19	2	8					(29/15,788; 0.18%)
M. phlei	16		7					(23/15,788; 0.14%)
M. vulneris	2		26					(28/15,788; 0.17%)
M. haemophilum	18		1		1			(20/15,788; 0.12%)

M. florentinum			1	5				(6/15,788; 0.038%)
M. nebraskense			11					(11/15,788; 0.069%)
M. neoaurum	5		4					(9/15,788; 0.057%)
M. insubricum			1					(1/15,788; 0.006%)
M. parascrofulaceum			4					(4/15,788; 0.025%)
M. alvei	7							(7/15,788; 0.025%)
M. porcinum	1			2				(3/15,788; 0.019%)
M. europaeum			3					(3/15,788; 0.019%)
M. shimoidei	3				1			(4/15,788; 0.025%)
M. paraffinicum	1		1					(2/15,788; 0.012%)
M. branderi	3		1					(4/15,788; 0.025%)
M. goodie	5		1					(6/15,788; 0.038%)
M. holsaticum	2		2		1			(5/15,788; 0.031%)

(*Continued*)

TABLE 10.9 (Continued)

Species	Country (Reference) (Period)							
	The Netherlands (Hoefsloot et al., 2013; Van der Werf et al., 2014) (2006–11)	France (Hoefsloot et al., 2013; Martin-Casabona et al., 2004; Blanc et al., 2016) (1991–2013)	Germany (Hoefsloot et al., 2013; Martin-Casabona et al., 2004; Van der Werf et al., 2014) (1991–2011)	Luxembourg (Van der Werf et al., 2014) (2009–11)	Belgium (Hoefsloot et al., 2013; Martin-Casabona et al., 2004) (1991–2008)	Austria (Hoefsloot et al., 2013) (2008)	Switzerland (Martin-Casabona et al., 2004) (2008)	Total Western Europe
M. smegmatis	2							(2/15,788; 0.012%)
M. cosmeticum	4	1						(5/15,788; 0.031%)
M. heckeshornense	1		1					(2/15,788; 0.012%)
M. agri	2							(2/15,788; 0.012%)
M. brumae	1							(1/15,788; 0.006%)
M. flavescens/M. novocastrense			3					(3/15,788; 0.019%)
M. flavescens		1						(1/15,788; 0.006%)
M. hassiacum	2		1					(3/15,788; 0.019%)

M. heidelbergense	2		1					(3/15,788; 0.019%)
M. lepraemurium			3					(3/15,788; 0.019%)
M. mantenii	2							(2/15,788; 0.012%)
M. noviomagense	2							(2/15,788; 0.012%)
M. wolinskyi	3		1					(4/15,788; 0.025%)
M. mageritense	3		1					(4/15,788; 0.025%)
M. novocastrense	2							(2/15,788; 0.012%)
M. pulveris	1							(1/15,788; 0.006%)
M. phocaicum			1					(1/15,788; 0.006%)
M. pulveris/elephantis complex			1					(1/15,788; 0.006%)
M. riyadhense			1					(1/15,788; 0.006%)
M. asiaticum	1							(1/15,788; 0.006%)
M. aubagnense			1		1			(2/15,788; 0.012%)
M. conspicuum	1							(1/15,788; 0.006%)

(Continued)

TABLE 10.9 (Continued)

Species	Country (Reference) (Period)							
	The Netherlands (Hoefsloot et al., 2013; Van der Werf et al., 2014) (2006–11)	France (Hoefsloot et al., 2013; Martin-Casabona et al., 2004; Blanc et al., 2016) (1991–2013)	Germany (Hoefsloot et al., 2013; Martin-Casabona et al., 2004; Van der Werf et al., 2014) (1991–2011)	Luxembourg (Van der Werf et al., 2014) (2009–11)	Belgium (Hoefsloot et al., 2013; Martin-Casabona et al., 2004) (1991–2008)	Austria (Hoefsloot et al., 2013) (2008)	Switzerland (Martin-Casabona et al., 2004) (2008)	Total Western Europe
M. fluoroanthenivorans			1					(1/15,788; 0.006%)
M. kyorinense			1					(1/15,788; 0.006%)
M. massiliense				1				(1/15,788; 0.006%)
M. timonense				1				(1/15,788; 0.006%)
Other	27	760	139		284	3	573	(1786/15,788; 11.31%)
Unidentified	200		41	2				(243/15,788; 1.53%)
Total NTM	4573	4359	2655	123	1620	117	2341	(15,788/ 15,788; 100%)

geographical distribution of NTM based on environmental specimens is shown in Fig. 10.8 by the higher frequency of species. In general, in Western Europe (1417/4275; 33.14%) NTM has been reported from environmental specimens. The highest prevalence of NTM is in Germany with (534/1417; 37.68%), followed by France (352/1417; 24.84%), and The Netherlands (418/1417; 29.49%).

Asia Geography

Asia is the largest and most populous continent. It covers 44,579,000 km^2 (17,212,000 mi^2), about 30% of Earth's total land area, and 8.7% of the Earth's total surface area. Asia is has an overall population of about 4.5

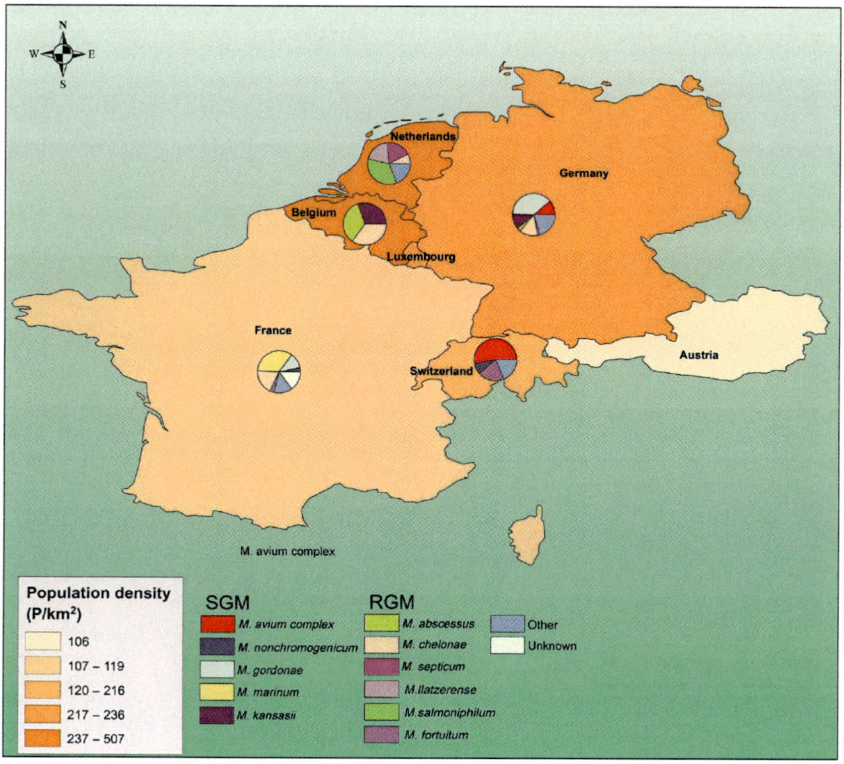

FIGURE 10.8 The distribution of nontuberculous mycobacteria species from environmental samples identified in Western Europe. *M. gordonae* (243/1417; 17.14%), *M. chelonae* (186/1417; 13.12%), and *M. salmoniphilum* (144/1417; 10.16%) were the most frequent species among different regions in Western Europe (Table 10.10).

TABLE 10.10 Species Distribution of Nontuberculous Mycobacteria From Environmental Samples in Western Europe

Kind of Species	Country (Reference) (Period)					
	France (Thorel et al., 2004; Le Dantec et al., 2002; Käser et al., 2009) (1996–2014)	The Netherlands (van Ingen et al., 2010) (2010)	Germany (Sajduda et al., 2012; Fischeder et al., 1991; Peters et al., 1995; Lahiri, 2014) (1991–2014)	Switzerland (Bono et al., 1995) (1994, 2007)	Belgium (Sajduda et al., 2012) (2001–12)	Total/ Western Europe
M. fortuitum	11 (2.15%)		9 (1.69%)	14 (20%)		(34/1417; 2.39%)
M. avium complex			60 (11.24%)	37 (52.86%)		(97/1417; 6.84%)
M. gordonae	40 (7.82%)		203 (38.02%)			(243/1417; 17.14%)
M. terrae				2 (2.86%)		(2/1417; 0.14%)
M. intracellulare	1 (0.20%)		3 (0.56%)	4 (5.72%)		(8/1417; 0.56%)
M. lentiflavum			5 (0.94%)			(5/1417; 0.35%)
M. smegmatis		16 (3.83%)	1 (0.19%)			(17/1417; 1.19%)
M. flavescens			45 (8.43%)			(45/1417; 3.17%)
M. peregrinum	6 (1.18%)	24 (5.74%)	14 (2.63%)			(44/1417; 3.10%)
M. kansasii	3 (0.59%)		63 (11.80%)		12 (27.91%)	(78/1417; 5.50%)
M. mucogenicum			5 (0.94%)			(5/1417; 0.35%)

M. marinum	120 (23.44%)					(120/1417; 8.46%)
M. gastri			12 (2.25%)			(12/1417; 0.84%)
M. abscessus			13 (2.44%)		3 (6.98%)	(16/1417; 1.12%)
M. szulgai	2 (0.39%)					(2/1417; 0.14%)
M. chelonae	66 (12.89%)	32 (7.66%)	75 (14.05%)		13 (30.24%)	(186/1417; 13.12%)
M. gadium	1 (0.20%)					(1/1417; 0.07%)
M. alvei		24 (5.75%)				(24/1417; 1.69%)
M. nonchromogenicum	12 (2.34%)		3 (0.56%)	6 (8.58%)		(21/1417; 1.48%)
M. porcinum			1 (0.19%)			(1/1417; 0.07%)
M. vaccae		16 (3.83%)				(16/1417; 1.12%)
M. septicum		79 (18.90%)				(79/1417; 5.57%)
M. ulcerans	24 (4.69%)					(24/1417; 1.69%)
M. xenopi			3 (0.56%)	2 (2.86%)		(5/1417; 0.35%)
M. fallax			2 (0.38%)			(2/1417; 0.14%)

(Continued)

TABLE 10.10 (Continued)

Kind of Species	Country (Reference) (Period)					
	France (Thorel et al., 2004; Le Dantec et al., 2002; Käser et al., 2009) (1996–2014)	The Netherlands (van Ingen et al., 2010) (2010)	Germany (Sajduda et al., 2012; Fischeder et al., 1991; Peters et al., 1995; Lahiri, 2014) (1991–2014)	Switzerland (Bono et al., 1995) (1994, 2007)	Belgium (Sajduda et al., 2012) (2001–12)	Total/ Western Europe
M. aurum	1 (0.20%)					(1/1417; 0.07%)
M. malmoense	13 (2.54%)		3 (0.56%)			(16/1417; 1.12%)
M. llatzerense		83 (19.86%)				(83/1417; 8.85%)
M. salmoniphilum		144 (34.45%)				(144/1417; 10.16%)
M. hiberniae			6 (1.13%)			(6/1417; 0.42%)
M. neglectum			8 (1.50%)			(8/1417; 0.56%)
M. genavense				5 (7.15%)		(5/1417; 0.35%)
M. immunogenum					13 (30.24%)	(13/1417; 0.91%)
Other					2 (4.66%)	(2/1417; 0.14%)
Unidentified	52 (10.16%)					(52/1417; 3.66%)
Total NTM	352	418	534	70	43	(1417/1417; 100%)

billion people, which constitutes roughly 60% of the world's population. Generally, the boundary starts from the east by the Pacific Ocean, on the south by the Indian Ocean, and on the north by the Arctic Ocean. China and India are the major economic countries in the region. There are 48 countries and three independable state countries (Taiwan, Hong Kong, Macao). According to the United Nations, the continent is divided into Eastern Asia, Southern Asia, Southeast Asia, Western Asia, and Central Asia. Central Asia itself has a population of 105 million, consisting of five republics: Kazakhstan (18 million), Kyrgyzstan (6 million), Tajikistan (9 million), Turkmenistan (6 million), and Uzbekistan (31 million). Afghanistan (35 million), is a part of South Asia but is also sometimes included in Central Asia. The Middle East, or near East, encompasses the territories of Turkey, Cyprus, Syria, Lebanon, Iraq, Iran, the West bank, Gaza strip, Jordan, Egypt, Sudan, and Libya and the various states and territories of Arabia proper (Saudi Arabia, Kuwait, Yemen, Oman, Bahrain, Qatar, and the Trucial States, or Trucial Oman [now United Arab Emirates]). Subsequent events have tended, in loose usage, to enlarge the number of countries included in the definition. Three North African countries, Tunisia, Algeria, and Morocco, are closely connected in sentiment and foreign policy with the Arab states. In addition, geographic factors often require statesmen and others to take account of Afghanistan and Pakistan in connection with the affairs of the Middle East.

Distribution of Nontuberculous Mycobacteria Based on Clinical and Environmental Specimens From Asia

This section is based on data compiled from information published on the distribution of NTM from 13 countries in different regions of the continent in Asia between 1969 and 2015 (Tables 10.11 and 10.12). Tables 10.11 and 10.12 show an overview of studies on NTM from clinical and environmental specimens from the entire Asian continent. In general, 82,138 and 2555 NTMs have been reported in clinical and environmental specimens from different regions of the Asia, respectively. Among Asian countries, Taiwan (36,923/82,138; 44.95%) reported the majority of NTM from clinical specimens, followed by Japan (31,042/82,138; 37.79%), Korea (11,774/82,138; 14.33%), and Iran (983/82,138; 1.19%). Iran (1107/2555; 43.32%) reported the majority of NTM from environmental specimens, followed by Korea (437/2555; 17.10%), Iraq (271/2555; 10.60%), and India (170/2555; 6.65%).

Different NTM species were identified from clinical and environmental specimens. *Mycobacterium avium*, *Mycobacterium avium* complex, *Mycobacterium gordonae*, *Mycobacterium xenopi*, *Mycobacterium kansasii*, *Mycobacterium intracellulare*, *Mycobacterium fortuitum*, *Mycobacterium mageritense*, *Mycobacterium chelonae*, and *Mycobacterium terrae* were identified as the most frequent species.

TABLE 10.11 Overview of Studies on Nontuberculous Mycobacteria From Clinical Specimens in Asia

Countries/Period	Types of Samples	Common Species (*n*)	Total NTM
China (Liu et al., 2016) (2005–12)	Clinical specimens	*M. intracellulare* (206), *M. avium* (106), *M. abscessus* (60)	488
India (Jesudason and Gladstone, 2005) (1999–2004)	Biopsy, sputum, CSF, urine, blood	*M. chelonae* (53), *M. fortuitum* (47), *M. terrae* (3)	173
India (Maurya et al., 2015) (2014–15)	Pulmonary and extra Pulmonary	*M. fortuitum* (17), *M. intracellulare* (13), *M. chelonae* (8)	62
Iran (Velayati et al., 2015) (1984–2014)	Clinical specimens	*M. fortuitum* (136), *M. simiae* (103), *M. kansasii* (57)	983
japan (Tsukamura et al., 1988b) (1971–84)	Sputum culture	*M. avium* complex (999), *M. kansasii* (193), *M. fortuitum* (22)	1248
japan (Namkoong et al., 2016) (1980–2014)	Pulmonary samples	*M. avium* (1433), *M. intracellulare* (679), *M. avium* complex (243)	2652
Japan (Morimoto et al., 2017) (2012–13)	Clinical specimens	*M. avium* (16,115), *M. intracellulare* (8112), *M. fortuitum* (277)	27,142
Korea, Seoul (Koh et al., 2006) (2002–03)	Respiratory specimens	*M. avium* complex (491), *M. abscessus* (442), *M. fortuitum* complex (268)	1548
Korea, Cheonan (Kim and Rheem, 2013) (2005–11)	AFB culture	*M. avium* complex (153), *M. gordonae* (12), *M. kansasii* (18)	232
Korean (Ryoo et al., 2008) (1993–2006)	Respiratory specimens	*M. avium* complex (5964), *M. fortuitum* complex (986), *M. abscessus* (957)	9994
Kuwait (Velayati et al., 2015) (1984–2014)	Pulmonary samples	*M. fortuitum* (7), *M. avium* complex (2), *M. abscessus* (2)	14
Lebanon (Velayati et al., 2015) (1984–2014)	Extra Pulmonary	*M. marinum* (22), *M. avium* complex (17), *M. chelonae* (2)	43
Oman (Velayati et al., 2015) (1984–2014)	Pulmonary samples	*M. avium* complex (9)	13

(*Continued*)

TABLE 10.11 (Continued)

Countries/Period	Types of Samples	Common Species (*n*)	Total NTM
Pakistan (Velayati et al., 2015) (1984–2014)	Pulmonary	*M. fortuitum* (23), *M. xenopi* (22), *M. kansasii* (16)	116
Saudi Arabia (Velayati et al., 2015; Al-Ghafli and Al-Hajoj, 2017) (1984–2014)	Pulmonary samples	*M. fortuitum* (49), *M. abscessus* (29), *M. avium* complex (18)	137
Taiwan (Ding et al., 2006) (1997–2003)	Clinical specimens	*M. avium* complex (161), *M. abscessus* (76), *M. fortuitum* (40)	386
Taiwan (Shu et al., 2008) (1999–2007)	Respiratory specimens	*M. avium* complex (1527), *M. abscessus* (898), *M. fortuitum* (646)	4779
Taiwan (Lai et al., 2010) (2000–08)	Clinical specimens	*M. avium* complex (2761), *M. abscessus* (1609)	9204
Taiwan (Chien et al., 2014) (2000–12)	Respiratory cultures	*M. avium* complex (2681), *M. fortuitum* (1462), *M. abscessus* (1348)	7774
Taiwan Northern (Chien et al., 2014) (2010–14)	Respiratory specimens	*M. avium* complex (3432), *M. abscessus* (1677), *M. kansasii* (592)	8427
Taiwan Sothern (Chien et al., 2014) (2010–14)	Respiratory specimens	*M. abscessus* (1497), *M. avium* complex (1372), *M. kansasii* (557)	6353
Thailand (Sriyabhaya and Wongwatana, 1981) (1969–78)	Sputum or laryngeal swab	*M. avium* complex (7), *M. avium* (6), *M. scrofulaceum* (6)	24
Turkey (Velayati et al., 2015) (1984–2014)	Clinical specimens	*M. avium* complex (54), *M. gordonae* (53), *M. fortuitum* (48)	280
Turkey (Martin-Casabona et al., 2004) (1991–96)	Clinical specimens	*M. fortuitum* (18), *M. gordonae* (12), *M. avium* complex (5)	53
Turkey (Hoefsloot et al., 2013) (2008)	Pulmonary and extra pulmonary samples	*M. szulgai* (4), *M. gordonae* (3), *M. chelonae* (3)	37

TABLE 10.12 Overview of Studies on Nontuberculous Mycobacteria From Environmental Specimens in Asia

Country (Period)	Type of Sample	Most Common Species	Total NTM
China (Hu et al., 2017) (2015–16)	Soil	*M. septicum* 26 (25.74%), *M. vulneris* 14 (12.72%), *M. fortuitum* 11 (10%)	110
South Korea (Lee et al., 2008) (2004–05)	Water	*M. lentiflavum* 40 (25.32%), *M. gordonae* 10 (6.32%), *M. terrae* 4 (2.53%)	158
South Korea (Shin et al., 2007) (2003–05)	Taps water in hospital	*M. gordonae* 44 (73.34%), *M. mucogenicum* 10 (16.66%), *M. kansasii* 2 (3.33%)	60
South Korea (Jin et al., 1984) (1980)	Soil	*M. fortuitum* 22 (39.28%), *M. gordonae* 11 (19.64%), *M. scrofulaceum* 9 (16.07%)	56
South Korea (Jin et al., 1984) (1980)	House dust	*M. fortuitum* 5 (62.5%), *M. avium* complex 1 (12.5%), *M. terrae* 1 (12.5%)	8
South Korea (Jin et al., 1984) (1980)	Well water	*M. fortuitum* 12 (25%), *M. terrae* 12 (25%), *M. gordonae* 6 (12.5%)	48
South Korea (Jin et al., 1984) (1980)	Sewage	*M. terrae* 37 (34.58%), *M. fortuitum* 27 (25.23%), *M. gordonae* 10 (9.34%)	107
Japan (Kobayashi et al., 2014) (2009–11)	Water	*M. avium* complex 12 (40%), *M. gordonae* 5 (16.66%), *M. nonchromogenicum* 3 (10%)	30
India (Mishra et al., 2018) (2018)	Soil	*M. fortuitum* 38 (30.89%), *M. gordonae* 26 (21.13%), *M. flavescens* 12 (9.75%)	123
India (Mishra et al., 2018) (2018)	Water	*M. gordonae* 13 (27.65%), *M. fortuitum* 10 (21.27%), *M. flavescens* 5 (10.63%)	47
Sri Lanka (Edirisinghe et al., 2014) (2014)	Aquatic sources	*M. fortuitum* 17 (37.78%), *M. phlei* 9 (20%), *M. marinum* 4 (8.88%)	45

(*Continued*)

TABLE 10.12 (Continued)

Country (Period)	Type of Sample	Most Common Species	Total NTM
Nepal (Jha et al., 2007) (2003)	Animal	*M. thermoresistibile* 6 (3.40%), *M. fortuitum* 2 (1.13%), *M. chelonae* 2 (1.13%)	163
Iran (Varahram et al., 2016) (2016)	Water and soil	*M. fortuitum* 72 (22.36%), *M. senegalense* 58 (18.01%), *M. parafortuitum* 44 (13.66%)	322
Turkey (Velayati et al., 2015) (1984–2014)	Environmental	*M. lentiflavum* 20 (32.26%), *M. gordonae* 10 (16.12%), *M. terrae* 6 (9.67%)	62
Iran (Velayati et al., 2015) (1984–2014)	Environment	*M. fortuitum* 93 (34.32%), *M. flavescens* 44 (16.23%), *M. chelonae* 38 (14.02%)	271
Iraq (Velayati et al., 2015) (1984–2014)	Environment	*M. chelonae* 50 (18.45%), *M. avium* complex 40 (14.76%), *M. fortuitum* 37 (13.65%)	271
Saudi Arabia (Velayati et al., 2015) (1984–2014)	Environment	*M. gordonae* 8 (53.34%), *M. fortuitum* 3 (20%), *M. avium* complex 2 (13.33%)	15
Saudi Arabia (Alqumber, 2014) (2013)	Water	*M. malmoense* 21 (14.48%), *M. chelonae* 18 (12.41%), *M. terrae* 16 (11.03%)	145
Iran (Livani and Ghaemi, 2014) (1976–2013)	Water	*M. fortuitum* 16 (24.62%), *M. chelonae* 16 (24.62%), *M. mucogenicum* 8 (12.30%)	65
Iran (Livani and Ghaemi, 2014) (1976–2013)	Soil	*M. fortuitum* 90 (19.66%), *M. flavescens* 61 (13.58%), *M. chelonae* 40 (8.90%)	449

Clinical Nontuberculous Mycobacteria Reported From Eastern Asia

The Eastern region includes data collected from four countries: Taiwan, Korean, Japan, and China. The geographical distribution of the eastern part of Asia is shown in Fig. 10.9 by the higher frequency of species. In general, in Eastern Asia (80,227/82,138; 97.67%) NTM has been reported from

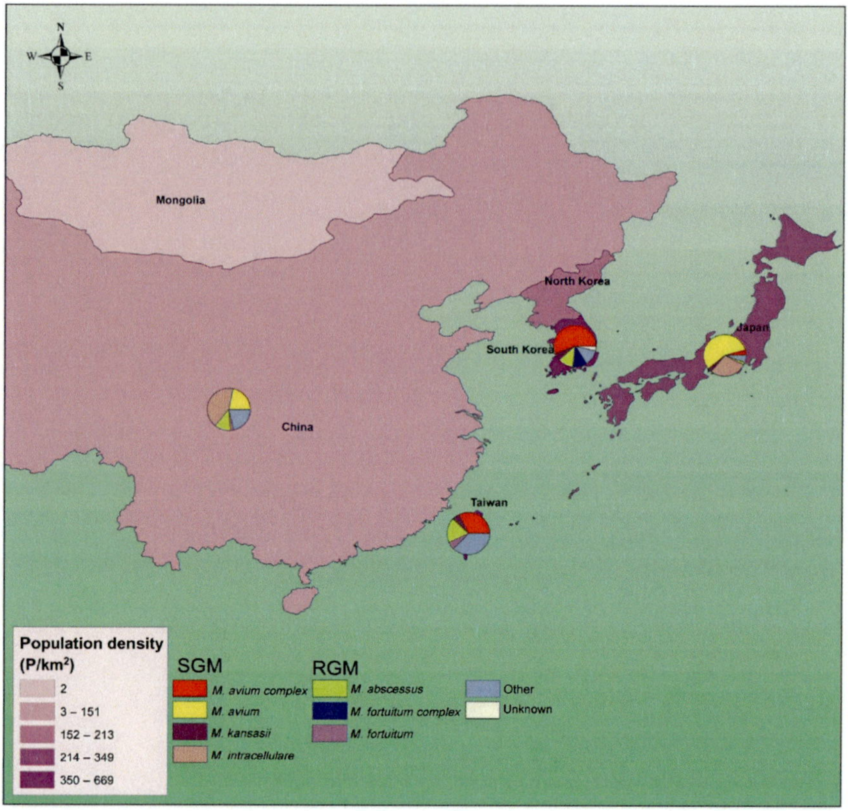

FIGURE 10.9 The distribution of nontuberculous mycobacteria species from clinical samples identified in Eastern Asia. *Mycobacterium avium* complex (19,784/80,227; 24.66%), *Mycobacterium avium* (17,654/80,227; 22%), *Mycobacterium abscessus* (9182/80,227; 11.44%), and *Mycobacterium intracellulare* (8997/80,227; 11.21%) were the most frequent species among different regions in Eastern Asia (Table 10.13). Table 10.13 shows the data recorded for eastern Asia by country, period of study and the frequency of different species.

clinical samples. The highest prevalence of NTM is in Taiwan with (36,923/80,227; 46.02%), followed by Japan (31,042/80,227; 38.69%), Korea (11,774/80,227; 14.67%), and China (488/80,227; 0.60%).

Distribution of Nontuberculous Mycobacteria Based on Environmental Specimens From Eastern Asia

The frequency of NTM in environmental samples includes data collected from China, Korea, and Japan. The geographical distribution of NTM based

TABLE 10.13 Species Distribution of Nontuberculous Mycobacteria From Clinical Samples in Eastern Asia

Species	Country (Reference) (Period)				
	Taiwan (Ding et al., 2006; Shu et al., 2008; Lai et al., 2010; Chien et al., 2014) (1997–2014)	Korean (Koh et al., 2006; Kim and Rheem, 2013; Ryoo et al., 2008) (1993–2011)	Japan (Namkoong et al., 2016; Morimoto et al., 2017; Tsukamura et al., 1988a) (1971–2014)	China (Liu et al., 2016) (2005–12)	Total East Asia
M. avium complex	11,934	6608	1242		(19,784/80,227; 24.66%)
M. avium			17,548	106	(17,654/80,227; 22%)
M. gordonae	882	200	1	34	(1117/80,227; 1.39%)
M. xenopi	37	1	28		(66/80,227; 0.08%)
M. intracellulare			8791	206	(8997/80,227; 11.21%)
M. fortuitum	2148	10	299	11	(2468/80,227; 3.07%)
M. fortuitum complex		1254			(1254/80,227; 1.56%)
M. kansasii	2018	629	866	2	(3515/80,227; 4.38%)
M. chelonae	1163	31	198	6	(1398/80,227; 1.74%)
M. chelonae complex		913			(913/80,227; 1.13%)
M. abscessus	7105	1412	605	60	(9182/80,227; 11.44%)
M. lentiflavum				6	(6/80,227; 0.007%)
M. marinum	18	10	3		(31/80,227; 0.03%)
M. terrae		8	92		(100/80,227; 0.12%)

(Continued)

TABLE 10.13 (Continued)

Species	Country (Reference) (Period)				
	Taiwan (Ding et al., 2006; Shu et al., 2008; Lai et al., 2010; Chien et al., 2014) (1997–2014)	Korean (Koh et al., 2006; Kim and Rheem, 2013; Ryoo et al., 2008) (1993–2011)	Japan (Namkoong et al., 2016; Morimoto et al., 2017; Tsukamura et al., 1988a) (1971–2014)	China (Liu et al., 2016) (2005–12)	Total East Asia
M. terrae complex	92	52			(144/80,227; 0.17%)
M. peregrinum			53		(53/80,227; 0.06%)
M. simiae	3		8		(11/80,227; 0.013%)
M. nonchromogenicum	1		42		(43/80,227; 0.053%)
M. celatum	2	57			(59/80,227; 0.07%)
M. scrofulaceum	54	16	43		(113/80,227; 0.14%)
M. szulgai	9	132	40	3	(184/80,227; 0.22%)
M. genavense				1	(1/80,227; 0.001%)
M. phlei	36				(36/80,227; 0.04%)
M. triviale	3		4		(7/80,227; 0.008%)
M. kumamotonense				2	(2/80,227; 0.002%)
M. neoaurum	1			2	(3/80,227; 0.003%)
M. vaccae	10				(10/80,227; 0.012%)
M. parascrofulaceum				3	(3/80,227; 0.003%)
M. shimoidei				2	(2/80,227; 0.002%)

M. triplex				2	(2/80,227; 0.002%)
M. holsaticum				1	(1/80,227; 0.001%)
M. heidelbergense				1	(1/80,227; 0.001%)
M. setense				1	(1/80,227; 0.001%)
M. immunogenum				1	(1/80,227; 0.001%)
M. gastri		2			(2/80,227; 0.002%)
M. phocaicum				1	(1/80,227; 0.001%)
M. colombiense				3	(3/80,227; 0.003%)
M. stomatepiae				1	(1/80,227; 0.001%)
M. seoulense				2	(2/80,227; 0.002%)
M. saskatchewanense				2	(2/80,227; 0.002%)
M. marseillense				5	(5/80,227; 0.006%)
M. septicum				2	(2/80,227; 0.002%)
M. massiliense				22	(22/80,227; 0.027%)
M. asiaticum	1				(1/80,227; 0.001%)
M. mageritense	1				(1/80,227; 0.001%)
M. smegmatis	41				(41/80,227; 0.051%)
M. flavescens	25				(25/80,227; 0.031%)
Other	11,339	7	1179		(12,525/80,227; 15.61%)
Unidentified		432			(432/80,227; 0.53%)
Total NTM	36,923	11,774	31,042	488	(80,227/80,227; 100%)

on environmental specimens is shown in Fig. 10.10 by the higher frequency of species. In general, in Eastern Asia (577/2555; 22.58%) NTM has been reported from environmental specimens. The highest prevalence of NTM is in Korea with (437/577; 75.73%), followed by China (110/577; 19.06%), and Japan (30/577; 5.19%).

Clinical Nontuberculous Mycobacteria Reported From Western Asia

The western region includes data collected from six countries: Iran, Lebanon, Saudi Arabia, Kuwait, Oman, and Turkey. The geographical distribution of

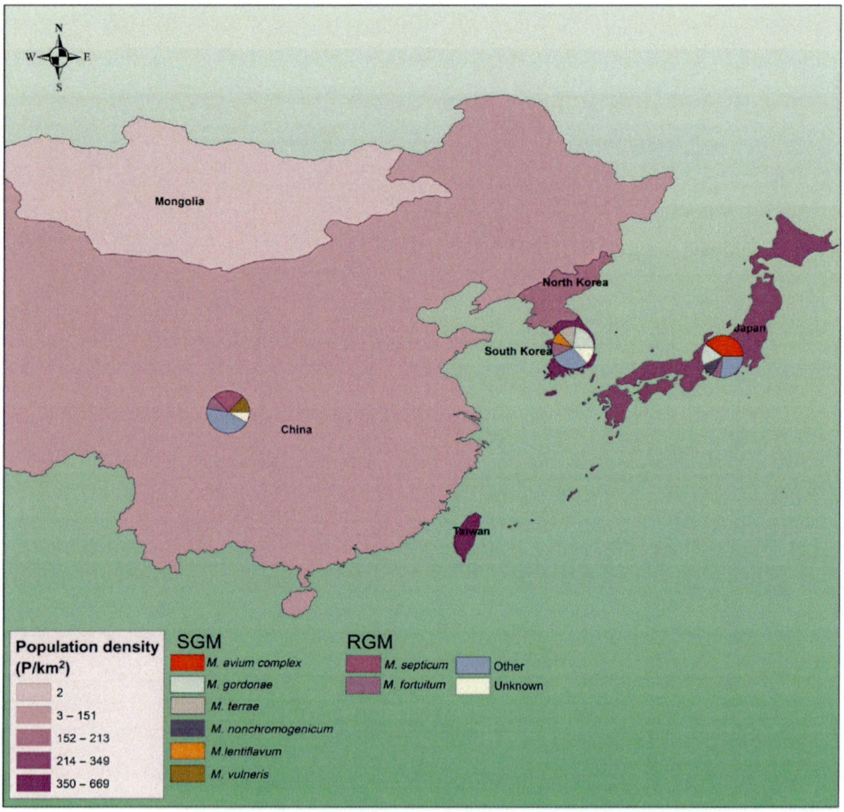

FIGURE 10.10 The distribution of nontuberculous mycobacteria species from environmental samples identified in Eastern Asia. *M. gordonae* (105/577; 18.19%), *M. terrae* (61/577; 10.57%), *M. fortuitum* (59/577; 10.22%), and *M. lentiflavum* (40/577; 6.93%) were the most frequent species among different regions in Eastern Asia (Table 10.14).

TABLE 10.14 Species Distribution of Nontuberculous Mycobacteria From Environmental Samples in Eastern Asia

Country (Reference) (Period)	China (Hu et al., 2017) (2015–16)	South Korea (Lee et al., 2008; Shin et al., 2007; Jin et al., 1984) (1980, 2004–05)	Japan (Kobayashi et al., 2014) (2009–14)	Total/Eastern Asia
M. fortuitum	11 (10%)	46 (10.53%)	2 (6.67%)	(59/577; 10.22%)
M. avium complex		4 (0.92%)	12 (40%)	(16/577; 2.77%)
M. simiae			1 (3.34%)	(1/577; 0.17%)
M. gordonae	1 (0.91%)	99 (22.65%)	5 (16.67%)	(105/577; 18.19%)
M. terrae		61 (13.96%)		(61/577;10.57%)
M. intracellulare	3 (2.73%)	2 (0.46%)	1 (3.34%)	(6/577; 1.03%)
M. arupense		1 (0.23%)		(1/577; 0.17%)
M. nonchromogenicum			3 (10%)	(3/577; 0.51%)
M. chubuense		1 (0.23%)		(1/577; 0.17%)
M. lentiflavum		40 (9.16%)		(40/577; 6.93%)
M. smegmatis		4 (0.92%)		(4/577; 0.69%)
M. flavescens		8 (1.84%)		(8/577; 1.38%)
M. gilvum		1 (0.23%)		(1/577; 0.17%)
M. kumamotonense		3 (0.69%)		(3/577; 0.51%)
M. peregrinum	1 (0.91%)	3 (0.69%)		(4/577; 0.69%)
M. scrofulaceum		18 (4.12%)	1 (3.34%)	(19/577; 3.29%)

(*Continued*)

TABLE 10.14 (Continued)

Country (Reference) (Period)	China (Hu et al., 2017) (2015–16)	South Korea (Lee et al., 2008; Shin et al., 2007; Jin et al., 1984) (1980, 2004–05)	Japan (Kobayashi et al., 2014) (2009–14)	Total/Eastern Asia
M. mucogenicum		10 (2.29%)		(10/577; 1.73%)
M. marinum			1 (3.34%)	(1/577; 0.17%)
M. gastri			1 (3.34%)	(1/577; 0.17%)
M. abscessus			1 (3.34%)	(1/577; 0.17%)
M. szulgai			1 (3.34%)	(1/577; 0.17%)
M. chelonae	4 (3.64%)	6 (1.38%)		(10/577; 1.73%)
M. farcinogenes	2 (1.82%)			(2/577; 0.34%)
M. triplex	2 (1.82%)	3 (0.69%)		(5/577; 0.86%)
M. alvei	10 (9.09%)			(10/577; 1.73%)
M. nonchromogenicum	1 (0.91%)			(1/577; 0.17%)
M. porcinum	3 (2.73%)			(3/577; 0.51%)
M. vaccae	2 (1.82%)	15 (3.44%)		(17/577; 2.94%)
M. vulneris	14 (12.73%)			(14/577; 2.42%)
M. timonense	9 (8.19%)			(9/577; 1.55%)
M. bouchedurhonense	4 (3.64%)			(4/577; 0.69%)
M. mantenii	3 (2.73%)			(3/577; 0.51%)

(Continued)

TABLE 10.14 (Continued)

Country (Reference) (Period)	China (Hu et al., 2017) (2015–16)	South Korea (Lee et al., 2008; Shin et al., 2007; Jin et al., 1984) (1980, 2004–05)	Japan (Kobayashi et al., 2014) (2009–14)	Total/Eastern Asia
M. colombiense	2 (1.82%)			(2/577; 0.34%)
M. virginiense	2 (1.82%)			(2/577; 0.34%)
M. florentinum	1 (0.91%)			(1/577; 0.17%)
M. aichiense		2 (0.46%)		(2/577; 0.34%)
M. brisbanense		1 (0.23%)		(1/577; 0.17%)
M. holsaticum		2 (0.46%)		(2/577; 0.34%)
M. phocaicum		2 (0.46%)		(2/577; 0.34%)
M. phlei		3 (0.69%)		(3/577; 0.51%)
M. triviale			1 (3.34%)	(1/577; 0.17%)
M. septicum	26 (23.64%)			(26/577; 4.50%)
M. gadium		1 (0.23%)		(1/577; 0.17%)
Unidentified	9 (8.19%)	59		(68/577; 11.78%)
Other		42		(42/577; 7.27%)
Total NTM	110	437	30	(577/577; 100%)

the western part of Asia is shown in Fig. 10.11 by the higher frequency of species. In general, in Western Asia (1560/82,138; 1.89%) NTM has been reported from clinical samples. The highest prevalence of NTM is in Iran with (983/1560; 63.01%), followed by Turkey (370/1560; 23.71%), Saudi Arabia (137/1560; 8.78%), and Lebanon (43/1560; 2.75%).

Distribution of Nontuberculous Mycobacteria Based on Environmental Specimens From Western Asia

The frequency of NTM in environmental samples includes data collected from Turkey, Iraq, Iran, and Saudi Arabia. The geographical distribution of

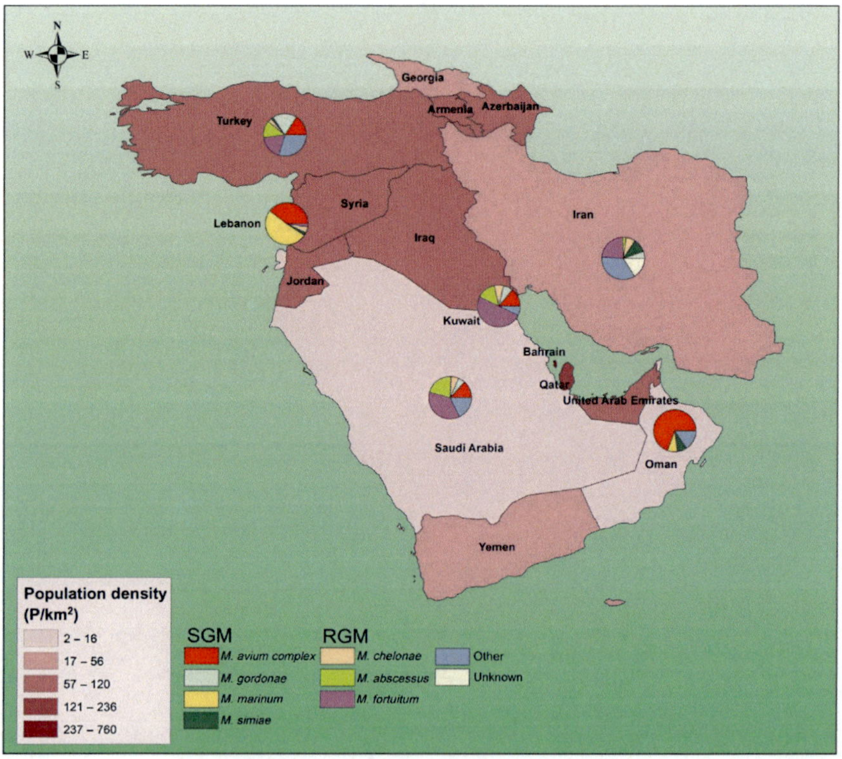

FIGURE 10.11 The distribution of nontuberculous mycobacteria species from clinical samples identified in Western Asia. *M. fortuitum* (353/1560; 22.62%), *M. gordonae* (128/1560; 8.20%), *M. simiae* (112/1560; 7.17%), and *M. avium* complex (105/1560; 6.73%) were the most frequent species among different regions in west Asia (Table 10.15).

TABLE 10.15 Species Distribution of Nontuberculous Mycobacteria From Clinical Samples in Western Asia

Species	Country (Reference) (Period)						
	Iran (Velayati et al., 2015) (1984–2014)	Lebanon (Velayati et al., 2015) (1984–2014)	Saudi Arabia (13) (Velayati et al., 2015; Al-Ghafli and Al-Hajoj, 2017) (1984–2014)	Kuwait (Velayati et al., 2015) (1984–2014)	Oman (Velayati et al., 2015) (1984–2014)	Turkey (Velayati et al., 2015) (1991–2014)	Total Western Asia
M. avium complex		17	18	2	9	59	(105/1560; 6.73%)
M. avium	13					1	(14/1560; 0.89%)
M. gordonae	52		7	1		68	(128/1560; 8.20%)
M. xenopi	9		4			7	(20/1560; 1.28%)
M. intracellulare	19					1	(20/1560; 1.28%)
M. fortuitum	229	1	49	7		67	(353/1560; 22.62%)
M. parafortuitum	5						(5/1560; 0.32%)
M. kansasii	60		7	1	1	18	(87/1560; 5.57%)
M. chelonae	70	2	8	1		12	(93/1560; 5.96%)

(Continued)

TABLE 10.15 (Continued)

Species	Country (Reference) (Period)						
	Iran (Velayati et al., 2015) (1984–2014)	Lebanon (Velayati et al., 2015) (1984–2014)	Saudi Arabia (13) (Velayati et al., 2015; Al-Ghafli and Al-Hajoj, 2017) (1984–2014)	Kuwait (Velayati et al., 2015) (1984–2014)	Oman (Velayati et al., 2015) (1984–2014)	Turkey (Velayati et al., 2015) (1991–2014)	Total Western Asia
M. abscessus	23		29	2		47	(101/1560; 6.47%)
M. malmoense	3						(3/1560; 0.19%)
M. lentiflavum	4		2			11	(17/1560; 1.08%)
M. marinum	6	22	1		1	1	(31/1560; 1.98%)
M. terrae	21		2			1	(24/1560; 1.53%)
M. peregrinum	14					5	(19/1560; 1.21%)
M. simiae	103	1	1		1	6	(112/1560; 7.17%)
M. interjectum						2	(2/1560; 0.12%)
M. fallax	5						(5/1560; 0.32%)

M. nonchromogenicum	2						(2/1560; 0.12%)
M. mucogenicum	13					1	(14/1560; 0.89%)
M. arupense	1						(1/1560; 0.06%)
M. scrofulaceum	8		2			11	(21/1560; 1.34%)
M. szulgai	3		1			13	(17/1560; 1.08%)
M. genavense	1					1	(2/1560; 0.12%)
M. phlei	18						(18/1560; 1.15%)
M. intermedium						1	(1/1560; 0.06%)
M. triviale	11						(11/1560; 0.70%)
M. nebraskense	1						(1/1560; 0.06%)
M. aurum	2						(2/1560; 0.12%)
M. neoaurum	2						(2/1560; 0.12%)
M. elephantis	1						(1/1560; 0.06%)
M. austroafricanum	1						(1/1560; 0.06%)

(*Continued*)

TABLE 10.15 (Continued)

Species	Country (Reference) (Period)						
	Iran (Velayati et al., 2015) (1984–2014)	Lebanon (Velayati et al., 2015) (1984–2014)	Saudi Arabia (13) (Velayati et al., 2015; Al-Ghafli and Al-Hajoj, 2017) (1984–2014)	Kuwait (Velayati et al., 2015) (1984–2014)	Oman (Velayati et al., 2015) (1984–2014)	Turkey (Velayati et al., 2015) (1991–2014)	Total Western Asia
M. parascrofulaceum	3						(3/1560; 0.19%)
M. sherrisii	1						(1/1560; 0.06%)
M. porcinum	3						(3/1560; 0.19%)
M. triplex	1						(1/1560; 0.06%)
M. branderi	3						(3/1560; 0.19%)
M. thermoresistibile	28						(28/1560; 1.79%)
M. gastri	6						(6/1560; 0.38%)
M. monacense	2						(2/1560; 0.12%)
M. montefiorense	1						(1/1560; 0.06%)

M. novocastrense	1						(1/1560; 0.06%)
M. conceptionense	9						(9/1560; 0.57%)
M. massiliense	3						(3/1560; 0.19%)
M. poriferae	1						(1/1560; 0.06%)
M. senegalense	11						(11/1560; 0.70%)
M. obuense	1						(1/1560; 0.06%)
M. haemophilum						2	(2/1560; 0.12%)
M. smegmatis	7		4			1	(12/1560; 0.76%)
M. flavescens	45		2		1	1	(49/1560; 3.14%)
Other						33	(33/1560; 2.11%)
Unidentified	157						(157/1560; 10.06%)
Total NTM	983	43	137	14	13	370	(1560/1560; 100%)

NTM based on environmental specimens is shown in Fig. 10.12 by the higher frequency of species. In general, in Western Asia (1600/2555; 62.63%) NTM has been reported from environmental specimens. The highest prevalence of NTM is in Iran with (1107/1600; 69.18%), followed by Iraq (271/1600; 16.93%), and Saudi Arabia (160/1600; 10%).

Clinical Nontuberculous Mycobacteria Reported From Southern Asia

The southern region of Asia includes data collected from two countries: India and Pakistan. The geographical distribution of the southern part of

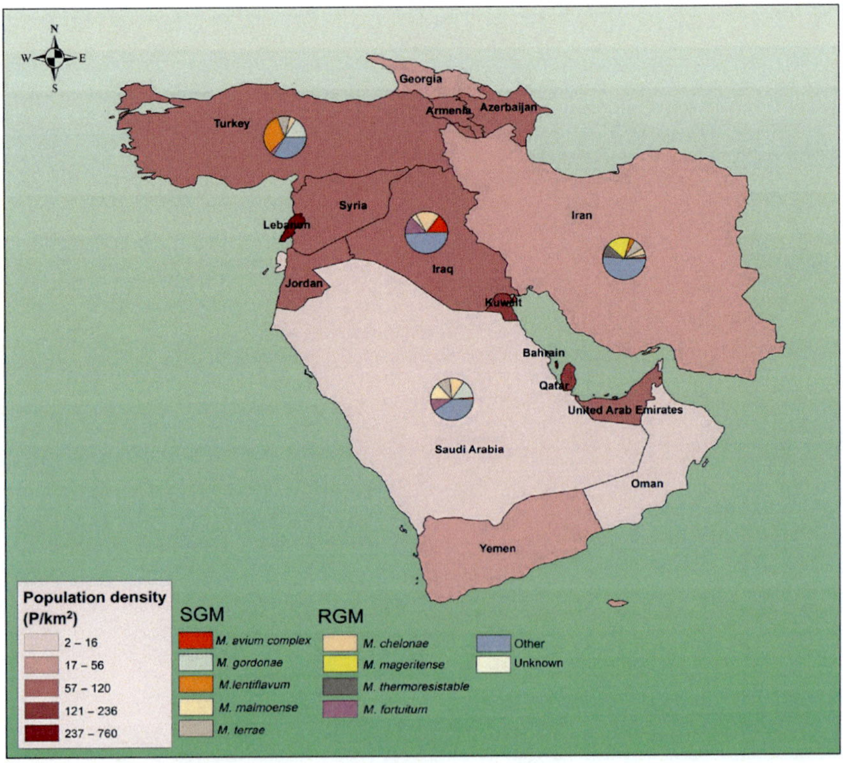

FIGURE 10.12 The distribution of nontuberculous mycobacteria species from environmental samples identified in Western Asia. *M. mageritense* (199/1600; 12.43%), *M. chelonae* (126/1600; 7.87%), *M. terrae* (116/1600; 7.24%), and *M. thermoresistable* (107/1600; 6.68%) were the most frequent species among different regions in Western Asia (Table 10.16).

TABLE 10.16 Species Distribution of Nontuberculous Mycobacteria From Environmental Samples in Western Asia

Kind of Species	Country (Reference) (Period)				
	Turkey (Velayati et al., 2015) (1984–2014)	Iraq (Velayati et al., 2015) (1984–2014)	Iran (Velayati et al., 2015; Varahram et al., 2016; Livani and Ghaemi, 2014) (1976–2013, 1984–2014, 2016)	Saudi Arabia (Velayati et al., 2015; Alqumber, 2014) (1984–2014)	Total/Western Asia
M. fortuitum	2 (3.23%)	37 (13.65%)	6 (0.54%)	15 (9.37%)	(60/1600; 3.75%)
M. avium complex		40 (14.76%)	14 (1.27%)	2 (1.25%)	(56/1600; 3.50%)
M. avium				12 (7.5%)	(12/1600; 0.75%)
M. mageritense			199 (17.98%)		(199/1600; 12.43%)
M. chubuense			26 (2.35%)		(26/1600; 1.62%)
M. gadium			3 (0.27%)		(3/1600; 0.18%)
M. brisbanense			2 (0.18%)		(2/1600; 0.12%)
M. neworleansense			3 (0.27%)		(3/1600; 0.18%)
M. sinense			2 (0.18%)		(2/1600; 0.12%)
M. bourgelatii			5 (0.45%)		(5/1600; 0.31%)
M. simiae	2 (3.23%)	20 (7.38%)	22 (1.98%)	15 (9.37%)	(59/1600; 3.68%)
M. gordonae	10 (16.13%)		20 (1.80%)	23 (14.38%)	(53/1600; 3.31%)
M. terrae	6 (9.68%)		94 (8.49%)	16 (10%)	(116/1600; 7.24%)
M. intracellulare				5 (3.13%)	(5/1600; 0.31%)
M. parafortuitum			38 (3.44%)		(38/1600; 2.37%)
M. neoaurum			26 (2.35%)		(26/1600; 1.62%)

(Continued)

TABLE 10.16 (Continued)

Kind of Species	Country (Reference) (Period)				
	Turkey (Velayati et al., 2015) (1984–2014)	Iraq (Velayati et al., 2015) (1984–2014)	Iran (Velayati et al., 2015; Varahram et al., 2016; Livani and Ghaemi, 2014) (1976–2013, 1984–2014, 2016)	Saudi Arabia (Velayati et al., 2015; Alqumber, 2014) (1984–2014)	Total/Western Asia
M. lentiflavum	20 (32.26%)		42 (3.79%)		(62/1600; 3.87%)
M. thermoresistable			107 (9.67%)		(107/1600; 6.68%)
M. smegmatis			3 (0.27%)		(3/1600; 0.18%)
M. peregrinum	3 (4.84%)		78 (7.05%)		(81/1600; 5.06%)
M. peregrinum			54 (4.88%)		(54/1600; 3.37%)
M. kansasii	5 (8.06%)	30 (11.07%)		16 (10%)	(51/1600; 3.18%)
M. mucogenicum			12 (1.09%)		(12/1600; 0.75%)
M. holsaticum			1 (0.09%)		(1/1600; 0.06%)
M. marinum		15 (5.54%)	4 (0.37%)		(19/1600; 1.18%)
M. abscessus		33 (12.17%)	20 (1.80%)	8 (5%)	(61/1600; 3.81%)
M. chelonae	3 (4.84%)	50 (18.45%)	55 (4.97%)	18 (11.25%)	(126/1600; 7.87%)
M. porcinum			11 (0.99%)		(11/1600; 0.68%)
M. vaccae			12 (1.09%)		(12/1600; 0.75%)
M. aichiense			2 (0.18%)		(2/1600; 0.12%)
M. phlei			37 (3.35%)		(37/1600; 2.31%)
M. triviale			20 (1.80%)		(20/1600; 1.25%)
M. xenopi		16 (5.90%)	10 (0.90%)		(26/1600; 1.62%)

M. fallax			10 (0.90%)		(10/1600; 0.62%)
M. conceptionence			58 (5.24%)		(58/1600; 3.62%)
M. aurum			8 (0.73%)		(8/1600; 0.5%)
M. poriferae			24 (2.17%)		(24/1600; 1.5%)
M. vanbaalenii			9 (0.82%)		(9/1600; 0.56%)
M. tilburgii			1 (0.09%)		(1/1600; 0.06%)
M. engbaekii			1 (0.09%)		(1/1600; 0.06%)
M. malmoense		11 (4.05%)		21 (13.13%)	(32/1600; 2%)
M. szulgai	2 (3.23%)	19 (7.01%)		9 (5.63%)	(30/1600; 1.87%)
M. genavense	6 (9.68%)				(6/1600; 0.37%)
M. haemophilum	3 (4.84%)		24 (2.17%)		(27/1600; 1.68%)
M. brumae			10 (0.90%)		(10/1600; 0.62%)
M. obuense			10 (0.90%)		(10/1600; 0.62%)
M. novocastrense			7 (0.64%)		(7/1600; 0.43%)
M. rhodesia			6 (0.55%)		(6/1600; 0.37%)
M. confluentis			3 (0.28%)		(3/1600; 0.18%)
M. gordonae-like			5 (0.46%)		(5/1600; 0.31%)
M. fluorantherivorans			3 (0.28%)		(3/1600; 0.18%)
Total NTM	62	271	1107	160	(1600/1600; 100%)

Asia is shown in Fig. 10.13 by the higher frequency of species. In general, in Southern Asia (351/82,138; 0.42%) NTM has been reported from clinical samples. The highest prevalence of NTM is in India with (235/351; 66.95%), followed by Pakistan (116/351; 33.04%).

Distribution of Nontuberculous Mycobacteria Based on Environmental Specimens From Southern Asia

The frequency of NTM in environmental samples includes data collected from India, Sri Lanka, and Nepal. The geographical distribution of NTM

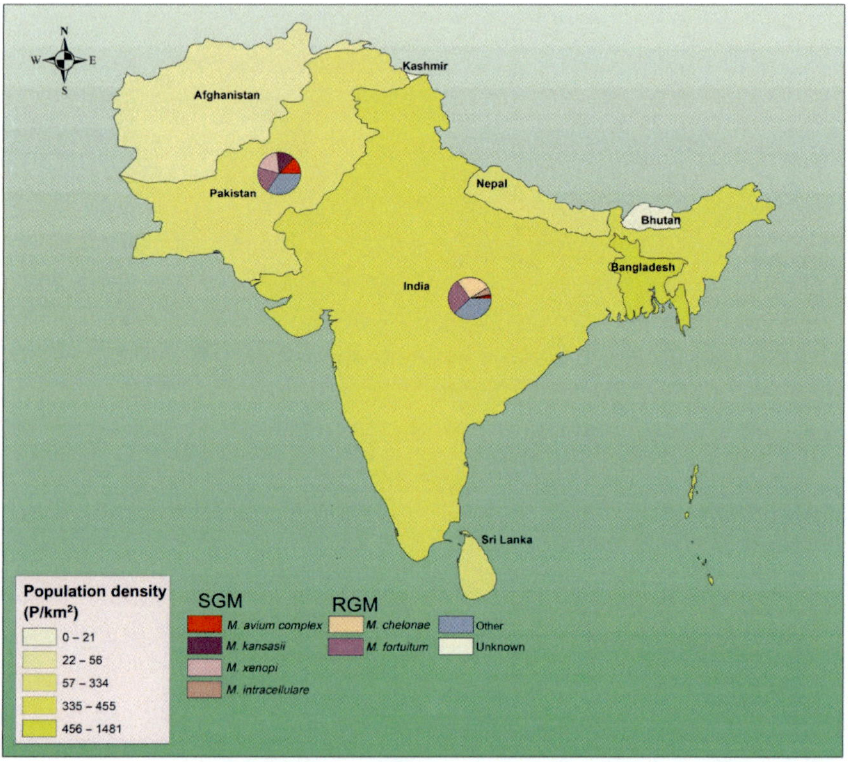

FIGURE 10.13 The distribution of nontuberculous mycobacteria species from clinical samples identified in Southern Asia. *M. fortuitum* (87/351; 24.7%), *M. chelonae* (61/351; 17.37%), *M. xenopi* (22/351; 6.26), and *M. avium* complex (20/351; 5.69%) were the most frequent species among different regions in Southern Asia (Table 10.17).

TABLE 10.17 Species Distribution of Nontuberculous Mycobacteria From Clinical Samples in Southern Asia

Species	Country (Reference) (Period)		
	India (Velayati et al., 2015; Maurya et al., 2015) (1999–2015)	Pakistan (Velayati et al., 2015) (1984–2014)	Total South Asia
M. avium complex	5	15	(20/351; 5.69%)
M. gordonae	3		(3/351; 0.85%)
M. xenopi		22	(22/351; 6.26%)
M. intracellulare	13		(13/351; 3.70%)
M. fortuitum	64	23	(87/351; 24.7%)
M. kansasii	3	16	(19/351; 5.41%)
M. chelonae	61		(61/351; 17.37%)
M. abscessus	9		(9/351; 2.56%)
M. marinum		1	(1/351; 0.28%)
M. terrae	3	1	(4/351; 1.13%)
M. simiae	1		(1/351; 0.28%)
M. interjectum	2		(1/351; 0.56%)
M. mucogenicum		12	(12/351; 3.41%)
M. scrofulaceum	1	2	(3/351; 0.85%)

(Continued)

TABLE 10.17 (Continued)

Species	Country (Reference) (Period)		
	India (Velayati et al., 2015; Maurya et al., 2015) (1999–2015)	Pakistan (Velayati et al., 2015) (1984–2014)	Total South Asia
M. szulgai	3		(3/351; 0.85%)
M. thermoresistibile		12	(12/351; 3.41%)
M. smegmatis	2	12	(14/351; 3.98%)
M. flavescens	1		(1/351; 0.28%)
Other	64		(64/351; 18.23%)
Unidentified			
Total NTM	235	116	(351/351; 100%)

based on environmental specimens is shown in Fig. 10.14 by the higher frequency of species. In general, in Western Asia (378/2555; 14.79%) NTM has been reported from environmental specimens. The highest prevalence of NTM is in India with (170/378; 44.97%), followed by Nepal (163/378; 43.12%), and Sri Lanka (45/378; 11.90%).

Africa Geography

Africa is the world's second largest and second most populous continent (after Asia). At about 30.3 million km^2 (11.7 million square miles) including adjacent islands, it covers 6% of Earth's total surface area and 20% of its land area. With 1.2 billion people as of 2016, it accounts for about 16% of the world's population. There are 54 countries in Africa, according to the United Nations. Additionally, there are four dependable state countries or territories, Reunion, Western Sahara, Mayotte, and Saint Helena.

Distribution of Nontuberculous Mycobacteria Based on Clinical and Environmental Specimens From Africa

This section is based on data compiled from information published on the distribution of NTM from nine countries in different regions of Africa

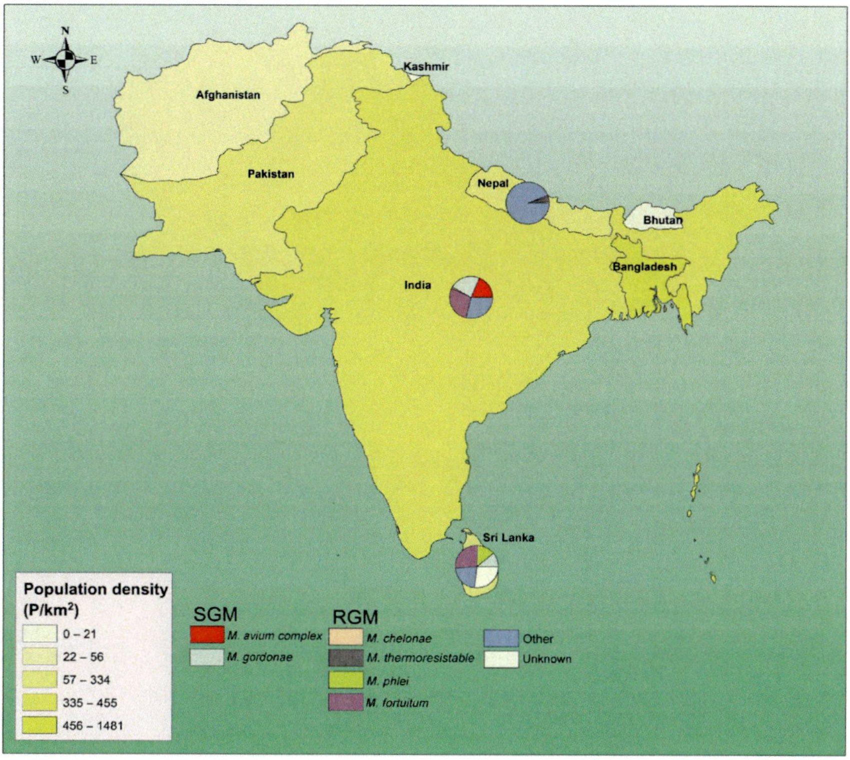

FIGURE 10.14 The distribution of nontuberculous mycobacteria species from environmental samples identified in Southern Asia. *M. fortuitum* (62/378; 16.40%), *M. gordonae* (45/378; 11.90%), and *M. avium* complex (32/378; 8.46%) were the most frequent species among different regions in Southern Asia (Table 10.18).

TABLE 10.18 The Distribution of Nontuberculous Mycobacteria Species From Environmental Samples Identified in Southern Asia

Kind of Species	Country (Reference) (Period)			
	India (Mishra et al., 2018) (2018)	Sri Lanka (Edirisinghe et al., 2014) (2014)	Nepal (Jha et al., 2007) (2003)	Total/ Southern Asia
M. fortuitum	48 (28.24%)	12 (26.67%)	2 (1.14%)	(62/378; 16.40%)
M. avium complex	32 (18.83%)			(32/378; 8.46%)
M. avium		1 (2.23%)		(1/378; 0.26%)
M. simiae	4 (2.36%)			(4/378; 1.05%)
M. gordonae	39 (22.95%)	5 (11.12%)	1 (0.57%)	(45/378; 11.90%)

(*Continued*)

TABLE 10.18 (Continued)

Kind of Species	Country (Reference) (Period)			
	India (Mishra et al., 2018) (2018)	Sri Lanka (Edirisinghe et al., 2014) (2014)	Nepal (Jha et al., 2007) (2003)	Total/ Southern Asia
M. terrae	1 (0.59%)	2 (4.45%)		(3/378; 0.79%)
M. vanbaalenii	1 (0.59%)			(1/378; 0.26%)
M. thermoresistable			6 (3.41%)	(6/378; 1.58%)
M. smegmatis	5 (2.95%)	1 (2.23%)		(6/378; 1.58%)
M. flavescens	17 (10%)			(17/378; 4.49%)
M. kansasii	3 (1.77%)			(3/378; 0.79%)
M. scrofulaceum	9 (5.30%)			(9/378; 2.38%)
M. marinum	4 (2.36%)	3 (6.67%)		(7/378; 1.85%)
M. chelonae	2 (1.18%)		2 (1.14%)	(4/378; 1.05%)
M. phlei		6 (13.34%)		(6/378; 1.58%)
M. xenopi	2(1.18%)			(2/378; 0.52%)
M. malmoense		2 (4.45%)		(2/378; 0.52%)
M. sherrisii	3 (1.77%)			(3/378; 0.79%)
M. celatum		1 (2.23%)		(1/378; 0.26%)
Other			152 (91.47%)	(152/378; 40.21%)
Unidentified		12 (26.67%)		(12/378; 3.17%)
Total NTM	170	45	163	(378/378; 100%)

between 1940 and 2016 (Tables 10.19 and 10.20). Tables 10.19 and 10.20 show an overview of studies that provided data for NTM from clinical and environmental specimens in Africa. In general, 8304 and 770 NTMs have been reported in clinical and environmental specimens from different regions of Africa, respectively. Among Egypt, Ghana, Kenya, Mali, Nigeria, South Arica, Tanzania, Uganda, Zambia, Ethiopia, Morocco, Cameroon, Zimbabwe, South Africa (6457/8304; 77.75%) and (420/770; 54.54%) had the majority of reported NTM from both clinical and environmental specimens, respectively. *M. avium*, *M. avium* complex, *M. kansasii*, *M. intracellulare*, *M. fortuitum*, *M. terrae*, and *M. nonchromogenicum* were identified as the most frequent species.

TABLE 10.19 Overview of Studies on Nontuberculous Mycobacteria From Clinical Specimens in Africa

Country (Reference) (Period)	Types of Samples	Common Species (*n*)	Total NTM
Egypt (Velayati et al., 2015) (1984–2014)	Extra pulmonary	*M. marinum* (39), *M. fortuitum* (5)	47
Ghana (Bjerrum et al., 2016; Okoi et al., 2017) (1940–2016)	Pulmonary samples	*M. avium* complex (9), *M. simiae* (3), *M. chelonae* complex (3)	38
Kenya (Okoi et al., 2017; Limo et al., 2015) (2007–09)	Sputum specimens	*M. intracellulare* (3), *M. fortuitum* (1)	15
Kenya (Okoi et al., 2017; Nyamogoba et al., 2012) (2014–15)	Sputum specimens	*M. intracellulare* (47), *M. abscessus* (12)	89
Mali (Okoi et al., 2017; Maiga et al., 2012) (2004–09)	Pulmonary samples	*M. avium* complex (11), *M. simiae* (2)	17
Nigeria (Okoi et al., 2017; Daniel et al., 2011; Aliyu and El-Kamary, 2014; Mawak et al., 2006; Pokam and Asuquo, 2012; Beer and Davis, 1965; Falodun et al.; Aliyu and El-Kamary, 2013) (1940–2016)	Pulmonary samples	*M. avium* complex (52), *M. fortuitum* (12), *M. gordonae* (8)	146
South Arica (Corbett et al., 1999a; Corbett et al., 1999c) (1993–98)	Sputum specimens	*M. kansasii* (419), *M. scrofulaceum* (85), *M. fortuitum* (17)	652
South Arica (Van Halsema et al., 2015; Katale et al., 2014; Corbett et al., 1999b) (2006–09)	Sputum specimens	*M. gordonae* (61), *M. kansasii* (50), *M. fortuitum* (21)	244

(*Continued*)

TABLE 10.19 (Continued)

Country (Reference) (Period)	Types of Samples	Common Species (*n*)	Total NTM
South Arica (Hoefsloot et al., 2013) (2008)	Pulmonary samples	*M. intracellulare* (2207), *M. avium* (593), *M. gordonae* (264)	5646
Tanzania (Okoi et al., 2017; Katale et al., 2014; Haraka et al., 2012; Hoza et al., 2016) (1940–2016)	Pulmonary samples	*M. gordonae* (7), *M. interjectum* (6), *M. intracellulare* (5)	49
Uganda (Okoi et al., 2017; Asiimwe et al., 2013; Bainomugisa et al., 2015) (2009–11)	Sputum specimens	*M. fortuitum* (40), *M. avium* (13)	109
Zambia (Malama et al., 2014; Chanda-Kapata et al., 2015; Buijtels et al., 2010; Buijtels et al., 2009; Buijtels et al., 2005; Mwikuma et al., 2015) (1940–2016)	Pulmonary samples	*M. avium* complex (327), *M. peregrinum* (59)	1250
Zambia (Buijtels et al., 2009) (2002–03)	Sputum or gargle specimens	*M. avium* complex (15), *M. intracellulare* (12)	72

TABLE 10.20 Overview of Studies on Nontuberculous Mycobacteria From Environmental Specimens in Africa

Country (Years)	Type of Sample	Most Common of Species	Total NTM
Uganda (Kankya et al., 2011) (2008–09)	Water	*M. gordonae* 5 (22.73%), *M. nonchromogenicum* 4 (18.18%), *M. fortuitum* 3 (13.63%)	22
Uganda (Kankya et al., 2011) (2008–09)	Soil	*M. fortuitum* 7 (30.43%), *M. intracellulare* 7 (30.43%), M. avium complex 3 (13.04%)	23
Uganda (Kankya et al., 2011) (2008–09)	Animal (fecal)	*M. fortuitum* 2 (66.67%), *M. engbaekii* 1 (33.33%)	3
Tanzania (Katale et al., 2014) (2012–13)	Wildlife	*M. intracellulare* 19 (46.34%), *M. lentiflavum* 11 (26.82%), *M. fortuitum* 3 (7.31%)	41

(Continued)

TABLE 10.20 (Continued)

Country (Years)	Type of Sample	Most Common of Species	Total NTM
Tanzania (Kazwala et al., 1998) (1987–96)	Cattle	*M. flavescens* 13 (41.94%), *M. terrae* 7 (22.58%), *M. smegmatis* 4 (12.90%)	31
Zambia (Monde et al., 2018) (2015–16)	Water	*M. gordonae* 5 (21.74%), *M. fortuitum* 3 (13.04%), *M. senegalense* 3 (13.04%)	23
Egypt (Mohamed et al., 2009) (2004–05)	Pig	*M. avium* complex 18 (50%), *M. scrofulaceum* 8 (22.22%), *M. chelonae* 5 (13.88%)	36
Ethiopia (Berg et al., 2009) (2006–08)	Cattle	*M. nonchromogenicum* 8 (30.77%), *M. gordonae* 5 (19.23%), *M. fortuitum* 3 (11.53%)	26
Morocco (Berrada, 1993) (1989–90)	Cattle	*M. avium* complex 73 (74.49%), *M. fortuitum* 14 (14.29%), Unknown NTM 11 (11.23%)	98
Cameroon (Egbe et al., 2016) (2012)	Cattle	*M. phlei* 11 (52.38%), *M. fortuitum* 5 (23.80%)	21
South Africa (Gcebe et al., 2013) (2010–12)	Soil	*M. terrae* 33 (20.63%), *M. nonchromogenicum* 16 (10%), *M. paraffinicum* 14 (8.75%)	160
South Africa (Gcebe et al., 2013) (2010–12)	Water	*M. nonchromogenicum* 11 (16.67%), M. terrae 7 (10.60%), *M. moriokaense*-like 6 (9.09%)	66
South Africa (Gcebe et al., 2013) (2010–12)	Swabs	*M. vaccae* 17 (16.50%), *M. moriokaense* 9 (8.73%), *M. acapulcensis* 9 (8.73%)	103
South Africa (Gcebe et al., 2013) (2010–12)	Animal source tissue	*M. avium* complex 16 (17.58%), *M. fortuitum* complex 8 (8.79%), *M. moriokaense*-like 9 (9.89%)	91
Zimbabwe (Padya et al., 2015) (2013–14)	Cattle	*M. neoaurum* 7 (26.92%), *M. fortuitum* 6 (23.07%), *M. terrae* 5 (19.23%)	26

Clinical Nontuberculous Mycobacteria Reported From Africa

This section includes data collected from nine countries: Uganda, Kenya, Tanzania, Zambia, South Arica, Mali, Ghana, Nigeria, and Egypt. The geographical distribution of NTM in African countries is shown in Fig. 10.15 by the higher frequency of species. In general, in Africa 8304 NTM has been

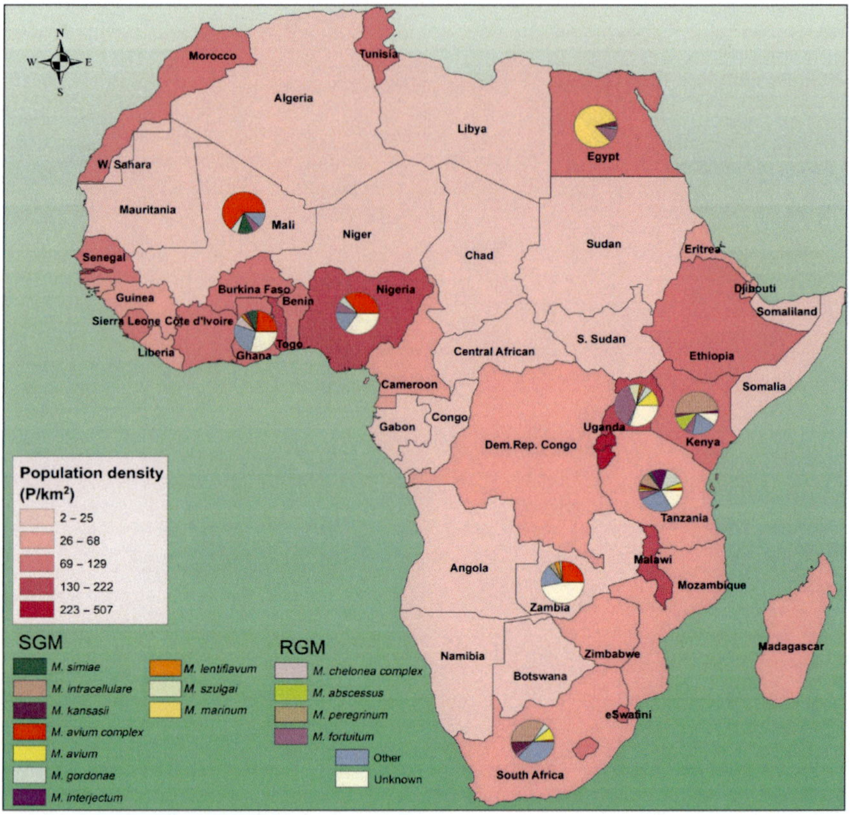

FIGURE 10.15 The distribution of nontuberculous mycobacteria species from clinical samples identified in Africa. *Mycobacterium kansasii* (636/8304; 7.65%), *Mycobacterium intracellulare* (2278/8304; 7.43%), *Mycobacterium avium* (611/8304; 7.35%), and *Mycobacterium avium* complex (517/8304; 6.22%) were the most frequent species among different countries in Africa (Table 10.21). Table 10.21 shows the data recorded for different countries in Africa, period of study, as well as the frequency of different species.

reported from clinical samples. The highest prevalence of NTM is in South Africa with (6457/8304; 77.75%), followed by Zambia (1337/8304; 16.10%), Uganda (109/8304; 1.31%), and Kenya (104/8304; 1.25%).

Distribution of Nontuberculous Mycobacteria Based on Environmental Specimens From Africa

The frequency of NTM in environmental samples includes data collected from Uganda, Tanzania, Zambia, Egypt, Morocco, Cameroon, Zimbabwe, Ethiopia, and South Africa. The geographical distribution of NTM based on

TABLE 10.21 Species Distribution of Nontuberculous Mycobacteria From Clinical Samples in Africa

Species	Country (Reference) (Period)									
	Uganda (Okoi et al., 2017; Asiimwe et al., 2013; Bainomugisa et al., 2015) (1940–2016)	Kenya (Okoi et al., 2017; Nyamogoba et al., 2012; Jacqueline et al.) (1940–2016)	Tanzania (Okoi et al., 2017; Katale et al., 2014; Hoza et al., 2016; Haraka et al., 2012) (1940–2016)	Zambia (Okoi et al., 2017; Malama et al., 2014; Chanda-Kapata et al., 2015; Buijtels et al., 2010; Buijtels et al., 2009; Buijtels et al., 2005; Mwikuma et al., 2015) (1940–2016)	South Arica (Hoefsloot et al., 2013; Okoi et al., 2017; Corbett et al., 1999a; Corbett et al., 1999c; Van Halsema et al., 2015; Katale et al., 2014; Corbett et al., 1999b) (1940–2016)	Mali (Okoi et al., 2017; Maiga et al., 2012) (1940–2016)	Ghana (Bjerrum et al., 2016; Okoi et al., 2017) (1940–2016)	Nigeria (Okoi et al., 2017; Daniel et al., 2011; Aliyu and El-Kamary, 2014; Mawak et al., 2006; Pokam and Asuquo, 2012; Beer and Davis, 1965; Aliyu and El-Kamary, 2013; Olutayo et al., 2016) (1940–2016)	Egypt (Velayati et al., 2015) (1984–2014)	Total NTM Africa
M. avium complex			1	342	102	11	9	52		(517/8304; 6.22%)
M. avium	13		2	3	593					(611/8304; 7.35%)
M. gordonae	6		7	23	325	1		8		(370/8304; 4.45%)
M. intracellulare	3	50	5	12	2208					(2278/8304; 7.43%)
M. fortuitum	40	7	3		209	1		12	5	(277/8304; 3.33%)

(*Continued*)

TABLE 10.21 (Continued)

Species	Country (Reference) (Period)								
	Uganda (Okoi et al., 2017; Asiimwe et al., 2013; Bainomugisa et al., 2015) (1940–2016)	Kenya (Okoi et al., 2017; Nyamogoba et al., 2012; Jacqueline et al.) (1940–2016)	Tanzania (Okoi et al., 2017; Katale et al., 2014; Hoza et al., 2016; Haraka et al., 2012) (1940–2016)	Zambia (Okoi et al., 2017; Malama et al., 2014; Chanda-Kapata et al., 2015; Buijtels et al., 2010; Buijtels et al., 2009; Buijtels et al., 2005; Mwikuma et al., 2015) (1940–2016)	South Arica (Hoefsloot et al., 2013; Okoi et al., 2017; Corbett et al., 1999a; Corbett et al., 1999c; Van Halsema et al., 2015; Katale et al., 2014; Corbett et al., 1999b) (1940–2016)	Mali (Okoi et al., 2017; Maiga et al., 2012) (1940–2016)	Ghana (Bjerrum et al., 2016; Okoi et al., 2017) (1940–2016)	Nigeria (Okoi et al., 2017; Daniel et al., 2011; Aliyu and El-Kamary, 2014; Mawak et al., 2006; Pokam and Asuquo, 2012; Beer and Davis, 1965; Aliyu and El-Kamary, 2013; Olutayo et al., 2016) (1940–2016)	Egypt (Velayati et al., 2015) (1984–2014)	Total NTM Africa
M. fortuitum complex							2			(2/8304; 0.024%)
M. kansasii	1	3	1		628		1		2	(636/8304; 7.65%)
M. chelonae/M. abscessus			3		3					(6/8304; 0.07%)
M. xenopi		1			41					(42/8304; 0.50%)
M. chelonae				8	196				1	(205/8304; 2.46%)

M. chelonae complex							3			(3/8304; 0.03%)
M. abscessus		12			18					(30/8304; 0.36%)
M. malmoense					43					(43/8304; 0.51%)
M. lentiflavum	2		1	49			1			(53/8304; 0.63%)
M. terrae complex					12					(12/8304; 0.14%)
M. marinum			1						39	(40/8304; 0.48%)
M. peregrinum	1	1		61	5					(68/8304; 0.81%)
M. chimaera				1						(1/8304; 0.012%)
M. terrae					37		1			(38/8304; 0.45%)
M. kumamotonense				2						(2/8304; 0.024%)
M. arupense							1			(1/8304; 0.012%)
M. neoaurum				1						(1/8304; 0.012%)
M. triviale					2					(2/8304; 0.024%)
M. parascrofulaceum					21					(21/8304; 0.25%)
M. elephantis				1						(1/8304; 0.012%)
M. flavescens					6					(6/8304; 0.072%)

(Continued)

TABLE 10.21 (Continued)

Species	Country (Reference) (Period)									
	Uganda (Okoi et al., 2017; Asiimwe et al., 2013; Bainomugisa et al., 2015) (1940–2016)	Kenya (Okoi et al., 2017; Nyamogoba et al., 2012; Jacqueline et al.) (1940–2016)	Tanzania (Okoi et al., 2017; Katale et al., 2014; Hoza et al., 2016; Haraka et al., 2012) (1940–2016)	Zambia (Okoi et al., 2017; Malama et al., 2014; Chanda-Kapata et al., 2015; Buijtels et al., 2010; Buijtels et al., 2009; Buijtels et al., 2005; Mwikuma et al., 2015) (1940–2016)	South Arica (Hoefsloot et al., 2013; Okoi et al., 2017; Corbett et al., 1999a; Corbett et al., 1999c; Van Halsema et al., 2015; Katale et al., 2014; Corbett et al., 1999b) (1940–2016)	Mali (Okoi et al., 2017; Maiga et al., 2012) (1940–2016)	Ghana (Bjerrum et al., 2016; Okoi et al., 2017) (1940–2016)	Nigeria (Okoi et al., 2017; Daniel et al., 2011; Aliyu and El-Kamary, 2014; Mawak et al., 2006; Pokam and Asuquo, 2012; Beer and Davis, 1965; Aliyu and El-Kamary, 2013; Olutayo et al., 2016) (1940–2016)	Egypt (Velayati et al., 2015) (1984–2014)	Total NTM Africa
M. porcinum				2						(2/8304; 0.024%)
M. europaeum				1						(1/8304; 0.012%)
M. goodii				1						(1/8304; 0.012%)
M. asiaticum				1	11					(12/8304; 0.014%)
M. bolletii				1						(1/8304; 0.012%)

M. bouchedurhonense				1						(1/8304; 0.012%)
M. simiae			1		5	2	3			(11/8304; 0.13%)
M. interjectum		2	6		1					(9/8304; 0.10%)
M. celatum			1							(1/8304; 0.012%)
M. scrofulaceum	2	4	3		468					(477/8304; 5.74%)
M. spaghni			2		2					(4/8304; 0.048%)
M. genavense			1		1					(2/8304; 0.024%)
M. gilvum			1		1					(2/8304; 0.024%)
M. gastri					19					(19/8304; 0.22%)
M. intermedium			1		1					(2/8304; 0.024%)
M. poriferae			1		1					(2/8304; 0.024%)
M. szulgai	9				5					(14/8304; 0.16%)
Other	32	14		208	1483	2	6	23		(1768/8304; 21.29%)
Unknown		10	8	619	10		11	51		(709/8304; 8.53%)
Total NTM	109	104	49	1337	6457	17	38	146	47	(8304/8304; 100%)

environmental specimens is shown in Fig. 10.16 by the higher frequency of species. In general, in Africa 770 NTM have been reported from environmental specimens. The highest prevalence of NTM is in South Africa with (420/770; 54.54%), followed by Morocco (98/770; 12.72%), Tanzania (72/770; 9.35%), and Uganda (48/770; 6.23%).

America Geography

The Americas comprise the continents of North and South America. Together, they make up most of the land in the Earth's western hemisphere and comprise the New World (America, Americas geography).

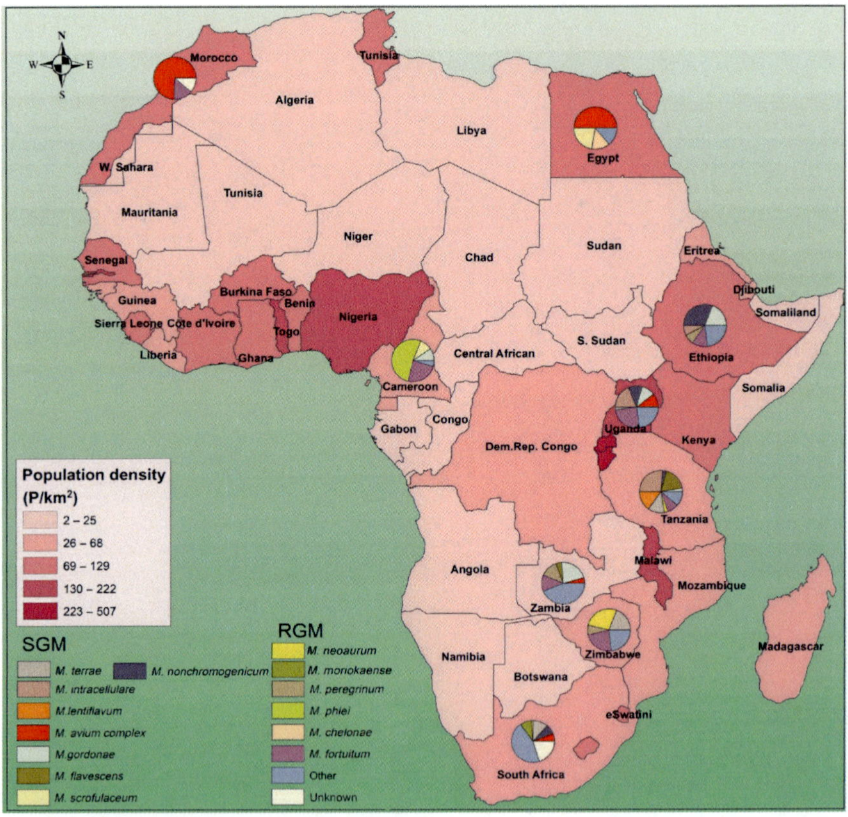

FIGURE 10.16 The distribution of nontuberculous mycobacteria species from environmental samples identified in Africa. *M. avium* complex (122/770; 15.84%), *M. terrae* (62/770; 8.05%), *M. fortuitum* (49/770; 6.36%), and *M. nonchromogenicum* (46/770; 5.97%) were the most frequent species among different countries in Africa (Table 10.22).

TABLE 10.22 Species Distribution of Nontuberculous Mycobacteria From Environmental Samples in Africa

Kind of Species	Country (Reference) (Period)									
	Uganda (Kankya et al., 2011) (2008–09)	Tanzania (Katale et al., 2014; Kazwala et al., 1998) (1987–96, 2012–13)	Zambia (Monde et al., 2018) (2015–16)	Egypt (Mohamed et al., 2009) (2004–05)	Morocco (Berrada, 1993) (1989–90)	Cameroon (Egbe et al., 2016) (2012)	Zimbabwe (Padya et al., 2015) (2013–14)	Ethiopia (Berg et al., 2009) (2006–08)	South Africa (Gcebe et al., 2013) (2010–12)	Total/ Africa
M. fortuitum	12 (25%)	6 (8.11%)	3 (13.05%)		14 (14.29%)	5 (23.81%)	6 (23.08%)	3 (11.54%)		(49/770; 6.36%)
M. avium complex	5 (10.42%)		1 (4.35%)	18 (37.5%)	73 (74.49%)				25 (5.96%)	(122/770; 15.84%)
M. simiae	1 (2.09%)	2 (2.71%)	1 (4.35%)	1 (2.09%)					10 (2.38%)	(15/770; 1.94%)
M. gordonae	5 (10.42%)	2 (2.71%)	5 (21.74%)			2 (9.53%)		5 (19.24%)	3 (0.72%)	(22/770; 2.85%)
M. terrae	1 (2.09%)	9 (12.17%)					5 (19.24%)		47 (11.19%)	(62/770; 8.05%)
M. intracellulare	9 (18.75%)	19 (25.68%)						2 (7.69%)		(30/770; 3.89%)
M. arupense	1 (2.09%)						2 (7.69%)		1 (0.24%)	(4/770; 0.51%)
M. nonchromogenicum	5 (10.42%)	2 (2.71%)						8 (30.77%)	31 (7.38%)	(46/770; 5.97%)
M. kubicae	2 (4.17%)									(2/770; 0.25%)
M. senuense	1 (2.09%)								1 (0.24%)	(2/770; 0.25%)
M. parafortuitum	1 (2.09%)								5 (1.19%)	(6/770; 0.77%)

(*Continued*)

TABLE 10.22 (Continued)

Kind of Species	Country (Reference) (Period)									
	Uganda (Kankya et al., 2011) (2008–09)	Tanzania (Katale et al., 2014; Kazwala et al., 1998) (1987–96, 2012–13)	Zambia (Monde et al., 2018) (2015–16)	Egypt (Mohamed et al., 2009) (2004–05)	Morocco (Berrada, 1993) (1989–90)	Cameroon (Egbe et al., 2016) (2012)	Zimbabwe (Padya et al., 2015) (2013–14)	Ethiopia (Berg et al., 2009) (2006–08)	South Africa (Gcebe et al., 2013) (2010–12)	Total/ Africa
M. hiberniae	1 (2.09%)									(1/770; 0.12%)
M. moriokaense-like									34 (8.09%)	(34/770; 4.41%)
M. szulgai									3 (0.72%)	(3/770; 0.38%)
M. engbaekii	2 (4.17%)							1 (3.85%)	5 (1.19%)	(8/770; 1.03%)
M. colombiense								1 (3.85%)		(1/770; 0.12%)
M. pulveris									4 (0.96%)	(4/770; 0.51%)
M. fortuitum complex									16 (3.81%)	(16/770; 2.07%)
M. vanbaalenii	1 (2.09%)									(1/770; 0.12%)
M. fluoranthenivorans									1 (0.24%)	(1/770; 0.12%)
M. chubuense	1 (2.09%)									(1/770; 0.12%)
M. lacticola									1 (0.24%)	(1/770; 0.12%)
M. acapulcensis								1 (3.85%)	19 (4.53%)	(20/770; 2.59%)

M. palustre									3 (0.72%)	(3/770; 0.38%)
M. neoaurum		2 (2.71%)					7 (26.93%)		6 (1.43%)	(15/770; 1.94%)
M. nebraskense									1 (0.24%)	(1/770; 0.12%)
M. goodii							3 (11.54%)		3 (0.72%)	(6/770; 0.77%)
M. lentiflavum		11 (14.87%)								(11/770; 1.42%)
M. austroafricanum									2 (0.48%)	(2/770; 0.25%)
M. thermoresistable		1 (1.36%)							1 (0.24%)	(2/770; 0.25%)
M. gallinarum				2 (4.17%)						(2/770; 0.25%)
M. duvalii									8 (1.91%)	(8/770; 1.03%)
M. interjectum									2 (0.48%)	(2/770; 0.25%)
M. confluentis		1 (1.36%)							3 (0.72%)	(4/770; 0.51%)
M. chelonae				5 (10.42%)					1 (0.24%)	(6/770; 0.77%)
M. elephantis							1 (3.85%)			(1/770; 0.12%)
M. smegmatis		4 (5.41%)	1 (4.35%)							(5/770; 0.64%)
M. monacense								1 (3.85%)	1 (0.24%)	(2/770; 0.25%)

(Continued)

TABLE 10.22 (Continued)

Kind of Species	Country (Reference) (Period)									
	Uganda (Kankya et al., 2011) (2008–09)	Tanzania (Katale et al., 2014; Kazwala et al., 1998) (1987–96, 2012–13)	Zambia (Monde et al., 2018) (2015–16)	Egypt (Mohamed et al., 2009) (2004–05)	Morocco (Berrada, 1993) (1989–90)	Cameroon (Egbe et al., 2016) (2012)	Zimbabwe (Padya et al., 2015) (2013–14)	Ethiopia (Berg et al., 2009) (2006–08)	South Africa (Gcebe et al., 2013) (2010–12)	Total/ Africa
M. wolinsky									1 (0.24%)	(1/770; 0.12%)
M. porcinum				2 (4.17%)						(2/770; 0.25%)
M. flavescens		13 (17.57%)	1 (4.35%)						4 (0.96%)	(18/770; 2.33%)
M. senegalense			3 (13.05%)							(3/770; 0.38%)
M. scrofulaceum				8 (16.67%)		2 (9.53%)				(10/770; 1.29%)
M. triviale									3 (0.72%)	(3/770; 0.38%)
M. triviplex									3 (0.72%)	(3/770; 0.38%)
M. asiaticum									6 (1.43%)	(6/770; 0.77%)
M. moriokaense									18 (4.29%)	(18/770; 2.33%)
M. gilvum			2 (8.69%)							(2/770; 0.25%)
M. madagascariense									1 (0.24%)	(1/770; 012%)

M. vaccae								1 (3.85%)	22 (5.24%)	(23/770; 2.98%)
M. intermedium									7 (1.67%)	(7/770; 0.90%)
M. kumamotonense			1 (4.35%)						3 (0.72%)	(4/770; 0.51%)
M. phlei						11 (57.39%)				(11/770; 1.42%)
M. septicum									9 (2.15%)	(9/770; 1.16%)
M. chitae									4 (0.96%)	(4/770; 0.51%)
M. holsaticum									2 (0.48%)	(2/770; 0.25%)
M. paraffinicum			1 (4.35%)						17 (4.05%)	(18/770; 2.33%)
M. mucogenicum						1 (4.77%)		1 (3.85%)		(2/770; 0.25%)
M. peregrinum			3 (13.05%)				2 (7.69%)	2 (7.69%)		(7/770; 0.90%)
M. housetonense			1 (4.35%)							(1/770; 0.12%)
Unidentified					11 (11.23%)				83 (19.77%)	(94/770; 12.20%)
Total NTM	48	72	23	36	98	21	26	26	420	(770/770; 100%)

North America is the third-largest continent, and is also a portion of the second largest supercontinent if North and South America are combined into the Americas and Africa, Europe, and Asia are considered to be part of the (largest) supercontinent Afro-Eurasia (Americas geography).

With an estimated population of 460 million and an area of 24,346,000 km^2 (9,400,000 mi^2), the northernmost of the two continents of the western hemisphere is bounded by the Pacific Ocean on the west; the Atlantic Ocean on the east; the Caribbean Sea, Atlantic, and Pacific oceans and South America on the south; and the Arctic Ocean on the north. The northern half of North America is sparsely populated and covered mostly by Canada, except for the Northeastern portion, which is occupied by Greenland, and the Northwestern portion, which is occupied by Alaska, the largest state of the United States. The central and southern portions of the continent are represented by the United States, Mexico, and numerous smaller states primarily in Central America and in the Caribbean (America, Americas geography).

The continent is delimited on the southeast by most geographers at the Darien watershed along the Colombia–Panama border, placing all of Panama within North America. Alternatively, a less common view would end North America at the Panama Canal. Islands generally associated with North America include Greenland, the world's largest island, and the archipelagos and islands in the Caribbean. The terminology of the Americas is complex, but "Anglo-America" can describe Canada and the United States, while "Latin America" comprises Mexico and the countries of Central America and the Caribbean, as well as South America (America, Americas geography).

Natural features of North America include the northern portion of the American Cordillera, represented by the geologically new Rocky Mountains in the west; and the considerably older Appalachian Mountains to the east. The north hosts an abundance of glacial lakes formed during the last glacial period, including the Great Lakes. North America's major continental divide is the Great Divide, which runs north and south down through the Rocky Mountains. The major watersheds all drain to the east: The Mississippi/Missouri and Rio Grande into the Gulf of Mexico, and St. Lawrence into the Atlantic Ocean (Americas geography).

Distribution of Nontuberculous Mycobacteria Based on Clinical and Environmental Specimens From America

Data compiled from information published on the distribution of NTM from four countries in different regions of America between 1990 and 2013 (Tables 10.23 and 10.24). Tables 10.23 and 10.24 shows the number of NTM detected and at least three species with the highest frequency and type

TABLE 10.23 Overview of Studies on Nontuberculous Mycobacteria From Clinical Specimens in America

Countries (Years)	Types of Samples	Common Species (*n*)	Total NTM
Argentina (Hoefsloot et al., 2013) (2008)	Pulmonary specimens	*M. avium* (11), *M. intracellulare* (6)	35
Brazil (Martin-Casabona et al., 2004) (1991–96)	Clinical specimens	*M. avium* complex (782), *M. gordonae* (106), *M. kansasii* (137)	1142
Brazil, Rio de Janeiro (Mello et al., 2013) (1993–2011)	Sputum samples	*M. kansasii* (59), *M. avium* complex (53), *M. abscessus* (23)	174
Canada, British Columbia (Hernandez-Garduno et al., 2009) (1990–2006)	Pulmonary specimens	*M. avium* complex (7398), *M. xenopi* (164), *M. fortuitum* (347)	9648
Canada, Ontario (Marras et al., 2007) (1997–2003)	Pulmonary, blood, urine, bone marrow	*M. avium* complex (6371), *M. xenopi* (2869), *M. kansasii* (183)	23,823
Canada, Ontario (Al Houqani et al., 2011) (2007)	Pulmonary specimens	*M. avium* complex (1521), *M. xenopi* (565), *M. kansasii* (40)	4364
United States (Cassidy et al., 2009) (2005–06)	Pulmonary and extra pulmonary	*M. avium* complex (901), *M. gordonae* (101), *M. abscessus* (19)	1301
United States (Henkle et al., 2017) (2007–12)	Extra pulmonary	*M. avium* complex (167), *M. fortuitum* complex (32), *M. chelonae/M. abscessus* (52)	334
United States, Washington (Ford et al., 2017) (1998–2011)	Pulmonary and extra pulmonary	*M. avium* complex (1620), *M. gordonae* (598), *M. abscessus* (155)	3689

of sample in each country. In general, 44,507 and 510 NTMs have been reported in clinical and environmental specimens from different regions of the America, respectively. Among American countries, Canada (37,835/44,510; 85%) reported the majority of NTM from pulmonary and nonpulmonary specimens, followed by United States (5324/44,510; 11.96%), Brazil (1316/44,510; 2.95%), and Argentina (35/44,510; 0.07%). Also, Unites

TABLE 10.24 Overview of Studies on Nontuberculous Mycobacteria From Environmental Specimens in America

Country (Years)	Type of Sample	Most Common of Species	Total NTM
Uruguay (Saenz and Errico, 1984) (1984)	Swin	*M. avium* complex 14 (46.67%), *M. terrae* 5 (16.66%), *M. kansasii* 3 (10%)	30
Argentina (Traversa et al., 2009) (2008)	Felis concolor	*M. gordonae* 2 (40%), *M. avium* complex 1 (20%), *M. simiae* 1 (20%)	5
Argentina (Barandiaran et al., 2015) (2007–09)	Swine	*M. avium* complex 36 (16.66%), *M. gordonae* 2 (1.29%), *M. scrofulaceum* 1 (0.64%)	154
Brazil (Sgarioni et al., 2014) (2012–13)	Milk	*M. nonchromogenicum* 5 (9.61%), *M. peregrinum* 4 (33.33%), *M. smegmatis* 3 (25%)	12
Brazil (Falcão et al., 1993) (1987)	Water	*M. fortuitum* 10 (32.26%), *M. nonchromogenicum* 4 (12.90%), *M. gordonae* 3 (9.67%)	31
Mexico (Cerna-Cortes et al., 2015) (2013)	Raw vegetables	*M. porcinum* 5 (26.31%), *M. fortuitum* 3 (15.79%), *M. mucogenicum* 3 (15.79%)	19
Canada (Cayer et al., 2007) (2000–03)	Peat	*M. intracellulare* 12 (24.49%), *M. interjectum* 7 (21.88%), *M. graecum* 2 (6.25%)	32
Ohio (Covert et al., 1999) (1999)	Drinking water	*M. mucogenicum* 13 (28.26%), *M. gordonae* 5 (10.41%), *M. intracellulare* 5 (10.41%)	48
California (Vugia et al., 2005) (2000)	Whirlpool footbaths	*M. fortuitum* 14 (48.27%), *M. mucogenicum* 7 (14.58%), *M. mageritense* 6 (12.5%)	48
Western United States (Hilborn et al., 2006) (2006)	Drinking water	*M. avium* complex 8 (21.05%), *M. kansasii* 3 (7.89%), *M. fortuitum* 2 (5.26%)	38
Kansas (Cox, 2012) (1990–2012)	Water	*M. gordonae* 9 (27.28%), *M. fortuitum* 7 (21.22%), M. chelonae 5 (15.16%)	33
Atlanta (Kline et al., 2004) (2004)	Water	*M. mucogenicum* 28 (46.67%), *M. gordonae* 15 (25%), *M. avium* complex 12 (20%)	60

States (227/510; 44.50%) had the majority reported NTM from environmental specimens, followed by Argentina (159/510; 31.7%).

Mycobacterium avium, *Mycobacterium avium* complex, *Mycobacterium gordonae*, *Mycobacterium xenopi*, *Mycobacterium kansasii*, *Mycobacterium chelonae*, *Mycobacterium intracellulare*, and *Mycobacterium fortuitum* were identified as the most frequent species.

Clinical Nontuberculous Mycobacteria Reported From America

The America study data collected from four countries: Argentina, Brazil, Canada, and United States. The geographical distribution of Northern and Southern America is shown in Fig. 10.17 by the higher frequency of species. In general, in America 44,507 NTM have been reported from clinical samples. The highest prevalence of NTM is in Canada with (37,835/44,510; 85%), followed by Unites states (5324/44,510; 11.96%), Brazil (1316/44,510; 2.95%), and Argentina (35/44,510; 0.07%).

Distribution of Nontuberculous Mycobacteria Based on Environmental Specimens From America

The frequency of NTM in environmental samples includes data collected from Uruguay, Argentina, Brazil, México, Canada, and the United States. The geographical distribution of NTM based on environmental specimens is

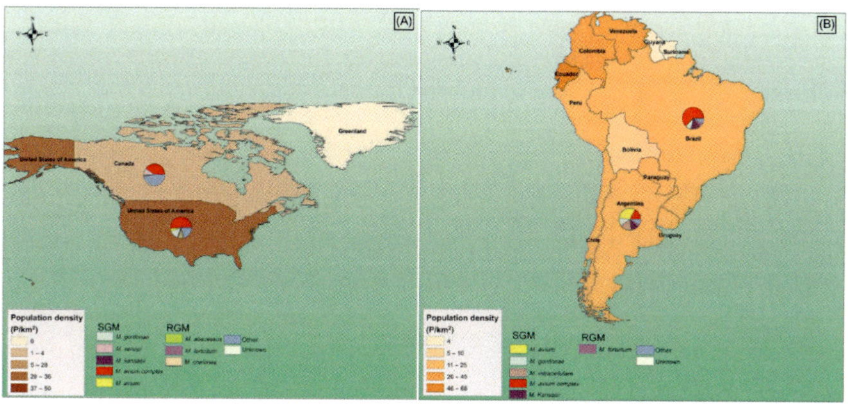

FIGURE 10.17 The distribution of nontuberculous mycobacteria species from clinical samples identified in (A) Northern and (B) Southern America. *M. avium* complex (18,816/44,510; 42.27%), *M. xenopi* (3647/44,510; 8.19%), *M. gordonae* (809/44,510; 1.81%), and *M. kansasii* (732/44,510; 1.64%) were the most frequent species among different regions in America (Table 10.25). Table 10.25 shows the data recorded for America by country, period of study, as well as the frequency of different species.

TABLE 10.25 Species Distribution of Nontuberculous Mycobacteria From Clinical Samples in America

Species	Country (Reference) (Period)				
	Argentina (Hoefsloot et al., 2013) (2008)	Brazil (Mello et al., 2013) (1991–2011)	Canada (Hernandez-Garduno et al., 2009; Marras et al., 2007; Al Houqani et al., 2011; Marras et al., 2017) (1990–2007)	United States (Cassidy et al., 2009; Henkle et al., 2017; Ford et al., 2017) (1998–2012)	Total America
M. avium complex	6	835	15,290	2685	(18,816/44,510; 42.27%)
M. avium	11			235	(246/44,510; 0.55%)
M. gordonae	4	106		699	(809/44,510; 1.81%)
M. xenopi		8	3598	41	(3647/44,510; 8.19%)
M. intracellulare	6			128	(134/44,510; 0.30%)
M. fortuitum group				10	(10/44,510; 0.022%)
M. fortuitum/chelonae group				16	(16/44,510; 0.035%)
M. fortuitum	1	64	347	82	(494/44,510; 1.10%)
M. fortuitum complex				53	(53/44,510; 0.11%)
M. fortuitum/peregrinum				28	(28/44,510; 0.06%)
M. kansasii	4	196	445	87	(732/44,510; 1.64%)
M. chelonae	1		482	190	(673/44,510; 1.51%)
M. chelonae/M. abscessus				92	(92/44,510; 0.20%)
M. abscessus		23		174	(197/44,510; 0.44%)
M. abscessus/massiliense				44	(44/44,510; 0.098%)
M. lentiflavum		1		66	(67/44,510; 0.15%)
M. sherrisii	1			1	(2/44,510; 0.004%)

M. marinum/M. ulcerans				5	(5/44,510; 0.011%)
M. ulcerans				1	(1/44,510; 0.002%)
M. marinum				88	(88/44,510; 0.19%)
M. terrae		1		8	(9/44,510; 0.020%)
M. terrae complex				3	(3/44,510; 0.006%)
M. peregrinum		2		7	(9/44,510; 0.020%)
M. simiae		1		23	(24/44,510; 0.053%)
M. simiae complex				4	(4/44,510; 0.008%)
M. nonchromogenicum				7	(7/44,510; 0.015%)
M. chimaera				3	(3/44,510; 0.006%)
M. interjectum				4	(4/44,510; 0.008%)
M. arupense				3	(3/44,510; 0.006%)
M. celatum		1		2	(3/44,510; 0.006%)
M. scrofulaceum				6	(6/44,510; 0.013%)
M. szulgai		1		17	(18/44,510; 0.040%)
M. mucogenicum/phocaicum				12	(12/44,510; 0.026%)
M. mucogenicum				21	(21/44,510; 0.047%)
M. genavense				8	(8/44,510; 0.017%)
M. phlei				2	(2/44,510; 0.004%)
M. haemophilum				12	(12/44,510; 0.026%)

(Continued)

TABLE 10.25 (Continued)

Species	Country (Reference) (Period)				
	Argentina (Hoefsloot et al., 2013) (2008)	Brazil (Mello et al., 2013) (1991–2011)	Canada (Hernandez-Garduno et al., 2009; Marras et al., 2007; Al Houqani et al., 2011; Marras et al., 2017) (1990–2007)	United States (Cassidy et al., 2009; Henkle et al., 2017; Ford et al., 2017) (1998–2012)	Total America
M. florentinum				1	(1/44,510; 0.002%)
M. nebraskense				26	(26/44,510; 0.058%)
M. obuense				10	(10/44,510; 0.022%)
M. doricum				6	(6/44,510; 0.013%)
M. bovis				20	(20/44,510; 0.044%)
M. kumamotonense				2	(2/44,510; 0.004%)
M. neoaurum				11	(11/44,510; 0.024%)
M. aurum				1	(1/44,510; 0.002%)
M. vaccae				3	(3/44,510; 0.006%)
M. elephantis				1	(1/44,510; 0.002%)
M. insubricum				1	(1/44,510; 0.002%)
M. porcinum				15	(15/44,510; 0.033%)
M. alvei				2	(2/44,510; 0.004%)
M. parascrofulaceum				5	(5/44,510; 0.011%)
M. shimoidei				2	(2/44,510; 0.004%)

M. paraffinicum				10	(10/44,510; 0.022%)
M. triplex				3	(3/44,510; 0.006%)
M. branderi				2	(2/44,510; 0.004%)
M. goodii				15	(15/44,510; 0.033%)
M. cosmeticum				1	(1/44,510; 0.002%)
M. heckeshornense				10	(10/44,510; 0.022%)
M. frederiksbergense				3	(3/44,510; 0.006%)
M. hassiacum				1	(1/44,510; 0.002%)
M. leprae				6	(6/44,510; 0.013%)
M. lepraemurium				1	(1/44,510; 0.002%)
M. immunogenum				7	(7/44,510; 0.015%)
M. gastri				1	(1/44,510; 0.002%)
M. ilatzerense				3	(3/44,510; 0.006%)
M. intermedium				1	(1/44,510; 0.002%)
M. seoulense				1	(1/44,510; 0.002%)
M. kyorinense				1	(1/44,510; 0.002%)
M. conceptionense				2	(2/44,510; 0.004%)
M. aubagnense				3	(3/44,510; 0.006%)
M. bolletii				7	(7/44,510; 0.015%)
M. brisbanense				1	(1/44,510; 0.002%)

(*Continued*)

TABLE 10.25 (Continued)

Species	Country (Reference) (Period)				
	Argentina (Hoefsloot et al., 2013) (2008)	Brazil (Mello et al., 2013) (1991–2011)	Canada (Hernandez-Garduno et al., 2009; Marras et al., 2007; Al Houqani et al., 2011; Marras et al., 2017) (1990–2007)	United States (Cassidy et al., 2009; Henkle et al., 2017; Ford et al., 2017) (1998–2012)	Total America
M. septicum				1	(1/44,510; 0.002%)
M. senegalense				1	(1/44,510; 0.002%)
M. parmense				1	(1/44,510; 0.002%)
M. pulveris				1	(1/44,510; 0.002%)
M. massiliense		3		6	(9/44,510; 0.020%)
M. asiaticum		1		1	(2/44,510; 0.004%)
M. mageritense				3	(3/44,510; 0.006%)
M. smegmatis				5	(5/44,510; 0.011%)
M. flavescens	1				(1/44,510; 0.002%)
Other		59	17,673	200	(17,932/44,510; 40.28%)
Unidentified		14		50	(64/44,507; 0.14%)
Total NTM	35	1316	37,835	5324	(44,510/44,510; 100%)

shown in Fig. 10.18 by the higher frequency of species. In general, in America 510 NTM have been reported from environmental specimens. The highest prevalence of NTM is in the United States with (227/510; 44.50%), followed by Argentina (159/510; 31.17%), Brazil (43/510; 8.43%), and Canada (32/510; 6.27%).

Australia Geography

Australia is a country, a continent, and an island. It is located in Oceania between the Indian Ocean and the South Pacific Ocean (Australia). It is the sixth largest country in the world with a total area of 7,686,850 km^2 (2,967,910 mi^2) (including Lord Howe Island and Macquarie Island), making

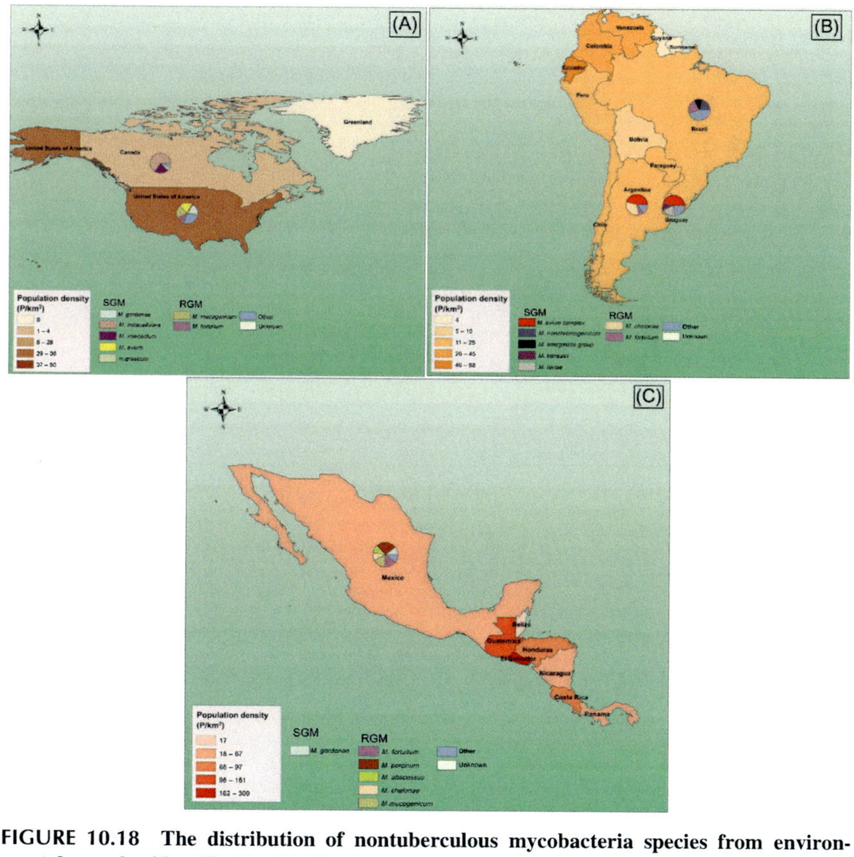

FIGURE 10.18 The distribution of nontuberculous mycobacteria species from environmental samples identified in (A) Northern), (B) Southern, and (C) Central of America. *M. avium* complex (89/510; 17.45%), *M. chelonae* (60/510; 11.76%), *M. mucogenicum* (56/510; 10.98%), and *M. fortuitum* (53/510; 10.39%) were the most frequent species among different regions in America.

TABLE 10.26 Species Distribution of Nontuberculous Mycobacteria From Environmental Samples in America

Kind of Species	Country (Reference) (Period)						
	Uruguay (Saenz and Errico, 1984) (1984)	Argentina (Traversa et al., 2009; Barandiaran et al., 2015) (2007–09)	Brazil (Sgarioni et al., 2014; Falcão et al., 1993) (1987, 2012–13)	México (Cerna-Cortes et al., 2015) (2013)	Canada (Cayer et al., 2007) (2000–03)	USA (Covert et al., 1999; Vugia et al., 2005; Cox, 2012; Kline et al., 2004) (1999–2012)	Total/America
M. fortuitum		12 (7.55%)	10 (23.26%)	3 (15.79%)		28 (12.34%)	(53/510; 10.39%)
M. fortuitum complex						2 (0.89%)	(2/510; 0.39%)
M. avium complex	14 (46.67%)	73 (45.92%)		1 (5.27%)		1 (0.45%)	(89/510; 17.45%)
M. avium						43 (18.95%)	(43/510; 8.43%)
M. simiae		1 (0.63%)				3 (1.33%)	(4/510; 0.78%)
M. gordonae		2 (1.26%)	3 (6.98%)	2 (10.53%)		35 (15.42%)	(42/510; 8.23%)
M. terrae	5 (16.67%)					4 (1.77%)	(9/510; 1.76%)
M. intracellulare					21 (65.63%)	5 (2.21%)	(26/510; 5.09%)
M. llatzerense		11 (6.92%)					(11/510; 2.15%)
M. interjectum					7 (21.88%)		(7/510; 1.37%)
M. immunogenum		3 (1.89%)					(3/510; 0.58%)

M. bohemicum					1 (3.13%)		(1/510; 0.19%)
M. nonchromogenicum			9 (20.93%)				(9/510; 1.76%)
M. graecum					2 (6.25%)		(2/510; 0.39%)
M. neoaurum						2 (0.89%)	(2/510; 0.39%)
M. lentiflavum						2 (0.89%)	(2/510; 0.39%)
M. angelicum						6 (2.65%)	(6/510; 1.17%)
M. smegmatis			5 (11.63%)		1 (3.13%)	4 (1.77%)	(10/510; 1.96%)
M. peregrinum			4 (9.31%)			2 (0.89%)	(6/510; 1.17%)
M. kansasii	3 (10%)					5 (2.21%)	(8/510; 1.56%)
M. scrofulaceum	2 (6.67%)	1 (0.63%)				2 (0.89%)	(5/510; 0.98%)
M. mucogenicum		5 (3.15%)		3 (15.79%)		48 (21.15%)	(56/510; 10.98%)
M. mageritense						6 (2.65%)	(6/510; 1.17%)
M. gastri	2 (6.67%)					3 (1.33%)	(5/510; 0.98%)
M. abscessus				2 (10.53%)			(2/510; 0.39%)
M. chelonae	1 (3.34%)	51 (32.08%)		2 (10.53%)		6 (2.65%)	(60/510; 11.76%)

(*Continued*)

TABLE 10.26 (Continued)

Kind of Species	Country (Reference) (Period)						
	Uruguay (Saenz and Errico, 1984) (1984)	Argentina (Traversa et al., 2009; Barandiaran et al., 2015) (2007–09)	Brazil (Sgarioni et al., 2014; Falcão et al., 1993) (1987, 2012–13)	México (Cerna-Cortes et al., 2015) (2013)	Canada (Cayer et al., 2007) (2000–03)	USA (Covert et al., 1999; Vugia et al., 2005; Cox, 2012; Kline et al., 2004) (1999–2012)	Total/America
M. porcinum				5 (26.32%)			(5/510; 0.98%)
M. vaccae	2 (6.67%)						(2/510; 0.39%)
M. phlei			1 (2.33%)				(1/510; 0.19%)
M. parafortuitum			1 (2.33%)				(1/510; 0.19%)
M. aurum	1 (3.34%)						(1/510; 0.19%)
Other			10 (23.26%)	1 (5.27%)		14 (6.17%)	(25/510; 4.90%)
Unidentified						6 (2.65%)	(6/510; 1.17%)
Total NTM	30	159	43	19	32	227	(510/510; 100%)

it slightly smaller than the 48 states of the contiguous United States and 31.5 times larger than the United Kingdom (Australia).

The Australian mainland has a total coastline length of 35,877 km (22,293 mi) with an additional 23,859 km (14,825 mi) of island coastlines. There are 758 estuaries around the country with most located in tropical and subtropical zones. Australia claims an extensive Exclusive Economic Zone of 8,148,250 km^2 (3,146,057 mi^2). This exclusive economic zone does not include the Australian Antarctic Territory (Australia; Australian Bureau of Statistics; Australia geography).

Australia has the largest area of ocean jurisdiction of any country on Earth. It has no land borders. The northernmost parts of the country are the Cape York Peninsula of Queensland and the Top End of the Northern Territory. The western half of Australia consists of the Western Plateau, which rises to mountain heights near the west coast and falls to lower elevations near the continental center. The Western Plateau region is generally flat, though broken by various mountain ranges, such as the Hamersley Range, the MacDonnell Range, and the Musgrave Range. Surface water is generally lacking in the Western Plateau, although there are several larger rivers in the west and north, such as the Murchison, Ash Burton, and Victoria Rivers. The Eastern Highlands, or Great Dividing Range, lie near the eastern coast of Australia, separating the relatively narrow eastern coastal plain from the rest of the continent. These Eastern Australian temperate forests have the greatest relief, the most rainfall, the most abundant and varied flora and fauna, and the densest human settlement (Australia; Australia geography).

Between the Eastern Highlands and the Western Plateau, lies the Central Lowlands, which are made up of the Great Artesian Basin and Australia's largest river systems, Murray-Darling Basin and Lake Eyre Basin (Australia).

Distribution of Nontuberculous Mycobacteria Based on Clinical and Environmental Specimens From Australia

This section is based on data compiled from information published on the distribution of NTM from different regions of Australia in 1989 and 2011 (Tables 10.27 and 10.28). Tables 10.27 and 10.28 show an overview of studies on NTM from clinical and environmental specimens in entire Australia. In general, 6658 and 538 NTMs have been reported in clinical and environmental specimens from different regions of the Australia, respectively. Among Australia, Queensland (6599/6658; 99.11%) had the reported major NTM from clinical specimens, followed by northern, Australia (59/6658; 0.88%). Also Brisbane 538 NTM have been reported in environmental specimens from different regions.

TABLE 10.27 Overview of Studies on Nontuberculous Mycobacteria From Clinical Specimens in Australia

Country (Reference) (Period)	Type of Sample	Most Common of Species	Total NTM
Queensland, Australia (Chou et al., 2014) (2001–11)	Clinical specimen	*M. intracellulare* (2306) 34.63%, *M. avium* (674) 10.1%, *M. fortuitum* (493) 7.40%	6599
Northern, Australia (O'brien et al., 2000) (1989–97)	Clinical specimen	*M. avium* complex (45) 76.27%, *M. fortuitum* (5) 8.47%, *M. abscessus* (2) 3.38%	59

TABLE 10.28 Overview of Studies on Nontuberculous Mycobacteria From Environmental Specimens in Australia

Country (Years)	Type of Sample	Most Common of Species	Total NTM
Australia (Thomson et al., 2013) (2013)	Water samples in Summer	*M. mucogenicum* (31) 23.85%, *M. gordonae* (24) 18.46%, *M. poriferae* (18) 13.84%	130
Australia (Thomson et al., 2013) (2013)	Water samples in Winter	*M. kansasii* (133) 32.60%, *M. gordonae* (120) 29.41%, *M. mucogenicum* (42) 10.29%	408

Different NTM species were identified from clinical and environmental specimens. *Mycobacterium avium*, *Mycobacterium avium* complex, *Mycobacterium gordonae*, *Mycobacterium xenopi*, *Mycobacterium kansasii*, *Mycobacterium intracellulare*, and *Mycobacterium fortuitum* identified most frequently species.

Clinical Nontuberculous Mycobacteria Reported From Australia

Data were collected from two different regions of Australia: Queensland and Northern Australia. The geographical distribution of Australia is shown in Fig. 10.19 by a higher frequency of species. In general, in Northern Australia (59/6658; 0.88%) NTM have been reported from clinical samples. The highest prevalence of NTM is in Queensland with (6599/6658; 99.11%).

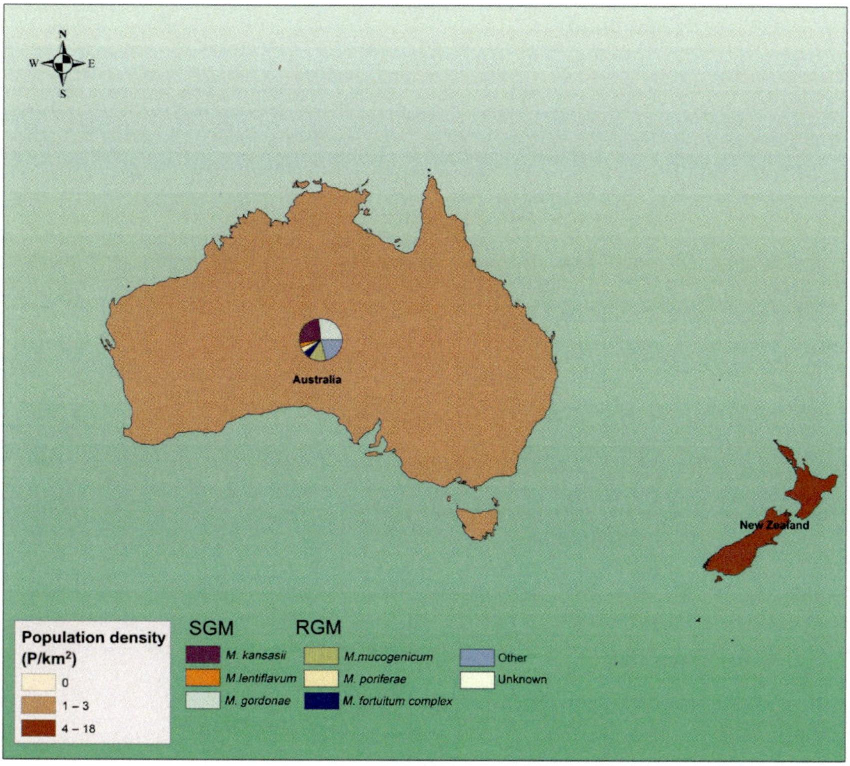

FIGURE 10.19 The distribution of nontuberculous mycobacteria species from clinical samples identified in Australia. *Mycobacterium intracellulare* (2306/6658; 34.63%), *Mycobacterium avium* (674/6658; 10.12%), and *Mycobacterium fortuitum* (498/6658; 7.47%) were the most frequent species among different regions in Australia (Table 10.29). Table 10.29 shows the data recorded for Australia by country, period of study, as well as the frequency of different species.

TABLE 10.29 Species Distribution of Nontuberculous Mycobacteria From Clinical Samples in Northern Australia

Species	Country (Reference) (Period)		
	Queensland, Australia (Chou et al., 2014) (2001–11)	Northern, Australia (O'brien et al., 2000) (1989–97)	Total Australia
M. avium complex		45	(45/6658; 0.67%)
M. avium	674		(674/6658; 10.12%)
M. gordonae	162	1	(163/6658; 2.44%)
M. intracellulare	2306		(2306/6658; 34.63%)
M. fortuitum	493	5	(498/6658; 7.47%)
M. kansasii	188		(188/6658; 2.82%)

(Continued)

TABLE 10.29 (Continued)

Species	Country (Reference) (Period)		
	Queensland, Australia (Chou et al., 2014) (2001–11)	**Northern, Australia (O'brien et al., 2000) (1989–97)**	**Total Australia**
M. chelonae	183		(183/6658; 2.74%)
M. abscessus	489	2	(491/6658; 7.37%)
M. terrae		1	(1/6658; 0.01%)
M. simiae		1	(1/6658; 0.01%)
M. scrofulaceum		2	(2/6658; 0.03%)
M. haemophilum		2	(2/6658; 0.03%)
Other	1357		(1357/6658; 20.38%)
Unknown	747		(747/6658; 11.21%)
Total NTM	6599	59	(6658/6658; 100%)

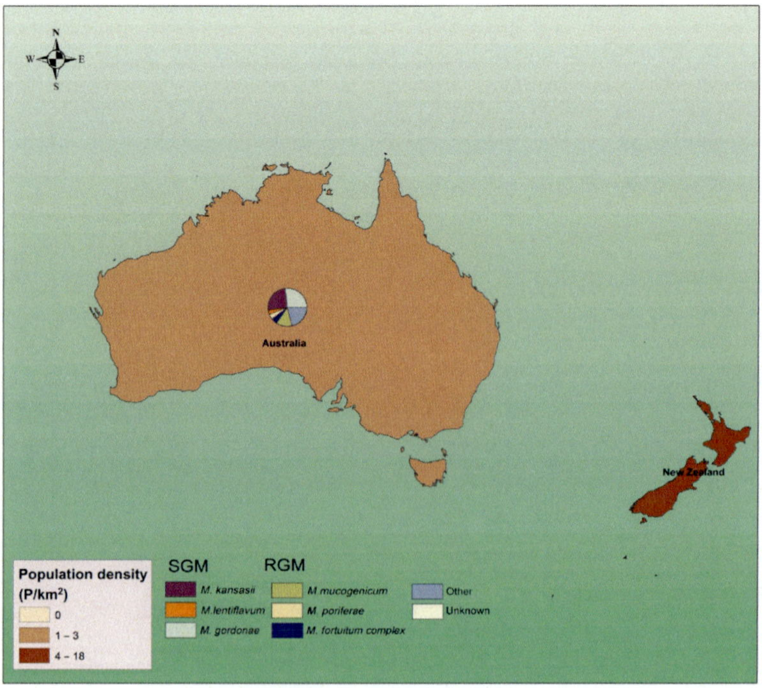

FIGURE 10.20 The distribution of nontuberculous mycobacteria species from environmental samples identified in Australia. *M. gordonae* (144/538; 26.76%), *M. kansasii* (138/538; 25.65%), *M. mucogenicum* (73/538; 13.56%), and *M. fortuitum* (27/538; 5.01%) were the most frequent species among different regions in Australia (Table 10.30).

TABLE 10.30 Species Distribution of Nontuberculous Mycobacteria From Environmental Samples in Australia

Kind of Species	Country (Reference) (Period)	
	Brisbane, Australia (Thomson et al., 2013) (2013)	Total Australia
M. fortuitum	27 (5.02%)	(27/538; 5.01%)
M. avium complex/ Avium	3 (0.56%)	(3/538; 0.55%)
M. simiae	2 (0.38%)	(2/538; 0.37%)
M. gordonae	144 (26.77%)	(144/538; 26.76%)
M. terrae	2 (0.38%)	(2/538; 0.37%)
M. intracellulare	2 (0.38%)	(2/538; 0.37%)
M. arupense	9 (1.68%)	(9/538; 1.67%)
M. lentiflavum	19 (3.54%)	(19/538; 3.53%)
M. flavescens	3 (0.56%)	(3/538; 0.55%)
M. senegalense	2 (0.38%)	(2/538; 0.37%)
M. gilvum	1 (0.19%)	(1/538; 0.18%)
M. angelicum	1 (0.19%)	(1/538; 0.18%)
M. bolletii	1 (0.19%)	(1/538; 0.18%)
M. kansasii	138 (25.66%)	(138/538; 25.65%)
M. mucogenicum	73 (13.57%)	(73/538; 13.56%)
M. mageritense	5 (0.93%)	(5/538; 0.92%)
M. abscessus	14 (2.61%)	(14/538; 2.60%)
M. chelonae	2 (0.38%)	(2/538; 0.37%)
M. farcinogenes	3 (0.56%)	(3/538; 0.55%)
M. gadium	5 (0.93%)	(5/538; 0.92%)
M. triplex	1 (0.19%)	(1/538; 0.18%)
M. austroafricanum	1 (0.19%)	(1/538; 0.18%)
M. diernhoferi	2 (0.38%)	(2/538; 0.37%)
M. interjectum	8 (1.49%)	(8/538; 1.48%)
M. cosmeticum	2 (0.38%)	(2/538; 0.37%)
M. moriokaense	1 (0.19%)	(1/538; 0.18%)

(*Continued*)

TABLE 10.30 (Continued)

Kind of Species	Country (Reference) (Period)	
	Brisbane, Australia (Thomson et al., 2013) (2013)	Total Australia
M. cookii	2 (0.38%)	(2/538; 0.37%)
M. wolinsky	1 (0.19%)	(2/538; 0.37%)
M. poriferae	24 (4.46%)	(24/538; 4.46%)
M. tilburgii	3 (0.56%)	(3/538; 0.55%)
M. rhodesia	1 (0.19%)	(1/538; 0.18%)
M. fluorantherivorans	15 (2.79%)	(15/538; 2.78%)
Other	21 (3.91%)	(21/538; 3.90%)
Total NTM	538	(538/538; 100%)

Distribution of Nontuberculous Mycobacteria Based on Environmental Specimens From Australia

The frequency of NTM in environmental samples includes data collected from Brisbane, Australia. The geographical distribution of NTM based on environmental specimens is shown in Fig. 10.20 by a higher frequency of species. In general, in Australia 538 NTM has been reported from environmental specimens.

REFERENCES

Agdestein, A., Olsen, I., Jørgensen, A., Djønne, B., Johansen, T.B., 2014. Novel insights into transmission routes of *Mycobacterium avium* in pigs and possible implications for human health. Vet. Res. 45 (1), 46.

Aghajani, J., Farnia, P., Velayati, A.A., 2017. Impact of geographical information system on public health sciences. Biomed. Biotechnol. Res J. 1 (2), 94.

Al Houqani, M., Jamieson, F., Chedore, P., Mehta, M., May, K., Marras, T.K., 2011. Isolation prevalence of pulmonary nontuberculous mycobacteria in Ontario in 2007. Can. Respir. J. 18 (1), 19–24.

Al-Ghafli, H., Al-Hajoj, S., 2017. Nontuberculous mycobacteria in Saudi Arabia and gulf countries: a review. Can. Respir. J. 2017.

Aliyu, G., El-Kamary, S.S., Abimiku, A.L., Brown, C., Tracy, K., Hungerford, L., et al., 2013. Prevalence of non-tuberculous mycobacterial infections among tuberculosis suspects in Nigeria. PLoS One. 8 (5), e63170.

Aliyu, G., El-Kamary, S.S., Abimiku, A.L., Hungerford, L., Obasanya, J., Blattner, W., 2014. Cost-effectiveness of point-of-care digital chest-x-ray in HIV patients with pulmonary mycobacterial infections in Nigeria. BMC Infect. Dis. 14 (1), 675.

Alqumber, M.A., 2014. Prevalence of mycobacteria in water reservoirs of Albaha, Saudi Arabia. Saudi Med. J. 35 (5), 466–471.

Americas Available from: <https://en.wikipedia.org/wiki/Americas>.

Americas geography. Geography of North America. Universal World Reference Encyclopedia. 11, book 1. V.S. Thatcher. 1964, 231–233. 64-12955.

Andréjak, C., Thomsen, V.Ø., Johansen, I.S., Riis, A., Benfield, T.L., Duhaut, P., et al., 2010. Nontuberculous pulmonary mycobacteriosis in Denmark: incidence and prognostic factors. Am. J. Respir. Crit. Care Med. 181 (5), 514–521.

Asiimwe, B.B., Bagyenzi, G.B., Ssengooba, W., Mumbowa, F., Mboowa, G., Wajja, A., et al., 2013. Species and genotypic diversity of non-tuberculous mycobacteria isolated from children investigated for pulmonary tuberculosis in rural Uganda. BMC Infect. Dis. 13 (1), 88.

Australia Available from: <https://en.wikipedia.org/wiki/Australia>.

Australian Bureau of Statistics. 2016 Census data reveals (no religion) is rising fast. Australian Bureau of Statistics. Archived from the original on June 27, 2017.

Australia geography. World of map. Geography of Australia Available from: <https://www.mapsofworld.com/australia/geography.html>.

Bainomugisa, A., Wampande, E., Muchwa, C., Akol, J., Mubiri, P., Ssenyungule, H., et al., 2015. Use of real time polymerase chain reaction for detection of *M. tuberculosis*, *M. avium* and *M. kansasii* from clinical specimens. BMC Infect. Dis. 15 (1), 181.

Barandiaran, S., Perez, A., Gioffre, A., Vivot, M.M., Cataldi, A., Zumarraga, M., 2015. Tuberculosis in swine co-infected with *Mycobacterium avium* subsp. hominissuis and *Mycobacterium bovis* in a cluster from Argentina. Epidemiol. Infect. 143 (5), 966–974.

Beer, A., Davis, G., 1965. 'Anonymous' mycobacteria isolated in Lagos, Nigeria. Tubercle 46 (1), 32–39.

Benson, C.A., Ellner, J.J., 1993. *Mycobacterium avium* complex infection and AIDS: advances in theory and practice. Clin. Infect. Dis. 17 (1), 7–20.

Beran, V., Matlova, L., Horvathova, A., Bartos, M., Moravkova, M., Pavlik, I., 2006. Mycobacteria in the animal's environment in the Czech Republic. Vet. Arh. 76 (Suppl.), 33–39.

Berg, S., Firdessa, R., Habtamu, M., Gadisa, E., Mengistu, A., Yamuah, L., et al., 2009. The burden of mycobacterial disease in Ethiopian cattle: implications for public health. PLoS One 4 (4), e5068.

Berrada, J., 1993. *Mycobacterium bovis* infection in cattle in Morocco: preparation and evaluation of chemical extracts for use in detection of immune responses.

Bjerrum, S., Oliver-Commey, J., Kenu, E., Lartey, M., Newman, M.J., Addo, K.K., et al., 2016. Tuberculosis and non-tuberculous mycobacteria among HIV-infected individuals in Ghana. Trop. Med. Int. Health 21 (9), 1181–1190.

Blanc, P., Dutronc, H., Peuchant, O., Dauchy, F.-A., Cazanave, C., Neau, D., et al., 2016. Nontuberculous mycobacterial infections in a French hospital: a 12-year retrospective study. PLoS One. 11 (12), e0168290.

Bono, M., Jemmi, T., Bernasconi, C., Burki, D., Telenti, A., Bodmer, T., 1995. Genotypic characterization of *Mycobacterium avium* strains recovered from animals and their comparison to human strains. Appl. Environ. Microbiol. 61 (1), 371–373.

Briancesco, R., Semproni, M., Paradiso, R., Bonadonna, L., 2014. Nontuberculous mycobacteria: an emerging risk in engineered environmental habitats. Ann. Microbiol. 64 (2), 735–740.

Buijtels, P.C., Petit, P.L., Verbrugh, H.A., van Belkum, A., van Soolingen, D., 2005. Isolation of nontuberculous mycobacteria in Zambia: eight case reports. J. Clin. Microbiol. 43 (12), 6020–6026.

Buijtels, P.C., van der Sande, M.A., de Graaff, C.S., Parkinson, S., Verbrugh, H.A., Petit, P.L., et al., 2009. Nontuberculous mycobacteria, Zambia. Emerging Infect. Dis. 15 (2), 242.

Buijtels, P.C.A.M., Van Der Sande, M., Parkinson, S., Verbrugh, H., Petit, P., van Soolingen, D., 2010. Isolation of non-tuberculous mycobacteria at three rural settings in Zambia; a pilot study. Clin. Microbiol. Infect. 16 (8), 1142–1148.

Bullin, C., Tanner, E.I., Collins, C., 1970. Isolation of *Mycobacterium xenopei* from water taps. Epidemiol. Infect. 68 (1), 97–100.

Cassidy, P.M., Hedberg, K., Saulson, A., McNelly, E., Winthrop, K.L., 2009. Nontuberculous mycobacterial disease prevalence and risk factors: a changing epidemiology. Clin. Infect. Dis. 49 (12), e124–e129.

Cayer, M.-P., Veillette, M., Pageau, P., Hamelin, R., Bergeron, M.-J., Mériaux, A., et al., 2007. Identification of mycobacteria in peat moss processing plants: application of molecular biology approaches. Can. J. Microbiol. 53 (1), 92–99.

Cerna-Cortes, J.F., Leon-Montes, N., Cortes-Cueto, A.L., Salas-Rangel, L.P., Helguera-Repetto, A.C., Lopez-Hernandez, D., et al., 2015. Microbiological quality of ready-to-eat vegetables collected in Mexico City: occurrence of aerobic-mesophilic bacteria, fecal coliforms, and potentially pathogenic nontuberculous mycobacteria. Biomed. Res. Int. 2015.

Chanda-Kapata, P., Kapata, N., Klinkenberg, E., Mulenga, L., Tembo, M., Katemangwe, P., et al., 2015. Non-tuberculous mycobacteria (NTM) in Zambia: prevalence, clinical, radiological and microbiological characteristics. BMC Infect. Dis. 15 (1), 500.

Chien, J.-Y., Lai, C.-C., Sheng, W.-H., Yu, C.-J., Hsueh, P.-R., 2014. Pulmonary infection and colonization with nontuberculous mycobacteria, Taiwan, 2000–2012. Emerging Infect. Dis. 20 (8), 1382.

Chou, M.P., Clements, A.C., Thomson, R.M., 2014. A spatial epidemiological analysis of nontuberculous mycobacterial infections in Queensland, Australia. BMC Infect. Dis. 14 (1), 279.

Corbett, E., Hay, M., Churchyard, G., Herselman, P., Clayton, T., Williams, B., et al., 1999a. *Mycobacterium kansasii* and *M. scrofulaceum* isolates from HIV-negative South African gold miners: incidence, clinical significance and radiology. Int. J. Tuberc. Lung Dis. 3 (6), 501–507.

Corbett, E.L., Churchyard, G.J., Clayton, T., Herselman, P., Williams, B., Hayes, R., et al., 1999b. Risk factors for pulmonary mycobacterial disease in South African gold miners: a case–control study. Am. J. Respir. Crit. Care Med. 159 (1), 94–99.

Corbett, E.L., Blumberg, L., Churchyard, G.J., Moloi, N., Mallory, K., Clayton, T., et al., 1999c. Nontuberculous mycobacteria: defining disease in a prospective cohort of South African miners. Am. J. Respir. Crit. Care Med. 160 (1), 15–21.

Covert, T.C., Rodgers, M.R., Reyes, A.L., Stelma, G.N., 1999. Occurrence of nontuberculous mycobacteria in environmental samples. Appl. Environ. Microbiol. 65 (6), 2492–2496.

Cox, L.K., 2012. Mycobacteria other than tuberculosis in eastern Kansas water supplies.

Daniel, O., Osman, E., Adebiyi, P., Mourad, G., Declarcq, E., Bakare, R., 2011. Non tuberculosis mycobacteria isolates among new and previously treated pulmonary tuberculosis patients in Nigeria. Asian Pac. J. Trop. Dis. 1 (2), 113–115.

Devana, J.V., Calambur, N., Reddy, B.R., 2018. Pacemaker site infection caused by rapidly growing nontuberculous mycobacteria (RGM). Biomed. Biotechnol. Res. J. 2 (1), 82.

Ding, L., Lai, C.-C., Lee, L., Hsueh, P., 2006. Disease caused by non-tuberculous mycobacteria in a university hospital in Taiwan, 1997–2003. Epidemiol. Infect. 134 (5), 1060–1067.

Ebani, V.V., Fratini, F., Bertelloni, F., Cerri, D., Tortoli, E., 2012. Isolation and identification of mycobacteria from captive reptiles. Res. Vet. Sci. 93 (3), 1136–1138.

Edirisinghe, E., Dissanayake, D., Abayasekera, C., Arulkanthan, A., 2014. Occurrence of nontuberculous mycobacteria in aquatic sources of Sri Lanka. Int. J. Mycobact. 3 (4), 242–246.

Egbe, N., Muwonge, A., Ndip, L., Kelly, R., Sander, M., Tanya, V., et al., 2016. Abattoir-based estimates of mycobacterial infections in Cameroon. Sci. Rep. 6, 24320.

Falcão, D.P., Valentini, S.R., Leite, C.Q.F., 1993. Pathogenic or potentially pathogenic bacteria as contaminants of fresh water from different sources in Araraquara, Brazil. Water Res. 27 (12), 1737–1741.

Falkinham III, J., 1996. Epidemiology of infection by nontuberculous mycobacteria. Clin. Microbiol. Rev. 9 (2), 177.

Falodun O., Fagade O., Cadmus S. Prevalence of Nontuberculous Mycobacteria Infections in Patients Diagnosed with Pulmonary Tuberculosis in Ibadan.

Fischeder, R., Schulze-Robbecke, R., Weber, A., 1991. Occurrence of mycobacteria in drinking water samples. Zentralbl. Hyg. Umweltmed. 192 (2), 154–158.

Fischer, O., Matlova, L., Dvorska, L., Švástová, P., Bartl, J., Melicharek, I., et al., 2001. Diptera as vectors of mycobacterial infections in cattle and pigs. Med. Vet. Entomol. 15 (2), 208–211.

Ford, E.S., Horne, D.J., Shah, J.A., Wallis, C.K., Fang, F.C., Hawn, T.R., 2017. Species-specific risk factors, treatment decisions, and clinical outcomes for laboratory isolates of less common nontuberculous mycobacteria in Washington State. Ann. Am. Thorac. Soc. 14 (7), 1129–1138.

Galassi, L., Donato, R., Tortoli, E., Burrini, D., Santianni, D., Dei, R., 2003. Nontuberculous mycobacteria in hospital water systems: application of HPLC for identification of environmental mycobacteria. J. Water Health 1 (3), 133–139.

Gcebe, N., Rutten, V., Gey van Pittius, N.C., Michel, A., 2013. Prevalence and distribution of non-tuberculous mycobacteria (NTM) in cattle, African buffaloes (*Syncerus caffer*) and their environments in South Africa. Transboundary Emerging Dis. 60, 74–84.

Griffith, D.E., Girard, W.M., Wallace Jr., R.J., 1993. Clinical features of pulmonary disease caused by rapidly growing mycobacteria. Am. Rev. Respir. Dis. 147, 1271–1278.

Griffith, D.E., Aksamit, T., Brown-Elliott, B.A., Catanzaro, A., Daley, C., Gordin, F., et al., 2007. An official ATS/IDSA statement: diagnosis, treatment, and prevention of nontuberculous mycobacterial diseases. Am. J. Respir. Crit. Care Med. 175 (4), 367–416.

Halstrom, S., Price, P., Thomson, R., 2015. Environmental mycobacteria as a cause of human infection. Int. J. Mycobact. 4 (2), 81–91.

Haraka, F., Rutaihwa, L.K., Battegay, M., Reither, K., 2012. Novel diagnostic procedure: *Mycobacterium intracellulare* infection in non-HIV infected patient in a region with a high burden of tuberculosis. BMJ Case Rep. 2012.

Henkle, E., Hedberg, K., Schafer, S.D., Winthrop, K.L., 2017. Surveillance of extrapulmonary nontuberculous mycobacteria infections, Oregon, USA, 2007–2012. Emerging Infect. Dis. 23 (10), 1627.

Hermansen, T.S., Ravn, P., Svensson, E., Lillebaek, T., 2017. Nontuberculous mycobacteria in Denmark, incidence and clinical importance during the last quarter-century. Sci. Rep. 7 (1), 6696.

Hernandez-Garduno, E., Rodrigues, M., Elwood, R., 2009. The incidence of pulmonary nontuberculous mycobacteria in British Columbia, Canada. Int. J. Tuberc. Lung Dis. 13 (9), 1086–1093.

Hilborn, E.D., Covert, T.C., Yakrus, M.A., Harris, S.I., Donnelly, S.F., Rice, E.W., et al., 2006. Persistence of nontuberculous mycobacteria in a drinking water system after addition of filtration treatment. Appl. Environ. Microbiol. 72 (9), 5864–5869.

Hoefsloot, W., Van Ingen, J., Andrejak, C., Ängeby, K., Bauriaud, R., Bemer, P., et al., 2013. The geographic diversity of nontuberculous mycobacteria isolated from pulmonary samples: an NTM-NET collaborative study. Eur. Respir. J. 42 (6), 1604–1613.

Honda, J.R., Hasan, N.A., Davidson, R.M., Williams, M.D., Epperson, L.E., Reynolds, P.R., et al., 2016. Environmental nontuberculous mycobacteria in the Hawaiian Islands. PLoS Neglected Trop. Dis. 10 (10), e0005068.

Hoza, A.S., Mfinanga, S.G., Rodloff, A.C., Moser, I., König, B., 2016. Increased isolation of nontuberculous mycobacteria among TB suspects in Northeastern, Tanzania: public health and diagnostic implications for control programmes. BMC Res. Notes 9 (1), 109.

Hu, Y., Yu, X., Zhao, D., Li, R., Liu, Y., Ge, M., et al., 2017. Isolation of nontuberculous mycobacteria from soil using Middlebrook 7H10 agar with increased malachite green concentration. AMB Express 7 (1), 69.

Jacqueline, L., Christine, B., Ngayo, M.O., Tura, G., Daniel, M., Rose, N., et al. Infection rates and correlates of Non-Tuberculous Mycobacteria Among Tuberculosis Retreatment Cases in Kenya.

Jankovic, M., Samarzija, M., Sabol, I., Jakopovic, M., Katalinic Jankovic, V., Zmak, L., et al., 2013. Geographical distribution and clinical relevance of non-tuberculous mycobacteria in Croatia. Int. J. Tuberc. Lung Dis. 17 (6), 836–841.

Jesudason, M., Gladstone, P., 2005. Non tuberculous mycobacteria isolated from clinical specimens at a tertiary care hospital in South India. Indian J. Med. Microbiol. 23 (3), 172.

Jha, V.C., Morita, Y., Dhakal, M., Besnet, B., Sato, T., Nagai, A., et al., 2007. Isolation of *Mycobacterium* spp. from milking buffaloes and cattle in Nepal. J. Vet. Med. Sci. 69 (8), 819–825.

Jin, B.W., Saito, H., Yoshii, Z., 1984. Environmental mycobacteria in Korea. Microbiol. Immunol. 28 (6), 667–677.

Joao, I., Cristovao, P., Antunes, L., Nunes, B., Jordao, L., 2014. Identification of nontuberculous mycobacteria by partial gene sequencing and public databases. Int. J. Mycobact. 3 (2), 144–151.

Kankya, C., Muwonge, A., Djønne, B., Munyeme, M., Opuda-Asibo, J., Skjerve, E., et al., 2011. Isolation of non-tuberculous mycobacteria from pastoral ecosystems of Uganda: public health significance. BMC Public Health 11 (1), 320.

Käser, M., Ruf, M.-T., Hauser, J., Marsollier, L., Pluschke, G., 2009. Optimized method for preparation of DNA from pathogenic and environmental mycobacteria. Appl. Environ. Microbiol. 75 (2), 414–418.

Katale, B.Z., Mbugi, E.V., Botha, L., Keyyu, J.D., Kendall, S., Dockrell, H.M., et al., 2014. Species diversity of non-tuberculous mycobacteria isolated from humans, livestock and wildlife in the Serengeti ecosystem, Tanzania. BMC Infect. Dis. 14 (1), 616.

Kazwala, R., Daborn, C., Kusiluka, L., Jiwa, S., Sharp, J., Kambarage, D., 1998. Isolation of *Mycobacterium* species from raw milk of pastoral cattle of the Southern Highlands of Tanzania. Trop. Anim. Health Prod. 30 (4), 233–239.

Kim, J.K., Rheem, I., 2013. Identification and distribution of nontuberculous mycobacteria from 2005 to 2011 in Cheonan, Korea. Tuberc. Respir. Dis. 74 (5), 215–221.

Kline, S., Cameron, S., Streifel, A., Yakrus, M.A., Kairis, F., Peacock, K., et al., 2004. An outbreak of bacteremias associated with *Mycobacterium mucogenicum* in a hospital water supply. Infect. Control Hosp. Epidemiol. 25 (12), 1042–1049.

Kobayashi, M., Oana, K., Kawakami, Y., 2014. Bath water contamination with *Legionella* and nontuberculous mycobacteria in 24-hour home baths, hot springs, and public bathhouses of Nagano Prefecture, Japan. Jpn. J. Infect. Dis. 67 (4), 276–281.

Koh, W.-J., Kwon, O.J., Jeon, K., Kim, T.S., Lee, K.S., Park, Y.K., et al., 2006. Clinical significance of nontuberculous mycobacteria isolated from respiratory specimens in Korea. Chest 129 (2), 341–348.

Kubalek, I., Mysak, J., 1996. The prevalence of environmental mycobacteria in drinking water supply systems in a demarcated region in Czech Republic, in the period 1984–1989. Eur. J. Epidemiol. 12 (5), 471–474.

Kušar, D., Zajc, U., Jenčič, V., Ocepek, M., Higgins, J., Žolnir-Dovč, M., et al., 2017. Mycobacteria in aquarium fish: results of a 3-year survey indicate caution required in handling pet-shop fish. J. Fish Dis. 40 (6), 773–784.

Lahiri, A., 2014. The Genetic Diversity of *Mycobacterium avium* subsp. *hominissuis*. Freie Universität, Berlin.

Lai, C.-C., Tan, C.-K., Chou, C.-H., Hsu, H.-L., Liao, C.-H., Huang, Y.-T., et al., 2010. Increasing incidence of nontuberculous mycobacteria, Taiwan, 2000–2008. Emerging Infect. Dis. 16 (2), 294.

Le Dantec, C., Duguet, J.-P., Montiel, A., Dumoutier, N., Dubrou, S., Vincent, V., 2002. Occurrence of mycobacteria in water treatment lines and in water distribution systems. Appl. Environ. Microbiol. 68 (11), 5318–5325.

Lecuona, M., Abreu, R., Rodríguez-Álvarez, C., Castro, B., Campos, S., Hernández-Porto, M., et al., 2016. First isolation of *Mycobacterium canariasense* from municipal water supplies in Tenerife, Canary Islands, Spain. Int. J. Hyg. Environ. Health 219 (1), 48–52.

Ledwon, A., Szeleszczuk, P., Zwolska, Z., Augustynowicz-Kopec, E., Kozak, M., 2008. Occurrence of *Mycobacterium* in the faeces of parrots kept in zoological gardens and private aviaries in Poland. Bull. Vet. Inst. Pulawy 52, 341–346.

Lee, E.-S., Lee, M.-Y., Han, S.-H., Ka, J.-O., 2008. Occurrence and molecular differentiation of environmental mycobacteria in surface waters. J. Microbiol. Biotechnol. 18 (7), 1207–1215.

Limo, J., Bii, C., Musa, O., Galgalo, T., Mutua, D., Nkirote, R., et al., 2015. Infection rates and correlates of non-tuberculous mycobacteria among tuberculosis retreatment cases in Kenya. Prime J. Social Sci. 4, 1128–1134.

Liu, H., Lian, L., Jiang, Y., Huang, M., Tan, Y., Zhao, X., et al., 2016. Identification of species of nontuberculous mycobacteria clinical isolates from 8 provinces of China. BioMed Res. Int. 2016.

Livani, S., Ghaemi, E., 2014. A review on environmental mycobacteria in Iran. Med. Lab. J. 8 (3), 114–128.

Maiga, M., Siddiqui, S., Diallo, S., Diarra, B., Traoré, B., Shea, Y.R., et al., 2012. Failure to recognize nontuberculous mycobacteria leads to misdiagnosis of chronic pulmonary tuberculosis. PLoS One 7 (5), e36902.

Makovcova, J., Slany, M., Babak, V., Slana, I., Kralik, P., 2014. The water environment as a source of potentially pathogenic mycobacteria. J. Water Health 12 (2), 254–263.

Malama, S., Munyeme, M., Mwanza, S., Muma, J.B., 2014. Isolation and characterization of non tuberculous mycobacteria from humans and animals in Namwala District of Zambia. BMC Res. Notes 7 (1), 622.

Marras, T.K., Chedore, P., Ying, A.M., Jamieson, F., 2007. Isolation prevalence of pulmonary non-tuberculous mycobacteria in Ontario, 1997–2003. Thorax 62 (8), 661–666.

Marras, T.K., Campitelli, M.A., Lu, H., Chung, H., Brode, S.K., Marchand-Austin, A., et al., 2017. Pulmonary nontuberculous mycobacteria–associated deaths, Ontario, Canada, 2001–2013. Emerging Infect. Dis. 23 (3), 468.

Martin-Casabona, N., Bahrmand, A., Bennedsen, J., Østergaard Thomsen, V., Curcio, M., Fauville-Dufaux, M., et al., 2004. Non-tuberculous mycobacteria: patterns of isolation. A multi-country retrospective survey. Int. J. Tuberc. Lung Dis. 8 (10), 1186–1193.

Matthews, P., Collins, P., Jones, P., 1976. Isolation of mycobacteria from dairy creamery effluent sludge. Epidemiol. Infect. 76 (3), 407–413.

Maurya, A., Nag, V., Kant, S., Kushwaha, R., Kumar, M., Singh, A., et al., 2015. Prevalence of nontuberculous mycobacteria among extrapulmonary tuberculosis cases in tertiary care centers in Northern India. BioMed Res. Int. 2015.

Maurya, A.K., Nag, V.L., Kant, S., Sharma, A., Gadepalli, R.S., Kushwaha, R.A.S., 2017. Recent methods for diagnosis of nontuberculous mycobacteria infections: relevance in clinical practice. Biomed. Biotechnol. Res. J. 1 (1), 14.

Mawak, J., Gomwalk, N., Bello, C., Kandakai-Olukemi, Y., 2006. Human pulmonary infections with bovine and environment (atypical) mycobacteria in Jos, Nigeria. Ghana Med. J. 40, 3.

McSwiggan, D., Collins, C., 1974. The isolation of *M. kansasii* and *M. xenopi* from water systems. Tubercle 55 (4), 291–297.

Mello, F., Borga, L., Rolla, V., Duarte, R., Sampaio, E., Holland, S., et al., 2013. Clinical and therapeutic features of pulmonary nontuberculous mycobacterial disease, Brazil, 1993–2011. Emerging Infect. Dis. 19 (3), 393–399.

Mencarini, J., Cresci, C., Simonetti, M., Truppa, C., Camiciottoli, G., Frilli, M., et al., 2017. Non-tuberculous mycobacteria: epidemiological pattern in a reference laboratory and risk factors associated with pulmonary disease. Epidemiol. Infect. 145 (3), 515–522.

Miguel, S., Irene, B., Pierre, M., Lucia, G.-L., Miguel, A.B., Josefina, S., et al., 2018. Pulmonary infections with nontuberculous mycobacteria, Catalonia, Spain, 1994–2014. Emerg. Infect. Dis. 24 (6).

Mirsaeidi, M.S., Tabarsi, P., Farnia, P., Ebrahimi, G., Morris, M.W., Masjedi, M.R., et al., 2007. Trends of drug resistant *Mycobacterium tuberculosis* in a tertiary tuberculosis center in Iran. Saudi Med. J. 28 (4), 544–550.

Mishra, P.S., Narang, P., Narang, R., Goswami, B., Mendiratta, D.K., 2018. Spatio-temporal study of environmental nontuberculous mycobacteria isolated from Wardha district in Central India. Antonie van Leeuwenhoek 111 (1), 73–87.

Mohamed, A.M., El-Ella, G.A.A., Nasr, E.A., 2009. Phenotypic and molecular typing of tuberculous and nontuberculous *Mycobacterium* species from slaughtered pigs in Egypt. J. Vet. Diagn. Invest. 21 (1), 48–52.

Monde, N., Munyeme, M., Muwonge, A., Muma, J.B., Malama, S., 2018. Characterization of non-tuberculous *Mycobacterium* from humans and water in an Agropastoral area in Zambia. BMC Infect. Dis. 18 (1), 20.

Moore, J.E., Kruijshaar, M.E., Ormerod, L.P., Drobniewski, F., Abubakar, I., 2010. Increasing reports of non-tuberculous mycobacteria in England, Wales and Northern Ireland, 1995–2006. BMC Public Health 10 (1), 612.

Morimoto, K., Hasegawa, N., Izumi, K., Namkoong, H., Uchimura, K., Yoshiyama, T., et al., 2017. A laboratory-based analysis of nontuberculous mycobacterial lung disease in Japan from 2012 to 2013. Ann. Am. Thorac. Soc. 14 (1), 49–56.

Mortaz, E., Adcock, I.M., Ricciardolo, F.L., Varahram, M., Jamaati, H., Velayati, A.A., et al., 2015. Anti-inflammatory effects of *Lactobacillus rhamnosus* and *Bifidobacterium breve* on cigarette smoke activated human macrophages. PLoS One 10 (8), e0136455.

Mwikuma, G., Kwenda, G., Hang'ombe, B.M., Simulundu, E., Kaile, T., Nzala, S., et al., 2015. Molecular identification of non-tuberculous mycobacteria isolated from clinical specimens in Zambia. Ann. Clin. Microbiol. Antimicrob. 14 (1), 1.

Namkoong, H., Kurashima, A., Morimoto, K., Hoshino, Y., Hasegawa, N., Ato, M., et al., 2016. Epidemiology of pulmonary nontuberculous mycobacterial disease, Japan (1). Emerging Infect. Dis. 22 (6), 1116–1117.

Petrini, B., 2006. Non-tuberculous mycobacterial infections. Scand. J. Infect. Dis. 38 (4), 246–255.

Nyamogoba, H., Mbuthia, G., Mining, S., Kikuvi, G., Kikuvi, R., Mpoke, S., et al., 2012. HIV co-infection with tuberculous and non-tuberculous mycobacteria in western Kenya: challenges in the diagnosis and management. Afr. Health Sci. 12 (3), 305–311.

Okoi, C., Anderson, S., Antonio, M., Mulwa, S.N., Gehre, F., Adetifa, I., 2017. Non-tuberculous mycobacteria isolated from pulmonary samples in sub-Saharan Africa—a systematic review and meta analyses. Sci. Rep. 7 (1), 12002.

Oliveira, M.J., Gaio, A.R., Gomes, M., Goncalves, A., Duarte, R., 2017. *Mycobacterium avium* infection in Portugal. Int. J. Tuberc. Lung Dis. 21 (2), 218–222.

O'brien, D.P., Currie, B.J., Krause, V.L., 2000. Nontuberculous mycobacterial disease in northern Australia: a case series and review of the literature. Clin. Infect. Dis. 31 (4), 958–967.

Olutayo, F., Fagade, O., Cadmus, S., 2016. Prevalence of nontuberculous mycobacteria infections in patients diagnosed with pulmonary tuberculosis in Ibadan. Int. J. Trop. Dis. Health 18, 1–8.

Padya, L., Chin'ombe, N., Magwenzi, M., Mbanga, J., Ruhanya, V., Nziramasanga, P., 2015. Molecular identification of mycobacterium species of public health importance in cattle in Zimbabwe by 16S rRNA gene sequencing. Open Microbiol. J. 9, 38.

Panagiotou, M., Papaioannou, A.I., Kostikas, K., Paraskeua, M., Velentza, E., Kanellopoulou, M., et al., 2014. The epidemiology of pulmonary nontuberculous mycobacteria: data from a general hospital in Athens, Greece, 2007–2013. Pulm. Med. 2014.

Pavlik, I., Trcka, I., Parmova, I., Svobodova, J., Melicharek, I., Nagy, G., et al., 2005. Detection of bovine and human tuberculosis in cattle and other animals in six Central European countries during the years 2000–2004. Vet. Med. 50 (7), 291.

Peters, M., Müller, C., Rüsch-Gerdes, S., Seidel, C., Göbel, U., Pohle, H., et al., 1995. Isolation of atypical mycobacteria from tap water in hospitals and homes: is this a possible source of disseminated MAC infection in AIDS patients? J. Infect. 31 (1), 39–44.

Pokam, B.T., Asuquo, A.E., 2012. Acid-fast bacilli other than mycobacteria in tuberculosis patients receiving directly observed therapy short course in cross river state, Nigeria. Tuberc. Res. Treat. 2012.

Portaels, F., 1995. Epidemiology of mycobacterial diseases. Clin. Dermatol. 13 (3), 207–222.

Pourahmad, F., Thompson, K.D., Adams, A., Richards, R.H., 2009. Comparative evaluation of polymerase chain reaction–restriction enzyme analysis (PRA) and sequencing of heat shock protein 65 (hsp65) gene for identification of aquatic mycobacteria. J. Microbiol. Methods 76 (2), 128–135.

Rajaei, E., Hadadi, M., Madadi, M., Aghajani, J., Ahmad, M.M., Farnia, P., et al., 2018. Outdoor air pollution affects tuberculosis development based on geographical information system modeling. Biomed. Biotechnol. Res. J. 2 (1), 39.

Rindi, L., Garzelli, C., 2015. Increase in non-tuberculous mycobacteria isolated from humans in Tuscany, Italy, from 2004 to 2014. BMC Infect. Dis. 16 (1), 44.

Rónai, Z., Eszterbauer, E., Csivincsik, A., Guti, C., Dencső, L., Jánosi, S., et al., 2016. Detection of wide genetic diversity and several novel strains among non-avium nontuberculous mycobacteria isolated from farmed and wild animals in Hungary. J. Appl. Microbiol. 121 (1), 41–54.

Ryoo, S.W., Shin, S., Shim, M.S., Park, Y.S., Lew, W.J., Park, S.N., et al., 2008. Spread of nontuberculous mycobacteria from 1993 to 2006 in Koreans. J. Clin. Lab. Anal. 22 (6), 415–420.

Saenz, A., Errico, F., 1984. Mycobacteria isolated from apparently normal swine lymph nodes in Uruguay.

Sajduda, A., Martin, A., Portaels, F., Palomino, J.C., 2012. hsp65 PCR-restriction analysis (PRA) with capillary electrophoresis for species identification and differentiation of *Mycobacterium kansasii* and *Mycobacterium chelonae–Mycobacterium abscessus* group. Int. J. Infect. Dis. 16 (3), e193–e197.

Santin, M., Barrabeig, I., Malchair, P., Gonzalez-Luquero, L., Benitez, M.A., Sabria, J., et al., 2018. Pulmonary infections with nontuberculous mycobacteria, Catalonia, Spain, 1994–2014. Emerging Infect. Dis. 24 (6), 1091.

Santos, R., Oliveira, F., Fernandes, J., Goncalves, S., Macieira, F., Cadete, M., 2005. Detection and identification of mycobacteria in the Lisbon water distribution system. Water Sci. Technol. 52 (8), 177–180.

Sebakova, H., Kozisek, F., Mudra, R., Kaustova, J., Fiedorova, M., Hanslikova, D., et al., 2008. Incidence of nontuberculous mycobacteria in four hot water systems using various types of disinfection. Can. J. Microbiol. 54 (11), 891–898.

Sgarioni, S.A., Hirata, R.D.C., Hiroyuki Hirata, M., Leite, C.Q.F., Prince, K.D., Leite, S.R.D.A., et al., 2014. Occurrence of *Mycobacterium bovis* and non-tuberculous mycobacteria (NTM) in raw and pasteurized milk in the northwestern region of Paraná, Brazil. Braz. J. Microbiol. 45 (2), 707–711.

Shin, J., Lee, E., Lee, H., Ryu, S., Kim, H., Chang, C., et al., 2007. Prevalence of non-tuberculous mycobacteria in a hospital environment. J. Hosp. Infect. 65 (2), 143–148.

Shu, C.-C., Lee, C.-H., Wang, J.-Y., Jerng, J.-S., Yu, C.-J., Hsueh, P.-R., et al., 2008. Nontuberculous mycobacteria pulmonary infection in medical intensive care unit: the incidence, patient characteristics, and clinical significance. Intensive Care Med. 34 (12), 2194.

Slany, M., Svobodova, J., Ettlova, A., Slana, I., Mrlik, V., Pavlik, I., 2010. *Mycobacterium arupense* among the isolates of non-tuberculous mycobacteria from human, animal and environmental samples. Vet. Med. 55 (8), 369–376.

Sriyabhaya, N., Wongwatana, S., 1981. Pulmonary infection caused by atypical mycobacteria: a report of 24 cases in Thailand. Rev. Infect. Dis. 3 (5), 1085–1089.

Thomson, R.M., Carter, R., Tolson, C., Coulter, C., Huygens, F., Hargreaves, M., 2013. Factors associated with the isolation of nontuberculous mycobacteria (NTM) from a large municipal water system in Brisbane, Australia. BMC Microbiol. 13 (1), 89.

Thorel, M.-F., Falkinham Iii, J.O., Moreau, R., 2004. Environmental mycobacteria from alpine and subalpine habitats. FEMS Microbiol. Ecol. 49 (3), 343–347.

Tortoli, E., 2003. Impact of genotypic studies on mycobacterial taxonomy: the new mycobacteria of the 1990s. Clin. Microbiol. Rev. 16 (2), 319–354.

Tortoli, E., Bartoloni, A., 1996. High-performance liquid chromatography and identification of mycobacteria. Rev. Med. Microbiol. 7 (4), 207–219.

Torvinen, E., Suomalainen, S., Lehtola, M.J., Miettinen, I.T., Zacheus, O., Paulin, L., et al., 2004. Mycobacteria in water and loose deposits of drinking water distribution systems in Finland. Appl. Environ. Microbiol. 70 (4), 1973–1981.

Torvinen, E., Suomalainen, S., Paulin, L., Kusnetsov, J., 2014. Mycobacteria in Finnish cooling tower waters. APMIS 122 (4), 353–358.

Traversa, M.J., Etchechoury, I., Jorge, M.C., Schettino, D.M., Bernadelli, A., Zumárraga, M., et al., 2009. Mycobacterial isolation from Felis concolor in captivity. Braz. J. Vet. Res. Anim. Sci. 46 (1), 25–31.

Trcka, I., Lamka, J., Suchy, R., Kopecna, M., Beran, V., Moravkova, M., et al., 2006. Mycobacterial infections in European wild boar (*Sus scrofa*) in the Czech Republic during the years 2002 to 2005. Vet. Med. 51 (5), 320–332.

Tsang, R.S., Seto, W., Leung, Y., Chow, K., Li, K., 1985. A pathogenic *Mycobacterium* isolated on simple agar media. Lab. Med. 16 (6), 374–376.

Tsukamura, M., Kita, N., Shimoide, H., Kurashima, A., Kuze, A., Shinoda, A., et al., 1988a. Studies on the nontuberculous lung mycobacteriosis in Japan (report of the study in the year 1986 of the mycobacteriosis research group of the Japanese National Chest Hospitals). Kekkaku 63 (7), 493–499.

Tsukamura, M., Kita, N., Shimoide, H., Arakawa, H., Kuze, A., 1988b. Studies on the epidemiology of nontuberculous mycobacteriosis in Japan. Am. Rev. Respir. Dis. 137 (6), 1280–1284.

van Ingen, J., Blaak, H., de Beer, J., de Roda Husman, A.M., van Soolingen, D., 2010. Rapidly growing nontuberculous mycobacteria cultured from home tap and shower water. Appl. Environ. Microbiol. 76 (17), 6017–6019.

Van der Werf, M.J., Ködmön, C., Katalinić-Janković, V., Kummik, T., Soini, H., Richter, E., et al., 2014. Inventory study of non-tuberculous mycobacteria in the European Union. BMC Infect. Dis. 14 (1), 62.

Van Halsema, C.L., Chihota, V.N., Gey van Pittius, N.C., Fielding, K.L., Lewis, J.J., Van Helden, P.D., et al., 2015. Clinical relevance of nontuberculous mycobacteria isolated from sputum in a gold mining workforce in South Africa: an observational, clinical study. BioMed Res. Int. 2015.

Varahram, M., Farnia, P., Saif, S., Marashian, M., Farnia, P., Ghanavi, J., et al., 2016. Identification of different subtypes of rapid growing atypical *Mycobacterium* from water and soil sources: using PCR-RFLP using hsp65 and rRNA 16s–23s genes. Int. J. Mycobact. 5, S212–S213.

Velayati, A.A., Farnia, P., Mozafari, M., Malekshahian, D., Seif, S., Rahideh, S., et al., 2014. Molecular epidemiology of nontuberculous mycobacteria isolates from clinical and environmental sources of a metropolitan city. PLoS One 9 (12), e114428.

Velayati, A.A., Rahideh, S., Nezhad, Z.D., Farnia, P., Mirsaeidi, M., 2015. Nontuberculous mycobacteria in Middle East: current situation and future challenges. Int. J. Mycobact. 4 (1), 7–17.

Vugia, D.J., Jang, Y., Zizek, C., Ely, J., Winthrop, K.L., Desmond, E., 2005. Mycobacteria in nail salon whirlpool footbaths, California. Emerging Infect. Dis. 11 (4), 616.

WHO, 2014. Available from: <www.WHO.int/gov/HIV/en/GlobalHealthObservatory(GHO)>.

Wayne, L.G., Sramek, H.A., 1992. Agents of newly recognized or infrequently encountered mycobacterial diseases. Clin. Microbiol. Rev. 5 (1), 1–25.

Weiss, C., Glassroth, J., 2012. Pulmonary disease caused by nontuberculous mycobacteria. Expert Rev. Respir. Med. 6, 597–613.

Zanoni, R., Florio, D., Fioravanti, M., Rossi, M., Prearo, M., 2008. Occurrence of *Mycobacterium* spp. in ornamental fish in Italy. J. Fish Dis. 31 (6), 433–441.

Zerihun, M., Hjortaas, M., Falk, K., Colquhoun, D., 2011. Immunohistochemical and Taqman real-time PCR detection of mycobacterial infections in fish. J. Fish Dis. 34 (3), 235–246.

Index

Note: Page numbers followed by "*f*", "*t*," and "*b*" refer to figures, tables, and boxes, respectively.

D

E

F

G

H

I

K

L

M

N

Made in the USA
Las Vegas, NV
24 May 2023

72514937R00207